Macaca mulatta

ENZYME HISTOCHEMISTRY OF THE NERVOUS SYSTEM

Macaca mulatta

ENZYME HISTOCHEMISTRY OF THE NERVOUS SYSTEM

Sohan L. Manocha and Totada R. Shantha
Yerkes Regional Primate Research Center
Emory University, Atlanta, Georgia

1970

ACADEMIC PRESS
New York and London

ACADEMIC PRESS, INC.
111 Fifth Avenue, New York, New York 10003

United Kingdom Edition published by
ACADEMIC PRESS, INC. (LONDON) LTD.
Berkeley Square House, London W1X 6BA

LIBRARY OF CONGRESS CATALOG CARD NUMBER: 77-108156

PRINTED IN THE UNITED STATES OF AMERICA

CONTENTS

FOREWORD

Although a number of studies of the enzyme biochemistry of mammalian brains have been carried out, very little has been done on primate brains, and no atlas of the distribution of multiple enzymes in the primate brain has been produced. There is an atlas of the distribution of some oxidative enzymes in the cat brain, and the brain of the squirrel monkey has been studied extensively from a histochemical viewpoint by Manocha, Shantha, Bourne, Iijima, Nakajima and others working in this institute.

This atlas of the enzyme histochemistry of the rhesus monkey (*Macaca mulatta*) brain, written by Drs. Manocha and Shantha, is unique because of the number of enzymes surveyed and because the brain of a rhesus monkey has been used. There is no doubt that *biochemical* studies of the brains of animals provide valuable quantitative information about a number of enzymes in various parts of the brain, but they do not tell us in what elements of the brain these enzymes are present. A biochemical study of the cerebellum, for example, does not tell us if the enzyme activities quantified by the biochemical techniques are present in the molecular layer, in the Purkinje cells, in the granule cells or the glomeruli, or just simply in the cerebellar capillaries. If there is a change of activity resulting from some experimental procedure, the biochemical method does not tell us whether any one or all of the elements have gained or lost in enzyme activity. In fact, if some elements lose and others gain—a result which could be demonstrated histochemically—there could be no biochemical change detectable. Thus there is a distinct place for enzyme histochemistry in the study of the brain in addition to what are, in many respects, the cruder techniques of biochemistry.

Dr. Manocha and Dr. Shantha have worked with me in my laboratory for almost 10 years and have proven themselves as first-class histochemists, especially in the area of nervous tissue. This book represents an enormous amount of work and tenacious adherance to their project, and I believe it will hold a most valuable position among scientific publications.

Atlanta, Georgia G. H. Bourne

PREFACE

This work is a general discussion of the distribution of certain oxidative and dephosphorylating enzymes and cholinesterases in the brain, spinal cord, olfactory bulb, and eye of the rhesus monkey (*Macaca mulatta*). It is intended to be a source of histological and histochemical data on the nervous system of the rhesus monkey, an animal which is used in laboratories throughout the world for all types of histological, neuroanatomical, neurophysiological, behavioral, and various other experimental studies. It is also hoped that this work will serve as a guide for future histochemical work, especially in the study of enzyme changes resulting from stimulation of certain selected areas of the brain and in the study of the effects of various drugs, microbia, viruses, toxins, etc., acting on the nervous system. Another function of this work is the investigation (under NASA grant NGR–11–001–016) of histological and histochemical changes in the nervous system of primates under zero gravity conditions and other hazards of prolonged space flights.

The various regions of the brain are so diverse histochemically that they pose a great challenge to correlate histological structure, histochemical nature, and function. Since the present study describes the cytoarchitectural anatomy as well as histochemical properties of individual areas of the brain with a brief comparison of numerous brain enzyme studies on other animals, it will be useful to students, teachers, and research workers in neuroanatomy, histochemistry, neurophysiology, neuropathology, animal behavior, and other related fields.

The authors* are very grateful to Dr. G. H. Bourne, Director, Yerkes Regional Primate Research Center, for his constant encouragement, informal discussions, and financial support from his grant (NGR–11–001–016) awarded by the National Aeronautics and Space Administration. The technical help, in experimental and library work, of Mrs. Mary G. Allison,

* Dr. T. R. Shantha is also named as Dr. T. R. Shanthaveerappa in previous publications.

Mrs. Patricia Brock, Miss Sandra Miller, Miss Marian Rogero, and Mrs. Mary J. Nimnicht is very much appreciated. Thanks are due to Mrs. Katherine Stowers and Mrs. Betsy Bradford for typing the manuscript. The authors also wish to thank Dr. S. P. Sharma of Kurukshetra University, India and Dr. Y. Nakajima, visiting scientist at Yerkes from Tokyo University, for their critique of the manuscript and valuable suggestions on its preparation.

This work has been carried out with the aid of grants FR–00165 from the Animal Resources Branch, National Institutes of Health and NGR–11–001–016 from the National Aeronautics and Space Administration.

Macaca mulatta

ENZYME HISTOCHEMISTRY OF THE NERVOUS SYSTEM

I

INTRODUCTION

Enzyme histochemistry makes use of histological preparations for the purpose of chemical characterization of enzymes in their natural locations. Instead of being colored with routine morphological stains, the sections are treated with complex chemical substrates in order to identify selectively specific enzyme activities. A basic requirement for the histochemical procedure, however, is that the reaction product be insoluble and demonstrable by staining the precipitate at the exact site of enzyme activity. Such a procedure greatly helps in correlating structure with function. This approach of *in situ* localization of chemical constituents without disturbing the environment is preferable under certain circumstances to the identification and quantitation of chemical constituents in the homogenates by biochemical means. In the latter, we can never be certain about the precise location of a particular enzyme, especially in the nervous system, because the homogenate may contain parts of adjacent areas. In biochemistry, profound difficulties are encountered in isolating pure samples out of heterogeneous material. For example, separation of gray matter from white matter is only an approximate procedure, as the former is always found mixed up with the latter (McNabb, 1951).

The biochemical study of the nervous system is further complicated by the intricacy of the nervous system itself, compounded by the large number of metabolic transformations by living cells (McDougal *et al.*, 1962). On the other hand, by maintaining the complete histological structure, not only is a detailed study of the chemoarchitectonics of the tissue possible, but also a distinction between the enzymes located in different *in situ* positions and acting on different substrates can be demonstrated; herein lies the chief advantage of histochemistry. Furthermore, in an analysis of the homogenates of whole tissue, an excess of a particular enzyme may prevent the detection of another closely related enzyme because of smaller quantity (Adams, 1965). For example, it may not be possible in the homogenates to detect the presence of acetylcholinesterase in fine motor or autonomic fibers because of excess amounts of nonspecific cholinesterases in the surrounding tissue. Koelle (1954) believed that the determination of cholinesterase activity by the use of homogenates leaves much to be desired. It cannot be ascertained biochemically whether a particular amount of enzyme activity in a brain area is located in the incoming axon, in the dendrites, or in the cytoplasmic or the nuclear area of neurons. A satisfactory answer to these questions can be provided only by a histochemical preparation which keeps an intact histological picture. Gerebtzoff (1959) stated that, in spite of its qualitative and even quantitative limitations, enzyme histochemistry holds an important position in the convergence of morphological, physiological, and biochemical research.

The validity of histochemical methods has been questioned by numerous biochemists and histochemists alike, and a great volume of literature is available on not only the specificity or nonspecificity of histochemical techniques, but also on the reliability and the reproducibility of the results. Adams (1965) has stressed that "histochemists have always been acutely aware of the need for stringent tests of validity for their methods—suitable histochemical tests for substances or substrates are considerably more difficult to devise than are corresponding biochemical tests. The biochemist can extract or elute the substance he wishes to estimate, so that it is relatively free of contaminating agents, before applying his color reaction. On the other hand, the histochemist has to devise a staining method which is specific for one substance when admixed with a heterogeneous collection of tissue components. On account of these difficulties and the resulting need for specificity, many histochemical reactions now tend to be more rigorous and more fully investigated than the corresponding methods used in biochemical chromatography." Glenner (1965, 1968) has made an excellent evaluation of the specificity of histochemical reactions, and it is clear from his illuminating articles that histochemistry and histochemical techniques must not be taken for granted. No attempt should be made to minimize the complexity of histochemical techniques. Glenner indicated that the assumption that the substrates are specific and identify the same enzyme activity all the time should not be made. Similarly, no conclusions should be drawn which are based on solubility or insolubility of an enzyme, effect of various histochemical steps on enzyme localization, or absence of enzyme activity in the presence of a specific substrate.

Quantitation in section or slide histochemistry cannot be exact because the intensity of stain or color (which generally is an indication of the enzyme content) can be varied by increasing or decreasing the reaction period. Still, a full account of the topographic distribution of enzymes belonging to the different metabolic cycles among the various brain nuclei can be as informative as a quantitative study. Such an estimate of the concentration and precise localization of enzyme activity can be related to the physiological activity of the tissues (since most of the chemical reactions in the living tissue are controlled by enzymes). Ishii and Friede (1967) gave an excellent example in this respect: a histochemical mapping of enzyme activity in the crisscrossing fiber bundles of the pons and its regional and cellular variation in the islands of griseum pontis will give more insight into the functional organization of the pons than would be expected of a biochemical assay of the pons or selected constituents. Friede (1961a, 1961b, 1961c) and Friede and Fleming (1963) have also successfully quantitated in relative figures the activity of some enzymes in various nuclei of the cat, guinea pig, and rhesus brains using a Welch densichron densitometer. Their estimates, when used to compare the enzyme content of the different areas of the brain, are as useful as a biochemical assay.

The present study is intended to investigate the enzyme architecture of the rhesus monkey (*Macaca mulatta*) brain, at both the gross and the microscopic levels. This approach provides a complete topographical map of the distribution of several enzymes with respect to the neuroanatomical structures and enables us to measure and compare the relative sites and concentrations of enzymes in different cytoarchitectural areas in the brain, cerebellum, spinal cord, dorsal root ganglia, olfactory bulb, and eyes. Particular attention has been paid to the distribution of a few hydrolytic and oxidative enzymes.

The hydrolytic group may be generally defined as those enzymes which catalyze the replacement by water of a peptide, ester, or hemiacetal bond. For a detailed analysis of the histochemical mechanism for the localization of the hydrolytic enzymes, the reader is referred to Holt and O'Sullivan (1958), Holt and Withers (1958), and Glenner (1965, 1968). In the localization of oxidative enzymes, the electrons liberated from the substrates are trapped by the tetrazolium salts with or without mediation by electron carriers. The binding of the tetrazolium salts to the tissue components allows electron transport *in situ* to form an insoluble colored product, the formazan, near the site of enzyme activity (Glenner, 1965). For details of the mechanism, the articles of Burtner *et al.* (1957), Slater *et al.* (1963), and Glenner (1965, 1968) are extremely valuable. Farber and Louviere (1956) explained that the tetrazolium salts can also accept electrons from a number of oxidation-reduction dyes in addition to enzymes. This makes the choice of a tetrazolium salt extremely important. Such a salt must be bound to the enzyme during the reaction process and the reduced intermediary product must not be displaced before the tetrazolium salt is reduced. The failure to reduce the tetrazole in an electron-transfer system can be attributed to two causes: (1) the enzyme may not be reacting with the tetrazolium salt and (2) the enzyme capable of reacting with a particular tetrazolium salt may be absent or lost by the solution (Glenner, 1965). In the present study Nitro BT [2,2′-di-*p*-nitrophenyl-5,5′-diphenyl-3,3′(3,3′-dimethoxy-4,4′-biphenyl) ditetrazolium chloride] as used by Nachlas *et al.* (1957), Glenner *et al.* (1957), and Matsui and Kobayashi (1965) is considered satisfactory. In a recent paper Tyrer *et al.* (1968) have, however, reported that in calibrating the reaction for MAO localization, it was found that the relation between proportional amounts of enzyme and the formazan product of enzyme activity was not linear.

The distribution of the oxidative enzymes in the various areas of the nervous system varies greatly, the most conspicuous difference being between the gray and white matter. The gray matter is rich in nerve cells and therefore has a much higher oxygen consumption (Holmes, 1930; Dixon and Meyer, 1936; McNabb, 1951) and glycolytic rate (Holmes, 1930) than the white matter. The gray areas of the brain also show higher quantities of cytochrome (Holmes, 1930), thiamine (Villela *et al.*, 1949), riboflavin (Gourévitch, 1937), and iron, which have intimate links within the metabolic processes. Strong enzyme activity of the citric acid and glycolytic pathway enzymes in our studies of the gray matter confirms the well-known fact that the brain uses mainly carbohydrates in its metabolic activities. Normally the brain oxidizes 85% of the glucose it utilizes; on the other hand, it metabolizes only 15% of the glucose by the glycolytic route to lactate (McIlwain, 1955).

It has been shown by some workers that the amount of oxidative enzyme activity is inversely proportional to the body size. For example, malic dehydrogenase activity in the mouse brain is nearly three times that in the cow brain (Fried and Tipton, 1953), and cytochrome oxidase activity in mouse cortex is about ten times that in the human cortex. But in spite of such large species variation in the enzyme content, the relative quantities of enzymes in the various areas in different species do not change to any great extent, according to Ishii and Friede (1967). These authors have discussed several examples: for instance, the cerebral cortex in man contains 96% of malate dehydrogenase activity in the nucleus caudatus (Tyler, 1960), 88% of lactic dehydrogenase activity (Tyler, 1960) and 106–122% of NAD-dehydrogenase activity (Friede and

Fleming, 1962); 86% of oxygen consumption in ox (Dixon and Meyer, 1936); 85% of oxygen consumption in dog (Himwich and Fazekas, 1941); and 88–115%, depending on layers, of the capillary density in the cat (Campbell, 1939). Thus, the ratio of activity in the cortex to that in the nucleus caudatus varies little among different species.

In addition, phosphatases (alkaline and acid phosphatase, adenosine triphosphatase) and carboxylesterases (simple esterases and cholinesterases) have been studied. There is reason to believe that phosphatases play an important role in various metabolic processes. As early as 1943, Sumner and Somers described phosphatases as being involved in the metabolism of nucleotides, phospholipids, and carbohydrates. The presence of alkaline phosphatase at the periphery of the spinal ganglion cells points toward a role in the exchange of ions (Shantha *et al.*, 1967; Thakar and Tewari, 1967). A similar role (active transport of sodium and potassium ions) has been attributed to adenosine triphosphatase (ATPase) (Skou, 1965; Albers, 1967; Fahn and Côté, 1968). The Na-K ATPase is present in highest quantities in the secretory nuclei and gray matter of the nervous tissue (Bonting *et al.*, 1961) and the nerve ending particles (Albers *et al.*, 1965; Hosie, 1965; Kurokawa *et al.*, 1965). Acid phosphatases (AC) are probably involved in the metabolism of nucleic acids (Shimizu, 1950) or in the synthesis of proteins (Bourne, 1958) or RNA (Meyer, 1963). The latter role is evidenced in their relation to the maturation of neuroblasts. Since the neuronal AC does not show as much variation as the oxidative enzymes, it may be closely related with the static, maintaining metabolism of cells rather than their dynamic, functional aspects. The lysosomes also show strong AC reaction, and increase in the lysosomal activity is indicated by increased AC activity. An increase in the lysosomal activity with the liberation of acid phosphatase appears to be responsible for some of the early changes observed in the chromatolytic neurons (Bodian and Mellors, 1945). Lysosomes are often associated with the lipofuscin pigment which explains the concentration of AC in those areas. Novikoff (1963) postulated that the larger lysosomes in neurons arise from the Golgi saccules where these are apposed to the endoplasmic reticulum. He believed that these bodies acquired their acid hydrolases from vesicles separating from the Golgi saccules. The saccules may concentrate the enzymes that are transported to them after formation in association with the ribosomes by way of the endoplasmic reticulum. Novikoff concluded that the neurons which show acid phosphatase activity in the Golgi vesicles show it also in the endoplasmic reticulum.

Simple esterases are found in locations similar to those of the acid phosphatases and are as abundant. The quantitative studies of Barron and Bernsohn (1965) showed that the phosphatases and esterases are actively involved in the degradation of myelin lipid which accompanies certain induced central and peripheral neuronal lesions of experimental animals. Simple esterases may also be involved in the hydrolysis of certain transmitter substances or substances other than acetylcholine.

The cholinesterases are specific and nonspecific (acetyl and butyryl, AChE and BChE) and these enzymes differ in the rate of hydrolysis of substrates; it decreases for AChE and increases for BChE with increasing length of the acetyl group, from acetyl to *n*-butyryl (Gerebtzoff, 1959). AChE is also known as acetylcholine hydrolase, AChE-1, specific, true, or aceto- or E-cholinesterase. In the present study the term acetylcholinesterase (AChE) is used, as recommended by the Enzyme Commission in 1964.

The major interest in this enzyme is centered on the fact that AChE hydrolyzes acetylcholine (ACh), and the ACh-cholinesterase system is of great significance in the metabolism of the brain because of its contribution to synaptic transmission (Bennett *et al.*, 1958). Those areas which show larger quantities of AChE do have higher ACh metabolism than those with lower AChE activity. There is sufficient evidence in numerous studies to suggest that ACh acts as a transmitter substance and AChE will be present on these sites. Furthermore, in contrast to the fleeting nature of ACh, AChE is a stable enzyme (Okinaka *et al.*, 1960). Dale, in 1933, first introduced the word "cholinergic" to designate cells which release ACh as a transmitter; hence, the presence of the latter is a more reliable indicator of cholinergic transmission (Silver, 1967), although the localization of AChE in various sites is also useful when related to the presence of ACh. It is well known that AChE activity is essential for the normal transmitter function of ACh at peripheral sites such as the neuromuscular junction. It is tempting, therefore, to conclude that the presence of AChE is an indication of cholinergic transmission in a particular area (Silver, 1967). AChE is consistently present in the nervous system, particularly in the cell bodies, axons, and terminals of all cholinergic neurons, and a good correlation exists between the AChE activity and ACh content in the brain (Feldberg and Vogt, 1948; Eränkö, 1967). Shute and Lewis (1963, 1966) and some other workers have shown that the cell bodies of neurons may show AChE and a negative reaction for ACh in their axons. This may mean that the presence of AChE on the axon membrane should be an essential criterion of a cholinergic neuron (Lewis and Shute, 1965).

Silver (1967) discussed the role of AChE in the glial cells. It is not likely that the Schwann cells or glia manufacture neuronal AChE, because the molecular weight of AChE is approximately 250,000 (Kremzner and Wilson, 1964), and this would make the AChE molecule too large to penetrate the membranes between the Schwann cells and the axons (Koenig and Koelle, 1961; Silver, 1967).

Ishii and Friede (1967) compared the distribution of AChE in different areas of the brain with some other factors involved in neuronal transmission—for example, norepinephrine, dopamine, serotonin and the like—and did not find any correlation. Norepinephrine has no consistency in the human brain. Both norepinephrine and AChE are low in cerebral and cerebellar cortices, but the hypothalamus shows larger amounts of norepinephrine than AChE (Sano *et al.*, 1959; Ehringer and Horneykiewicz, 1960; Bertler, 1961). The putamen and substantia nigra seem to show large amounts of dopamine and AChE, yet other centers with high AChE do not show a high concentration of dopamine (Carlsson, 1959; Sano *et al.*, 1959; Ishii and Friede, 1967).

The following pages include observations on some of the oxidative and hydrolytic enzymes in the rhesus brain and their interpretation with respect to some recent literature. Besides neurons and other familiar terms, the word "neuropil" points to the intercellular ground substance which consists of neuronal processes, the neuroglial cells and their processes, and small extracellular spaces made up of very thin (100–200 Å) gaps between the adjacent membranes (Hydén, 1960; Bondareff, 1965; Singh, 1968). This extracellular space may contain an electron lucid matrix material (Robertson *et al.*, 1963; Bondareff, 1965).

Light microscopy has made some very important contributions to enzyme histochemistry, but the ultimate goal of histochemical localization in the cell components of different enzymes mediating the metabolic processes will be reached only when these histochemical techniques are perfected and used at the ultrastructural level. Some progress has been made in this direction, and the functional organization of different tissues is understood better today than yesterday.

REFERENCES

Adams, C. W. M. (ed.) Histochemistry of the cells in the nervous system. In: *Neurohistochemistry*, pp. 253–331. Elsevier, Amsterdam, 1965.

Albers, R. W. Biochemical aspects of active transport. *Ann. Rev. Biochem.* **36**:727–756 (1967).

Albers, R. W., Arnaiz, G. R. de Lores, and De Robertis, E. Sodium potassium-activated ATPase and potassium-activated *p*-nitrophenyl-phosphatase: a comparison of their subcellular localizations in rat brain. *Proc. Nat. Acad. Sci. U.S.* **53**:557–564 (1965).

Barron, K. D., and Bernsohn, J. Brain esterases and phosphatases in multiple sclerosis. *Ann. New York Acad. Sci.* **122**:369–399 (1965).

Bennett, E. L., Krech, D., Rosenzweig, M. R., Karlsson, H., Dye, N., and Ohlander, A. Cholinesterase and lactic dehydrogenase activity in the rat brain. *J. Neurochem.* **3**:153–160 (1958).

Bertler, A. Occurrence and localization of catecholamines in the human brain. *Acta Physiol. Scand.* **51**:97–107 (1961).

Bodian, D., and Mellors, R. C. The regenerative cycle of motoneurons with special reference to phosphatase activity. *J. Exp. Med.* **81**:469–488 (1945).

Bondareff, W. The extracellular compartment of the cerebral cortex. *Anat. Rec.* **152**:119–127 (1965).

Bonting, S. L., Simon, K. A., and Hawkins, N. M. Studies on sodium-potassium activated adenosine triphosphatase I. Quantitative distribution in several tissues of the cat. *Arch. Biochem. Biophys.* **95**:416–423 (1961).

Bourne, G. H. Histochemical demonstration of phosphatases in the central nervous system of the rat. *Exptl. Cell Res., Suppl.* **5**:101–117 (1958).

Burtner, H. J., Bahn, R. C., and Longley, J. B. Observations on the reduction and quantitation of neotetrazolium. *J. Histochem. Cytochem.* **5**:127–134 (1957).

Campbell, A. C. P. Variation in vascular and oxidase content in different regions of the brain of the cat. *Arch. Neurol. Psychiat.* **41**:223–242 (1939).

Carlsson, A. The occurrence, distribution and physiological role of catecholamines in the nervous system. *Physiol. Rev.* **11**:490–493 (1959).

Dale, H. H. Nomenclature of fibers in autonomic system and their effects. *J. Physiol. (London)* **80**:10–11 (1933).

Dixon, T. F., and Meyer, A. Respiration of brain. *Biochem. J.* **30**:1577–1582 (1936).

Ehringer, H., and Horneykiewicz, O. Verteilung von Noradrenalin and Dopamin (13-Hydroxyturamin) im Gehirn des Menschen und ihr Verhalten bei Erkrankungen des extrapyramidalen Systems. *Klin. Wochrschr.* **38**:1236–1239 (1960).

Eränkö, O. Histochemistry of nervous tissues: catecholamines and cholinesterases. *Ann. Rev. Pharmocal.* **7**:203–222 (1967).

Fahn, S., and Côté, L. J. Regional distribution of sodium-potassium activated adenosine triphosphatase in the brain of the rhesus monkey. *J. Neurochem.* **15**:433–436 (1968).

Farber, E., and Louviere, C. D. Histochemical localization of specific oxidative enzymes. IV. Soluble oxidation reduction dyes as aids in the histochemical localization of oxidative enzymes with tetrazolium salts. *J. Histochem. Cytochem.* **4**:347–362 (1956).

Feldberg, W., and Vogt, M. Acetylcholine synthesis in different regions of the central nervous system. *J. Physiol. (London)* **107**:372–381 (1948).

Fried, G. H., and Tipton, S. R. Comparison of respiratory enzyme levels in tissues of mammals of different sizes. *Proc. Soc. Exptl. Biol. Med.* **82**:531–532 (1953).

Friede, R. L. A histochemical study of DPN diaphorase in human white matter; with some notes on myelination. *J. Neurochem.* **8**:17–30 (1961a).

Friede, R. L. Surface structures of the aqueduct and the ventricular walls; a morphologic, comparative and histochemical study. *J. Comp. Neurol.* **116**:229–297 (1961b).

Friede, R. L. *A Histochemical Atlas of Tissue Oxidation in the Brain Stem of the Cat.* Karger, Basel and Stechert-Hafner, New York, 1961c.

Friede, R. L., and Fleming, L. M. A mapping of oxidative enzymes in the human brain. *J. Neurochem.* **9**:179–198 (1962).

Friede, R. L., and Fleming, L. M. A mapping of the distribution of lactic dehydrogenase in the brain of the rhesus monkey. *Am. J. Anat.* **113**:215–234 (1963).

Gerebtzoff, M. A. *Cholinesterases*. International Series of Monographs on Pure and Applied Biology (Division: Modern Trends in Physiological Sciences), (P. Alexander, and Z. M. Bacq, eds.). Pergamon Press, New York, 1959.

Glenner, G. G. Enzyme histochemistry. In: *Neurohistochemistry* (C. W. M. Adams, ed.), pp. 109–160. Elsevier, Amsterdam, 1965.

Glenner, G. G. Evaluation of the specificity of enzyme histochemical reactions. *J. Histochem. Cytochem.* **16**:519–529 (1968).

Glenner, G. G., Burtner, H. J., and Brown, G. W., Jr. The histochemical demonstration of monoamine oxidase activity by tetrazolium salts. *J. Histochem. Cytochem.* **5**:591–600 (1957).

Gourévitch, A. La distribution de la flavine dans les tissus des mammiferes en relation avec leur respiration residuelle en presence des cyanures. *Bull. Soc. Chim. Biol. (Paris)* **19**:527–554 (1937).

Himwich, H. E., and Fazekas, B. J. F. Comparative studies of the metabolism of the brain of infant and adult dogs. *Am. J. Physiol.* **132**:454–459 (1941).

Holmes, E. G. Oxidations in central and peripheral nervous tissue. *Biochem. J.* **24**:914–925 (1930).

Holt, S. J., and O'Sullivan, D. G. Studies in enzyme cytochemistry. I. Principles of cytochemical staining methods. *Proc. Roy Soc.* [*Biol.*] **148**:465–480 (1958).

Holt, S. J., and Withers, R. F. J. Studies in enzyme cytochemistry. V. An appraisal of indigogenic reactions for esterase localization. *Proc. Roy. Soc.* [*Biol.*] **148**: 520–532 (1958).

Hosie, R. J. A. The localization of adenosine triphosphatases in morphologically characterized subcellular fractions of guinea pig brain. *Biochem. J.* **96**:404–412 (1965).

Hydén, H. The neuron. In: *The Cell* (J. Brachet and A. E. Mirsky, eds.), Vol. 4, Chap. 5, pp. 215–323. Academic Press, New York, 1960.

Ishii, T., and Friede, R. L. A comparative histochemical mapping of the distribution of acetylcholinesterase and nicotinamide adenine-dinucleotide-diaphorase activities in the human brain. *Intern. Rev. Neurobiol.* **10**:231–275 (1967).

Koelle, G. B. The histochemical localization of cholinesterases in the central nervous system of the rat. *J. Comp. Neurol.* **100**:211–228 (1954).

Koenig, E., and Koelle, G. B. Mode of regeneration of acetylcholinesterase in cholinergic neurons following irreversible inactivation. *J. Neurochem.* **8**:169–188 (1961).

Kremzner, L. T., and Wilson, I. B. A partial characterization of acetylcholinesterase. *Biochemistry* **3**:1902–1905, (1964).

Kurokawa, M., Sakamoto, T., and Kato, M. Distribution of sodium-plus-potassium stimulated adenosine triphosphatase activity in isolated nerve-ending particles. *Biochem. J.* **97**:833–844 (1965).

Lewis, P. R., and Shute, C. C. D. Fine localization of acetylcholinesterase in the optic nerve and retina of the rat. *J. Physiol (London)* **180**:8P–10P (1965).

Matsui, T., and Kobayashi, H. Histochemical demonstration of monoamine oxidase in hypothalamo-hypophysial system of the tree sparrow and the rat. *Z. Zellforsch.* **68**:172–182 (1965).

McDougal, D. B., Jr., Jones, E. M., and Sila, U. I. Distribution of enzymes of the tricarboxylic acid cycle in white matter. *Res. Publ. Assoc. Res. Nervous Mental Disease* **40**:182–188 (1962).

McIlwain, H. *Biochemistry and the Central Nervous System*. Churchill, London, 1955.

McNabb, A. R. Enzymes of gray matter and white matter of dog brain: the distribution of certain monoxidative enzymes. *Can. J. Med. Sci.* **29**:208–215 (1951).

Meyer, P. Histochemistry of the developing human brain. I. Alkaline phosphatase, acid phosphatase and AS-esterase in the cerebellum. *Acta Neurol. Scand.* **39**: 123–138 (1963).

Nachlas, M. M., Tsou, K., DeSouza, E., Cheng, C., and Seligman, A. M. Cytochemical demonstration of succinic dehydrogenase by the use of a new *p*-nitrophenyl substituted ditetrazole. *J. Histochem. Cytochem.* **5**:420–436 (1957).

Novikoff, A. B. Lysosomes in the physiology and pathology of cells: contributions of staining methods. *Ciba Found. Symp., Lysosomes,* pp. 36–77 (1963).

Okinaka, S., Yoshikawa, M., Uono, M., Mozai, T., Toyota, M., Muro, T., Igata, A., Tanabe, H., and Ueda, T. Histochemical study on cholinesterase of the human hypothalamus. *Acta Neuroveget.* (*Vienna*) **22**:53–62 (1960).

Robertson, J., Bodenheimer, D., and Stage, D. E. The ultrastructure of mauthner cell synapses and nodes in goldfish brains. *J. Cell. Biol.* **19**:159–199 (1963).

Sano, I., Gamo, T., Kakimoto, Y., Taniguchi, K., Takesada, M., and Nishinuma, K. Distribution of catechol compounds in human brain. *Biochim. Biophys. Acta* **32**:586–587 (1959).

Shantha T. R., Manocha, S. L., and Bourne, G. H. Enzyme histochemistry of the mesenteric and dorsal root ganglion cells of cat and squirrel monkey. *Histochem.* **10**:234–245 (1967).

Shimizu, N. Histochemical studies on the phosphatase of the nervous system. *J. Comp. Neurol.* **93**:201–218 (1950).

Shute, C. C. D., and Lewis, P. R. Cholinesterase-containing systems of the brain of the rat. *Nature* **199**:1160–1164 (1963).

Shute, C. C. D., and Lewis, P. R. Electron microscopy of cholinergic terminals and acetylcholinesterase containing neurons in the hippocampal formation of the rat. *Z. Zellforsch.* **69**:334–343 (1966).

Silver, A. Cholinesterases of the central nervous system with special reference to the cerebellum. *Intern. Rev. Neurobiol.* **10**:57–109 (1967).

Singh, R. Some observations on the histochemistry of the neuropile tissue of the brain of parrot (*Psittacula krameri*). *Acta Anatom.* **68**: 567–577 (1968).

Skou, J. C. Enzymatic basis for active transport of Na^+ and K^+ across cell membranes. *Physiol. Rev.* **45**:596–617 (1965).

Slater, T. F., Sawyer, B., and Strauli, U. Studies on succinate-tetrazolium reductase systems. III. Points of coupling of four different tetrazolium systems. *Biochim. Biophys. Acta* **77**:383–393 (1963).

Summer, J. B., and Somers, G. F. *Chemistry and Methods of Enzymes.* Academic Press, New York, 1943.

Thakar, D. S., and Tewari, H. B. Histochemical studies on the distribution of alkaline and acid phosphatases among the neurons of the cerebellum, spinal, and trigeminal ganglia of bat. *Acta Histochem.* **28**:359–367 (1967).

Tyler, H. R. Enzyme distribution in human brain. Lactic and malic dehydrogenases. *Proc. Soc. Exptl. Biol. Med.* **104**:79–83 (1960).

Tyrer, J. H., Eadie, M. J., and Kukums, J. R. Histochemical measurements of relative concentrations of monoamine oxidase in various regions of rabbit brain. *Brain* **91**:507–519 (1968).

Villela, G. G., Diaz, M. V., and Queriroga, S. T. Distribution of thiamine in the brain. *Arch. Biochem. Biophys.* **23**:81–84 (1949).

II

MATERIAL and METHODS

MATERIAL

Twelve rhesus monkeys of both sexes were used in the present study. All the animals had healthy coats and weighed from 10.5 to 12 pounds. Prior to being used for histochemical study, the animals were examined for parasitic infestations and other infections. They were well fed and watered regularly and were in good health before being sacrificed for this study. After decapitation the skull cap was removed within the shortest possible time with an electric saw. The dura mater was slit open, and powdered dry ice was sprinkled all over the surface in order to harden it to some extent. Then the skull was broken from the sides, and the brain was removed. Two or three blocks of the brain were cut without distorting its shape. The blocks were then quickly frozen in dry ice (−78°C).

Some earlier reports indicated that expediting the freezing of tissues is of minor importance. For example, Lazarus *et al.* (1962) showed that tissues from animals left at room temperature for four hours after death displayed minimal damage to cell organelles with few other apparent alterations. Similarly, Chason *et al.* (1963) showed that there was no change in the localization of malate and lactic dehydrogenase activity in the brain layers examined during postmortem intervals of four to six hours. Our experience with various histochemical techniques has shown, however, that the quick freezing of small blocks of tissue immediately after removal is an extremely important step for enzyme studies at the cellular level. This eliminates the formation of the large ice crystals which show up if the tissue is frozen slowly in a cryostat (—10°C to —25°C). Such ice crystals displace the enzymatically active cytoplasmic organelle, which may lead to the misinterpretation of the enzyme reactions. This procedure may also lead to the formation of large perinuclear or peripheral clump-like masses. The frozen block of tissue was mounted on a cryostat chuck, and a series of 15 μ thick sections were cut at —20° to —25°C. The sections were taken from the blade on cold cover glasses and dried for a few minutes at room temperature before being used for incubation and the demonstration of various enzymes. The sections were always serially numbered, and an attempt was made to use all the sections except those badly torn in the process of cutting.

Fresh frozen sections were used for the present study. Most histochemists agree that fresh unfixed tissue is best for the demonstration of most enzymes, in spite of the fact that routine cryostat sectioning has been recognized as a cause of serious physical damage to the tissue (Eckner *et al.*, 1968). Chason *et al.* (1963) reported that formalin fixation offers no advantage in the maintenance of a more accurate localization and has the disadvantage of a variable loss of enzyme activity (Friede, 1961). According to Neumann (1958) and Barka and Anderson (1963), freezing and drying of tissue followed by paraffin embedding gives the "best morphologic preservation with the least chemical alteration of the tissue"; nevertheless, it is believed to destroy the activity of many oxidative enzymes such as succinic dehydrogenase and cytochrome oxidase (Barka and Anderson, 1963). The physical damage caused by cryostat sectioning of fresh frozen tissue may be necessary and preferable to obtain adequate results. We have not used fixation and paraffin embedding, also, for fear of contraction caused by fixation and subsequent processing, which would lead to erroneous calculation of magnification and would introduce some uncertain elements in enzyme localization. Deane (1963) reported that those persons preferring paraffin sections to frozen sections are the ones still reporting nuclear enzymatic activity of phosphatases.

HISTOCHEMICAL PROCEDURES

The histochemical techniques are based on the identification of the product of an enzyme reaction with a physiologic substrate. Most substrates are synthetic and are prepared to resemble closely the physiologic substrate in order to satisfy the major criteria of enzyme-substrate specificity and also to yield a product from which an insoluble colored pigment can be formed in a short time (Ravin *et al.*, 1953). These authors have described the qualities of such a reaction thus: the substrate should be hydrolyzed rapidly to keep the incubation time of the tissue short; the coupling agent should not inhibit enzyme activity; and the conditions of pH, temperature, and ionic concentration optimal for enzymatic activity should not be different from those suitable for the smooth occurrence of the reaction. Most techniques used in the present study are believed to fulfill these criteria adequately.

The localization and distribution of phosphatases have been studied by Gomori salt methods (Gomori, 1939; Wolf *et al.*, 1943; Shimizu, 1950; Chiquoine, 1954; Bourne, 1957, 1958) and by the simultaneous coupling methods applying naphthol-AS-phosphate or other derivatives (Burstone, 1958a, 1958b, 1958c, 1960, 1961; Samorajski and McCloud, 1961; Anderson and Song, 1962) used with various diazonium salts including Fast Blue B, Fast Garnet GBC, Fast Red Violet TR, Fast Red Violet RC, and even basic fuchsin. The comparison between the salt-metal technique and azo dye methods is not an easy one. The results obtained with these two groups of techniques are greatly debated without much agreement since numerous influencing factors have been evidenced (Schiffer *et al.*, 1962). Pearse (1961) believes that the lead-acid phosphatase technique gives only an approximate distribution of the enzyme activity in various tissues. On the other hand, the azo dye methods present the problem of a choice of substrates in relation to their possible linkage with tissue structures (Defendi, 1957) and the choice of diazonium salts and their favorable or unfavorable influence on enzyme activity (Schiffer *et al.*, 1962). The major difference which has aroused a controversy about these two approaches is the nuclear staining for alkaline phosphatases by Gomori's method. Pearse and a number of other workers reported it as an artifact. We have relied more on the azo dye methods in the present study, due to biochemical data in favor of these methods (Schiffer *et al.*, 1962).

There are a number of methods for detection of ATPase. We have used the Padykula and Herman method (1955a, 1955b), which is a modified Gomori's calcium-cobalt method at pH 9.4 and the Wachstein and Meisel (1957) method at pH 7.2. The former workers demonstrated that sulfhydryl compounds such as BAL (2,3-dimercapto-1-propanol) or cysteine, which inhibit alkaline phosphatase, activate the site of ATPase activity. On the other hand, compounds like PCMB (P-chloromercuribenzoate), a sulfhydryl inhibitor, inhibit ATPase staining, which can be reversed by SH reagents. Wachstein and Meisel (1957), using a lead salt ATPase technique at pH 7.2 (tris buffer), demonstrated activity in cytoplasm, bile canaliculi, and mitochondria. These studies show clearly the pH variability for this enzyme localization. It is interesting to note that Bonting *et al.* (1962) showed the complete inhibition of Na-K-activated (membrane) ATPase by low concentrations of lead. Great care was, therefore, exercised in the interpretation of results obtained by the two methods.

Most of the workers have used fresh frozen sections, but fixed tissues have also been successfully used for the demonstration of phosphatases. Landow *et al.* (1942) used ethanol-fixed, paraffin-embedded tissue for Gomori's alkaline phosphatase method; Naidoo and Pratt (1951) used freeze-dried, paraffin-embedded tissue for Gomori's acid phosphatase and lead method for ATPase, and Bourne (1958) used acetone-fixed, paraffin-embedded brain tissue for alkaline phosphatase. Such examples can easily be multiplied.

Simple esterases are the enzymes which hydrolyze short-chain aliphatic esters. The method to detect simple esterase by using azo dye methods appears to be specific in most cases. Studies of Chessick (1954) and Gomori (1955) indicate that naphtholic substrates are capable of demonstrating a spectrum of simple esterases with considerable overlapping of substrate specificity.

The cholinesterases have been demonstrated by Koelle and Friedenwald's (1949) method in which acetyl or butyryl thiocholine iodide is used as the substrate, while the incubation medium consists of copper and sulfate ions. The enzymatically liberated thiocholine is precipitated as copper thiocholine sulfate, which is colorless and is rendered brown by the yellow ammonium sulfide solution. Although several substrates other than the thiocholine esters have been used (Pearse, 1960; Burstone, 1962; Eränkö *et al.*, 1964; Härkönen, 1964; Kokko, 1965), the above method remains ideal for the localization of cholinesterases. Several modifications by different workers are available and are widely used (Coupland and Holmes, 1957; Gerebtzoff, 1959; Lewis and Shute, 1959; Karnovsky and Roots, 1964; Joo *et al.*, 1965; Koelle and Gromadzki, 1966; Eränkö *et al.*, 1967). Lewis and Shute (1959) showed that the sensitivity of Koelle's technique can be controlled by varying the pH of the medium. At an acid pH, the method is highly selective for sites of high enzyme activity. To the present authors, the important factor in the localization of AChE is the selection of appropriate inhibitors, since acetylthiocholine is split not only by AChE but also by nonspecific cholinesterase, which must be selectively inhibited.

The specificity of the inhibitors has been discussed by a number of workers (Holmstedt and Sjögvist, 1961; Richardson, 1962; Koelle, 1963; Diegenbach, 1965; Koelle and Gromadzki, 1966). According to Ravin *et al.* (1953) eserine, in low concentration, is a selective inhibitor to allow clear differentiation between "cholinesterases on the one hand and aliesterases, lipases, or esterolytic function of the proteolytic enzymes on the other." Silver (1967) also proposed that appropriate combinations of substrates and inhibitors, pH and temperature of medium, and duration of incubation period are the important factors in the correct identification of cholinesterases. Prolonged incubation, although desirable in some instances to reveal all the cholinesterase sites, is inevitably followed by diffusion from areas of high AChE activity. A number of workers believe that prefixation in formalin for a short period is the best procedure for a correct localization (Coupland and Holmes, 1957; Karnovsky and Roots, 1964), although some others state as emphatically that the enzyme activity is inhibited to a marked degree by formaldehyde. Eränkö *et al.* (1967) suggested a formalin-Krebs–Ringer-calcium solution fixation for two to four hours at 0°C as ideal for the purpose. Austin and Phillis (1965) calculated that 70–80% of the original enzyme activity of cerebellar cholinesterases remained in the tissue fixed in formalin for four to six hours. Hardwick and Palmer (1961) believed that pseudocholinesterases in the sheep brain are less sensitive to formaldehyde than is AChE.

Oxidative enzymes are localized in the fresh frozen sections, although Samorajski (1960) reported that "brief fixation of dorsal ganglia and spinal cord segments in either chloraldehydrate-formalin or 2% calcium formalin prior to histochemical demonstration of DPN-diaphorase, not only succeeds in preserving enzyme activity, but actually causes an intense fixation of the reaction without any significant alteration in localization. Cold acetone causes a decrease in diaphorase activity." Potanos *et al.* (1959) studied the distribution of SDH by incubating fresh blocks of tissue, following this by paraffin embedding and sectioning. Wolfgram and Rose (1959) studied several dehydrogenases in the rat and cat brains by using unfixed frozen sections dried *in vacuo* at room temperature. Nitro BT has proven the most satisfactory tetrazolium salt for both succinic dehydrogenase and diaphorase systems. Samorajski (1960) showed that distribution of diaphorase activity with the MTT cobalt method is more restricted than what is revealed by the Nitro BT method.

The histochemical methods used in our investigations have been those considered standard by a number of workers over the years and are believed to give consistent results. The following histochemical techniques were employed for the demonstration of phosphatases, esterases, and oxidative enzymes.

Alkaline phosphatase

Calcium cobalt method (Gomori, 1939, 1952)

The incubation medium consists of 10 ml of 3% sodium β-glycerophosphate, 10 ml of 2% sodium diethylbarbiturate, 5 ml of distilled water, 20 ml of 2% calcium chloride, and 1 ml of 5% magnesium sulfate. The sections are incubated for four to six hours at 37°C, rinsed in running water, treated for a short time in 2% cobalt nitrate, and after washing in distilled water are treated with dilute yellow ammonium sulfide. The sections are thoroughly washed and mounted in aquamount.

α-Naphthyl phosphate method (Gomori, 1951, 1952)

The sections are incubated for 10 to 30 minutes at room temperature in an incubating medium containing 10 mg sodium α-naphthyl phosphate dissolved in 20 ml distilled water to which is added 5 ml of 0.2 *M* tris buffer (pH 8.9–9.3) and 6 to 8 drops of 10% magnesium sulfate. Immediately before use, 50 mg Red Violet LB salt is added to the mixture, which is shaken continuously and filtered. The sections are washed thoroughly in running tap water and mounted in aquamount.

Naphthol AS-BI phosphate azo dye method (Burstone, 1958a, 1958c, 1961)

The incubation medium consists of a stock solution containing 5 mg naphthol AS-BI phosphate dissolved in 0.25 ml dimethylformamide to which are added 25 ml distilled water and 25 ml of 0.2 *M* tris buffer (brought to pH 8.8–9.3 by 1 *M* Na_2CO_3). The stock solution can be stored for several weeks. Immediately before use, 50 mg diazonium salt (Fast Red RC or Fast Red TR proved satisfactory) is mixed in 50 ml of stock solution and filtered. The sections are incubated for 15 to 30 minutes at room temperature, washed thoroughly in tap water and mounted in aquamount.

Acid phosphatase

Lead nitrate method (Gomori, 1950, 1952)

The sections are incubated for about four hours at 37°C in a medium containing 12 mg lead nitrate, 30 mg sodium β-glycerophosphate, 2.5 ml of 0.05 *M* acetate buffer (pH 5.0), and 7.5 ml water. The solution is filtered before use. The sections are subsequently washed in water and developed in dilute yellow ammonium sulfide, then washed in water and mounted in aquamount.

Simultaneous coupling azo dye method (Grogg and Pearse, 1952; Barka and Anderson, 1963)

The incubation medium consists of 20 mg sodium α-naphthyl phosphate, 20 ml of 0.1 *M* phosphate buffer (pH 5.0) and 20 mg Garnet GBC. The solution is filtered, and the sections are incubated for 5 to 20 minutes at room temperature, washed in water, and mounted. Grogg and Pearse (1952) tested the qualities of 22 coupling agents at pH 5.0 for the demonstration of acid phosphatase and reported that Garnet GBC gave the most satisfactory results.

Simultaneous coupling azo dye method (Barka, 1960)

The incubation medium is prepared by adding 5 ml sodium α-naphthyl acid phosphate solution (4 mg/ml dissolved in Michaelis veronal acetate buffer stock solution) in 13 ml distilled water followed by 1.6 ml of the hexazonium salt solution. The latter is prepared by mixing equal amounts (0.8 ml) of 4% aqueous sodium nitrite and pararosanilin-HCl (2 gm added to 50 ml of 2 *N* HCl and heated, cooled, and filtered). The pH of the medium is adjusted to 6.5 with *N* NaOH. The sections are incubated for 5 to 20 minutes, rinsed in water and mounted in aquamount or dehydrated in an ascending series of alcohol concentrations and mounted in Canada balsam.

Adenosine triphosphatase (ATPase)

Lead method (Wachstein and Meisel, 1957)

This method is based on Gomori's lead nitrate medium at pH 7.2. The sections are incubated for 5 to 60 minutes at 37°C in a solution containing 25 mg adenosine triphosphate, disodium salt; 20 ml of 0.2 *M* tris buffer (pH 7.2); 87.5 magnesium sulfate; 60 mg lead nitrate; and 30 ml distilled water. The sections are washed in water after incubation, developed in yellow ammonium sulfide, and mounted in aquamount.

Calcium method (Padykula and Herman, 1955a, 1955b)

This method is based on reducing the concentration of salts and the substrate in the previously used calcium method after the Maengwyn–Davies *et al.* (1952) method. This reduction, according to Padykula and Herman, produced better cytoplasmic localization and less nuclear staining. The medium consists of 20 ml of 0.1 *M* sodium barbiturate (2.062 gm/100 ml); 10 ml of 0.18 $CaCl_2$ (1.998 gm/100 ml); 70 ml distilled water; and 152 mg adenosine triphosphate, disodium salt. The pH of the medium is adjusted to 9.6 with 0.1 *M* NaOH. The solution is filtered before use. The sections are incubated at 37°C for 30 to 60 minutes, washed in water, treated with dilute yellow ammonium sulfide, and mounted in aquamount.

Simple esterases

α-Naphthyl acetate method (Barka and Anderson, 1963)

α-Naphthyl acetate as a substrate for the simultaneous coupling azo dye methods was first introduced by Gomori because of its favorable characteristics compared to β-naphthyl acetate as earlier used by Nachlas and Seligman (1949a, 1949b) and Seligman *et al.* (1949). The incubation medium consists of 50 mg α-naphthyl acetate dissolved in 0.5 ml acetone to which is added 20 ml of 0.1 *M* phosphate buffer. Immediately before use, 100 mg Fast Red RC is added and the medium filtered. The sections are incubated for 5 to 20 minutes, washed in tap water, and mounted in aquamount.

Naphthol AS acetate method

Gomori (1952) substituted naphthol AS acetate for naphthyl acetate in a simultaneous coupling azo dye method for esterases. The technique employed in this study is after Burstone (1957) and Barka and Anderson (1963). The incubation medium consists of 20 to 30 mg naphthol AS-LC acetate dissolved in 2.5 ml dimethylformamide to which is added 7.5 ml ethylene glycol monoethyl ether, 10 ml of 0.2 *M* tris-maleate buffer (pH 7.1), and 30 ml water. Immediately before use, 20 to 30 mg Fast Garnet GBC is added and filtered. The sections are incubated for 5 to 20 minutes, washed in running water, and mounted in aquamount.

Cholinesterases

Direct coloring thiocholine method (Karnovsky and Roots, 1964)

The sections are incubated at 37°C for 30 to 60 minutes in a solution containing 9.5 mg acetylthiocholine iodide; 10 ml of 0.1 *M* phosphate buffer (pH 6.0); 1.5 ml of 0.1 *M* sodium citrate solution (2.9312 gm in 100 ml water); 1.5 ml of 30 m*M* cupric sulfate (477 mg in 100 ml water); 1.5 ml of 5 m*M* potassium ferricyanide (164.5 mg in 100 ml water); and 1.5 ml distilled water.

The advantage of this technique, according to these authors, lies in the fact that the color is produced directly at the site of enzymatic activity, eliminating the estimation of optimum incubation time. Because of the finely granular precipitate of the reaction products, needle-like deposits are not observed.

Koelle and Friedenwald's (1949) modification (Coupland and Holmes, 1957)

The incubation medium contains 1.2 ml copper glycine (glycine, 3.75 gm; $CuSO_4$–$5H_2O$, 2.5 gm; water, 100 ml); 1.2 ml $MgCl_2$ ($MgCl_2$–$6H_2O$, 9.52 gm; water, 100 ml); 10 ml of 0.2 *M* acetate buffer (pH 5.0); and 15.2 ml of 40% sodium sulfate. To this medium is added 3.2 ml acetyl or butyryl thiocholine iodide-$CuSO_4$ mixture (acetyl or butyryl thiocholine iodide 46 mg in 2.4 ml water and 0.8 ml of 0.1 *M* cupric sulfate centrifuged for 5 minutes at 300–350 rpm). The sections are incubated at 37°C for four to sixteen hours. Sections were taken out of the incubating mixture at varying intervals to determine the optimum time. The sections are washed in water, treated with a dilute solution (1%) of yellow ammonium sulfide, and mounted in aquamount.

Monoamine oxidase

Tetrazolium method (Glenner et al., 1957; Weissbach et al., 1957)

The substrate consists of tryptamine hydrochloride, 25 mg; sodium sulfate, 4 mg; Nitro BT, 5 mg; 0.1 *M* phosphate buffer (pH 7.6), 5 ml; and distilled water, 15 ml. The sections are incubated at 37°C for 30 to 35 minutes, washed in water for two minutes, fixed in 10% neutral formaldehyde from two hours to overnight, and mounted in aquamount.

Modified tetrazolium method (Matsui and Kobayashi, 1965)

This method is a modification of the Glenner *et al.* (1957) technique. A 20 ml solution of the incubation medium contains 5 ml of 0.1 *M* phosphate buffer (pH 7.6) and 15 ml saline solution containing 25 mg tryptamine hydrochloride, 1.67 gm anhydrous sodium sulfate, and 5 mg Nitro BT. According to Matsui and Kobayashi, the high concentration of sodium sulfate in the incubation mixture resulted in a decrease in the size of the formazan crystals and improvement in enzyme location. The present authors, however, did not observe any significant difference in MAO localization with either of the techniques.

Succinic dehydrogenase

Nitro BT succinate method (Nachlas et al., 1957)

The substrate consists of 5 ml of 0.2 *M* phosphate buffer (pH 7.6), 5 ml of 0.2 *M* sodium succinate solution, and 10 ml aqueous solution of Nitro BT (1 mg/1 ml). The sections are incubated for 20 to 30 minutes at 37°C, washed in saline, fixed subsequently in formal-saline for 10 minutes, rinsed in 15% alcohol for four to five minutes, and mounted in aquamount.

Lactic dehydrogenase

Nitro BT method (Hess et al., 1958)

The sections are incubated at 37°C for 30 to 40 minutes in a substrate consisting of sodium lactate, 112.07 mg; diphosphopyridine nucleotide, 66.344 mg; sodium cyanide, 4.9 mg; magnesium chloride, 10.05 mg; 0.1 *M* phosphate buffer (pH 7.4), 2.5 ml (out of 10 ml buffer plus 8 ml water); Nitro BT, 2.5 mg; polyvinylpyrrolidone (PVP), 750 mg. The sections are subsequently fixed in 10% formal-calcium for 10 minutes, washed briefly in distilled water, and mounted in aquamount.

Great attention has been paid to the proper execution of these techniques so that they give consistent results on repeated trials. The histochemical methods were accompanied by appropriate controls, and great caution was observed in the interpretation of enzyme activity to avoid misleading conclusions. The effects of freezing, thawing, fixation, osmotic change, and optimal pH of the enzyme system, as well as the duration of incubation, have been considered extremely important. Prolonged incubation often results in nonspecific reaction and the production of artifacts, making the isolation of negative sites particularly difficult (Gerebtzoff, 1959). Generalizations about a positive or negative reaction of an unusual type in some cells or a tissue must always be interpreted with great caution. Such generalizations, from derived fragmentary proof of an unusual reaction, can be extremely misleading. As stressed by Silver (1967), the most important factor in the enzyme localization is that the end-product must be accurately localized at the site of enzyme activity and there should be no diffusion of either the enzyme or the reaction product. Although fixation in formalin avoids diffusion of the end product, it can inhibit in varying degrees the enzyme activity (Taxi, 1952; Fukuda and Koelle, 1959; Hardwick and Palmer, 1961). Glenner (1965,

1968) has given an excellent evaluation of the specificity of enzyme reactions. For details, the reader is referred to these articles.

A list of enzymes to which we have referred in abbreviated form in the text may be found in the Appendix.

PHOTOGRAPHY

The photographs included in this monograph have been taken at the macroscopic and microscopic levels. At the macroscopic or gross level, the sections were photographed by our enlarger method (Shanthaveerappa *et al.*, 1965). The mounted slides were inserted in the negative carrier plate and mounted into the photographic enlarger. A 4 × 5 inch film was exposed in the manner of printing paper for half a second. The exposed films were processed as usual and the prints made from these negatives. For ascertaining the magnification, a small strip of the scale was placed along with the section while photographing and the final magnification calculated from the prints by measuring this scale. At the microscopic levels, a Carl–Zeiss microscope with an automatic built-in camera was used. The magnification in this system is easy to calculate, based on eyepiece, objective lens, and magnification while printing.

A list of abbreviations used in the figures may be found in the Appendix.

REFERENCES

Anderson, P. J., and Song, S. K. Acid phosphatase in the nervous system. *J. Neuropathol. Exptl. Neurol.* **21**:263–283 (1962).

Austin, L., and Phillis, J. W. The distribution of cerebellar cholinesterases in several species. *J. Neurochem.* **12**:709–727 (1965).

Barka, T. A simple azo-dye method for histochemical demonstration of acid phosphatase. *Nature* **187**:248–249 (1960).

Barka, T., and Anderson, P. J. *Histochemistry: Theory, Practice, and Bibliography.* Harper (Hoeber), New York, 1963.

Bonting, S. L., Caravaggio, L. L., and Hawkins, N. M. Studies on sodium-potassium-activated adenosine triphosphatase. IV. Correlation with cation transport sensitive to cardiac glycosides. *Arch. Biochem. Biophys.* **98**:413–419 (1962).

Bourne, G. H. Phosphatases in the central nervous system. *Nature* **179**:1247–1248 (1957).

Bourne, G. H. Histochemical demonstrations of phosphatases in the central nervous system of the rat. *Exptl. Cell Res.* **5**:101–117 (1958).

Burstone, M. S. The cytochemical localization of esterase. *J. Nat. Cancer Inst.* **18**:167–172 (1957).

Burstone, M. S. Comparison of naphthol-AS-phosphates for demonstration of phosphatases. *J. Nat. Cancer Inst.* **20**:601–614 (1958a).

Burstone, M. S. Histochemical demonstration of acid phosphatases with naphthol-AS-phosphatases. *J. Nat. Cancer Inst.* **21**:523–539 (1958b).

Burstone, M. S. The relationship between fixation and technique for the histochemical localization of hydrolytic enzymes. *J. Histochem. Cytochem.* **6**:322–339 (1958c).

Burstone, M. S. *Calcification in Biological Systems* (R. F. Sogmaess, ed.). Amer. Assoc. Adv. Sci., Washington, D.C., 1960.

Burstone, M. S. Histochemical demonstration of phosphatases in frozen sections with naphthol-AS-phosphatases. *J. Histochem. Cytochem.* **9**:146–153 (1961).

Burstone, M. S. *Enzyme Histochemistry.* Academic Press, New York and London, 1962.

Chason, J. L., Gonzalez, J. E., and Landers, J. W. Respiratory enzyme activity and distribution in the post mortem central nervous system. *J. Neuropathol. Exptl. Neurol.* **22**:248–254 (1963).

Chessick, R. D. The histochemical specificity of cholinesterases. *J. Histochem. Cytochem.* **2**:258–273 (1954).

Chiquoine, D. A. Distribution of alkaline phosphomonoesterase in the central nervous system. *J. Comp. Biol.* **100**:415–439 (1954).

Coupland, R. E., and Holmes, R. L. The use of cholinesterase techniques for demonstration of peripheral nervous structures. *Quart. J. Microscop. Sci.* **98**: 327–330 (1957).

Deane, H. W. Nuclear location of phosphatase activity: fact or artifact? *J. Histochem. Cytochem.* **11**:443–444 (1963).

Defendi, V. Observations on naphthol staining and the histochemical localization of enzymes by naphthol-azo dye technique. *J. Histochem. Cytochem.* **5**:1–10 (1957).

Diegenbach, P. C. Use of inhibitors in cholinesterase histochemistry. *Nature* **207**:308 (1965).

Eckner, F. A., Riele, B. H., Moulder, P. V., and Blackstone, E. H. Histochemical study of enzyme systems in frozen dried tissue. *Histochemie* **13**:283–288 (1968).

Eränkö, O., Härkönen, M., Kokko, A., and Räisänen, L. Histochemical and starch gel electrophoretic characterization of desmo- and lyo-esterases in the sympathetic and spinal ganglia of the rat. *J. Histochem. Cytochem.* **12**:570–581 (1964).

Eränkö, O., Koelle, G. B., and Räisänen, L. A thiocholine-lead ferrocyanide method for acetylcholinesterase. *J. Histochem. Cytochem.* **15**:674–680 (1967).

Friede, R. L. *A Histochemical Atlas of Tissue Oxidation in the Brain Stem of the Cat.* Karger, Basel, and Stechert-Hafner, New York, 1961.

Fukuda, T., and Koelle, G. B. The cytological localization of intracellular neuronal acetylcholinesterase. *J. Biochem. Cytol.* **5**:433–440 (1959).

Gerebtzoff, M. A. *Cholinesterases.* International Series of Monographs on Pure and Applied Biology (Division: Modern Trends in Physiological Sciences), (P. Alexander and Z. M. Bacq, eds.). Pergamon Press, New York, 1959.

Glenner, G. G. Enzyme histochemistry. In: *Neurohistochemistry* (C. W. M. Adams, ed.). Elsevier, Amsterdam, 1965.

Glenner, G. G. Evaluation of the specificity of enzyme histochemical reactions. *J. Histochem. Cytochem.* **16**:519–529 (1968).

Glenner, G. G., Burtner, H. J., and Brown, G. W., Jr. The histochemical demonstration of monoamine oxidase activity by tetrazolium salts. *J. Histochem. Cytochem.* **5**:591–600 (1957).

Gomori, G. Microtechnical demonstration of phosphatase in tissue sections. *Proc. Soc. Exptl. Biol. New York* **42**:23–26 (1939).

Gomori, G. An improved histochemical technique for acid phosphatase. *Stain Technol.* **25**:81–85 (1950).

Gomori, G. Alkaline phosphatase of cell nuclei. *J. Lab. Clin. Med.* **37**:526–531 (1951).

Gomori, G. *Microscopic Histochemistry; Principles and Practice.* Univ. of Chicago Press, Chicago, Illinois, 1952.

Gomori, G. Histochemistry of human esterases. *J. Histochem. Cytochem.* **3**:479–484 (1955).

Grogg, E., and Pearse, A. G. E. A critical study of the histochemical techniques for acid phosphatase. *J. Pathol. Bacteriol.* **64**:627–636 (1952).

Hardwick, D. C., and Palmer, A. C. Effect of formalin fixation on cholinesterase activity in sheep brain. *Quart. J. Exptl. Physiol.* **46**:350–352 (1961).

Härkönen, M. Carboxylic esterases, oxidative enzymes and catecholamines in the superior cervical ganglion of the rat and the effect of pre- and post-ganglionic nerve division. *Acta Physiol. Scand.* Suppl. **63**:1–94 (1964).

Hess, R., Scarpelli, D. G., and Pearse, A. G. E. Cytochemical localization of pyridine nucleotide linked dehydrogenases, *Nature* **181**:1531–1532 (1958).

Holmstedt, B., and Sjöqvist, F. Some principles about histochemistry of cholinesterase with special reference to thiocholine method. *Bibliotheca. Anat.* **2**:1-10 (1961).

Joo, F., Savay, G., and Csillik, B. A new modification of the Koelle-Friedenwald method for the histochemical demonstration of cholinesterase activity. *Acta Histochem.* **22**:40–45 (1965).

Karnovșky, M. J., and Roots, L. A "direct-coloring" thiocholine method for cholinesterases. *J. Histochem. Cytochem.* **12**:219–221 (1964).

Koelle, G. B. (ed.) Cytological distribution and physiological functions of cholinesterases (chap. 6). In: *Handbuch der Experimente Pharmakologie,* pp. 187–298. Springer-Verlag, Berlin, 1963.

Koelle, G. B., and Friedenwald, J. S. Histochemical method for localizing cholinesterase activity. *Proc. Soc. Exptl. Biol. Med.* **70**:617–622 (1949).

Koelle, G. B., and Gromadzki, C. G. Comparison of the gold-thiocholine and gold-thiolacetic acid methods for the histochemical localization of acetylcholinesterases and cholinesterases. *J. Histochem. Cytochem.* **14**:443–454 (1966).

Kokko, A. Histochemical and cytophotometric observations on esterases in the spinal ganglion of the rat. *Acta Physiol. Scand.* **66**:1–76 (1965).

Landow, H., Kabat, E. A., and Newman, W. Distribution of alkaline phosphatase in normal and neoplastic tissues of the nervous system. *Arch. Neurol. Psychiat.* **48**:518–530 (1942).

Lazarus, S. S., Wallace, B. J., Edgar, G. W. F., and Volk, B. W. Enzyme localization in rabbit cerebellum and effect of postmortem autolysis. *J. Neurochem.* **9**:227–232 (1962).

Lewis, P. R., and Shute, C. C. D. Selective staining of visceral efferents in the rat brain stem by a modified Koelle technique. *Nature* **183**:1743–1744 (1959).

Maengwyn-Davies, G. D., Friedenwald, J. S., and White, R. T. Histochemical studies of alkaline phosphatases in the tissues of the rat using frozen sections. II. Substrate specificity of enzymes hydrolyzing adenosine-triphosphate, muscle- and yeast-adenylic acids, and creatine phosphate at high pH; the histochemical demonstration of myosin ATPase. *J. Cellular. Comp. Physiol.* **39**:395–434 (1952).

Matsui, T., and Kobayashi, H. Histochemical demonstration of monoamine oxidase in hypothalamo-hypophysial system of the tree sparrow and the rat. *Z. Zellforsch.* **68**:172–182 (1965).

Nachlas, M. M., and Seligman, A. M. Evidence for the specificity of esterase and lipase by the use of three chromogenic substrates. *J. Biol. Chem.* **181**:343–355 (1949a).

Nachlas, M. M., and Seligman, A. M. The histochemical demonstration of esterase. *J. Nat. Cancer Inst.* **9**:415–425 (1949b).

Nachlas, M. M., Tsou, K., DeSouza, E., Cheng, C., and Seligman, A. M. Cytochemical demonstration of succinic dehydrogenase by the use of a new *p*-nitrophenyl substituted ditetrazole. *J. Histochem. Cytochem.* **5**:420–436 (1957).

Naidoo, D., and Pratt, O. E. The localization of some acid phosphatases in brain tissue. *J. Neurol. Neurosurg. Psychiatry* **14**:287–294 (1951).

Neumann, K. Anwendung der Gefriertrocknung für Histochemische Untersuchungen. In: *Handbuch de Histochemie* (W. Graumann and K. Neumann, eds.). Gustav Fischer, Stuttgart, 1958.

Padykula, H. A., and Herman, E. Factors affecting the activity of adenosine triphosphatase and other phosphatases as measured by histochemical techniques. *J. Histochem. Cytochem.* **3**:161–169 (1955a).

Padykula, H. A., and Herman, E. The specificity of the histochemical method for adenosine triphosphatase. *J. Histochem. Cytochem.* **3**:170–195 (1955b).

Pearse, A. G. E. *Histochemistry: Theoretical and Applied.* Churchill, London, 1960.

Pearse, A. G. E. *Histochemistry: Theoretical and Applied.* Little, Brown, Boston, 1961.

Potanos, J. N., Wolf, A., and Cowen, D. Cytochemical localization of oxidative enzymes in human nerve cells and neuroglia. *J. Neuropathol. Exptl. Neurol.* **18**:627–635 (1959).

Ravin, H. A., Zacks, S. I., and Seligman, A. M. The histochemical localization of acetylcholinesterase in nervous tissue. *J. Pharmacol. Exptl. Therap.* **107**:37–53 (1953).

Richardson, K. C. The fine structure of autonomic nerve endings in smooth muscle of the rat vas deferens. *J. Anat. (London)* **96**:427–442 (1962).

Samorajski, T. The application of diphosphopyridine nucleotide diaphorase methods in a study of dorsal ganglia and spinal cord. *J. Neurochem.* **5**:349–353 (1960).

Samorajski, T., and McCloud, J. Alkaline phosphomonoesterase and blood-brain permeability. *Lab. Invest.* **10**:492–501 (1961).

Schiffer, D., Vesco, C., and Piazza, L. Contribution to the histochemical demonstration and distribution of phosphatases and non-specific esterase in human nervous tissue. *Psychiat. Neurol. (Basel)* **144**:34–47 (1962).

Seligman, A. M., Nachlas, M. M., Manheimer, L. H., Friedman, O. M., and Wolf, G. Development of new methods for the histochemical demonstration of hydrolytic intracellular enzymes in a program of cancer research. *Ann. Surg.* **130**:333–341 (1949).

Shanthaveerappa, T. R., Manocha, S. L., and Bourne, G. H. Macrophotography of thick sections and gels with a regular photographic enlarger. *Stain Tech.* **40**:309 (1965).

Shimizu, N. Histochemical studies on the phosphatase of the nervous system. *J. Comp. Neurol.* **93**:201–218 (1950).

Silver, A. Cholinesterases of the central nervous system with special reference to the cerebellum. *Intern. Rev. Neurobiol.* **10**:57–109 (1967).

Taxi, J. Action du formol sur l'activite de diverses preparations de cholinesterases. *J. Physiol. (Paris)* **44**:595–599 (1952).

Wachstein, M., and Meisel, E. Histochemistry of hepatic phosphatases at a physiologic pH. *Am. J. Clin. Pathol.* **27**:13–23 (1957).

Weissbach, H., Redfield, B. G., Glenner, G. G., and Mitoma, C. Tetrazolium reduction as a measure of monoamine oxidase activity *in vitro*. *J. Histochem. Cytochem.* **5**:601–605 (1957).

Wolf, A., Kabat, E. A., and Newman, W. Histochemical studies in tissue enzymes. III. A study of the distribution of acid phosphatases with special reference to the nervous tissue. *Am. J. Pathol.* **19**:423–440 (1943).

Wolfgram, F., and Rose, A. S. The histochemical demonstration of dehydrogenases in neuroglia. *Exptl. Cell. Res.* **17**:526–530 (1959).

III

CORTICAL AREAS and RELATED FIBERS

In this study, the following cortical areas have been described: precentral and postcentral gyri, visual cortex, hippocampus, amygdaloid complex, tuberculum olfactorium, claustrum, and external and extreme capsule and capsula interna. A general discussion of the histochemistry of these areas follows the cytoarchitechtonic and histochemical observations.

PRECENTRAL AND POSTCENTRAL GYRI

The precentral and postcentral gyri, like many other areas of the cerebral cortex, have six layers, easily distinguished from each other in Nissl and Weil stained slides as well as histochemical preparations.

Layer I is the molecular or plexiform layer and contains mainly fibers and a few cells, most of which are horizontal.

Layer II is the external granular layer and is made up of small pyramidal cells in addition to granule cells. The Stripe of Kaes–Bechterew courses through the lower part of this layer and the upper part of Layer III.

Layer III is the external pyramidal layer and is composed mostly of pyramidal cells.

Layer IV is the internal granular layer, the outer part containing predominantly star pyramids, and the inner part small pyramids. The external Stripe of Baillarger is found within this layer.

Layer V is the internal pyramidal or ganglionic layer, which contains mostly small, medium, and large pyramidal cells. It is further divisible into three sublayers, the middle sublayer containing large pyramidal cells along with the granule cells. The internal Stripe of Baillarger occupies the deeper part of this layer.

Layer VI is the layer of fusiform or spindle-shaped cells; these are small, medium, and large in size. This layer also contains small star-shaped cells in addition to occasional large pyramidal cells.

The precentral gyrus contains pyramidal cells of Betz in Layer V. As we trace this layer into the postcentral gyrus, the Betz cells disappear, and the well-demarcated internal granular layer (Layer IV) appears. Layers III and V of the precentral gyrus contain a large number of pyramidal cells. Layer IV in this gyrus is also poorly developed and contains more pyramidal cells than the same layer in the postcentral gyrus.

The following histochemical description applies to both precentral and postcentral gyri unless otherwise mentioned.

AC. The neuropil in Layer I shows negligible activity, whereas the rest of the layers are mild. The neurons in Layer I are mild, Layers II and III moderate, IV moderately strong, Layer V strong to very strong in large and giant pyramidal cells, and moderate in small neurons. Betz cells show the strongest positive reaction. Layer IV in the postcentral gyrus shows much stronger positive cells than the same layer in the precentral gyrus.

AK. Only the blood vessels are positive. Layers I and II have the least number of blood vessels, and Layers IV and V have many more than the other four layers. The neuropil and the neurons show negative activity.

ATPase. The neuropil in Layer I shows stronger (moderately strong) activity than the other layers (mild to moderate). The positive activity gradually decreases as we trace this enzyme from Layer I to Layer VI. All the neurons are mild to moderate, and the positive activity is more evident in the cell membranes than in the cytoplasm. Betz cells are moderate. The nucleoli in all these neurons give positive reactions. The blood vessels also react strongly. In general, the neurons in Layers II, III, and IV of the postcentral gyrus show slightly stronger ATPase activity than the neurons in these layers of the precentral gyrus (Fig 4).

SE. The neuropil in Layer I shows mild activity compared to moderate in other layers. Betz cells in Layer V of the precentral gyrus give a strong reaction compared to moderate to moderately strong in other

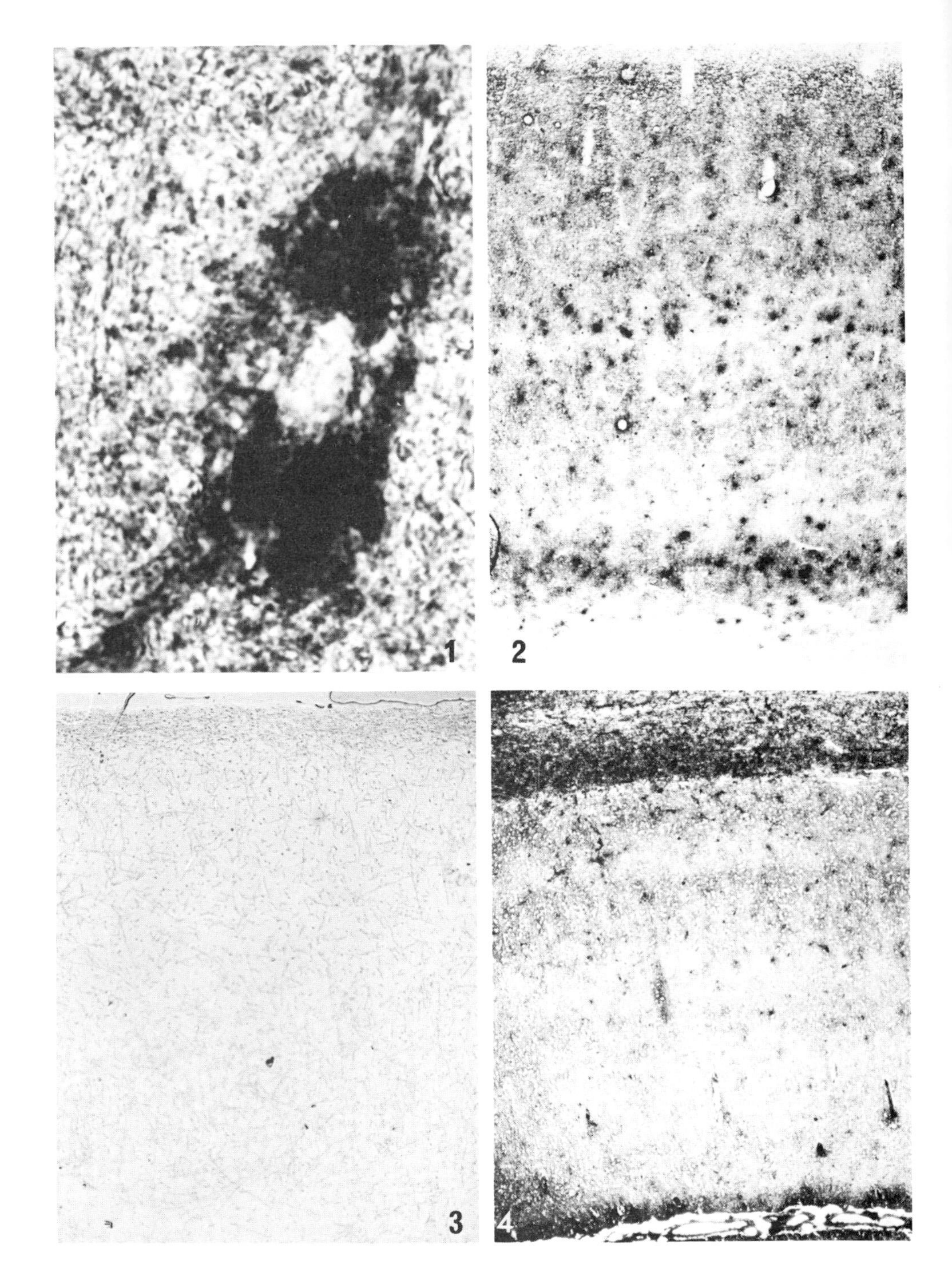
1
2
3
4

FIGURE 1. A pyramidal cell from Layer V of the precentral cortex showing SDH reaction. The enzyme activity can be seen in the cell processes for a considerable distance. ×1480.

FIGURE 2. AC reaction in the striate area of the cerebral cortex. The neurons in the fifth layer show the strongest enzyme reaction compared to the other layers. The nerve cells in the first layer are mild, the second layer moderate and the third, fourth, and sixth layers moderately strong. The neuropil is mild to moderate and the glial cells are mildly positive. ×37.

FIGURE 3. AChE reaction in the striate area. The first layer shows a number of closely packed horizontally running AChE positive fibers. The majority of the neurons, including the large ones, give a negligible to mild reaction. The large pyramidal cells of the fifth layer (lamina ganglionaris) are somewhat stronger and show mild to moderate positive activity. A microscopic examination reveals that the AChE activity is concentrated in the cell membranes and the processes. However, all the layers show a large number of moderately positive fibers. ×37.

FIGURE 4. ATPase preparation of the postcentral cortex showing varying degrees of enzyme activity in the different layers. The neuropil in the molecular layer is stronger (moderately strong) than the other layers. The positive activity gradually decreases from Layer I to VI. The neurons show a mild to moderate activity, and the positive reaction is more evident in the cell membrane than in the cytoplasm. ×37.

neurons. This positive enzymatic activity extends into cell processes for a considerable distance. The glial cells are generally moderate.

AChE. Many nerve fibers in various layers react moderately in AChE. Layer I shows a number of positive fibers running in a horizontal direction. The neuropil is negative, as are the majority of neurons. Large pyramidal cells and Betz cells show mild positive activity, while the capillaries give a moderate reaction.

MAO. The neuropil in all the layers shows moderate activity. Layer I shows much more enzyme activity than the other layers. The neurons show mild to moderate activity in their cytoplasm.

SDH. The neuropil shows mild activity in Layer I, moderate in Layers II and III, mild to moderate in Layers IV and V, and mild in Layer VI. Neurons in Layer I show negligible to mild, Layers II and III mild to moderate, Layer IV moderately strong, Layer V moderately strong to strong (Fig. 1), and Layer VI moderate SDH-positive activity. The Betz cells show the strongest reaction. Layer IV of the postcentral gyrus also has strong positive cells.

LDH. The neuropil in Layer I is mild, Layers II, III, and IV moderate, and Layers V and VI mild to moderately positive. Neurons in Layer I are mild, Layer II moderate, Layer III moderately strong, Layer IV strong, Layer V very strong, and Layer VI moderately strong. Layer IV in the postcentral gyrus and the Betz cells show strong positive activity, which extends into cell processes for a considerable distance. The glial cells show moderately strong activity.

THE VISUAL CORTEX OR STRIATE AREA

The visual cortex is characterized by a broad internal granular layer (Layer IV) separated into inner (more cellular) and outer (less cellular) sublayers by a large Stripe of Gennari. The whole striate area is divisible into the following layers:

Layer I. The *stratum moleculare,* or plexiform layer, is the superficial part, containing a large number of nerve fibers along with a few small granule cells and horizontal cells.

II. The *lamina granularis externa* contains small pyramidal cells with short axons.

Layer III. The *lamina pyramidalis* consists of outer, lightly stained pyramidal cells and an inner, darkly stained band of pyramidal cells mixed with granule cells.

Layer IV. The *lamina granularis interna* consists of giant stellate cells with oval or triangular cell bodies, solitary cells of Meynert, star pyramids, and granule cells.

Layer V. The *lamina ganglionaris* is not as rich in cells. It contains pyramidal cells of various sizes, including giant pyramidal cells in the deeper part of the layer.

Layer VI. The *lamina multiformis* contains medium-sized fusiform or triangular cells. The outer part is more cellular and darkly stained, whereas the inner part is less cellular and lightly stained.

The histochemical reactions are summarized below:

AC. The neuropil is mildly positive in the first layer and moderately positive in the rest of the layers. The neurons in Layer V show the strongest positive activity compared to the neurons in the other layers. The nerve cells in the first layer are mild, second layer moderate, and the third, fourth, and sixth layers moderately strong. The glial cells are mildly positive (Fig. 2).

AK. Only the blood vessels appear to react positively, except in Layer V which shows mild activity in the neuropil and in some neurons. Layers I, II, and VI have fewer blood vessels than the other layers.

ATPase. The neuropil in Layer I shows moderately strong activity, Layer II moderate, and the positive activity gradually decreases as it is traced toward the sixth layer. Neurons are mild to moderately positive, and the large neurons in Layer V are moderate to moderately strong.

The nucleoli of the neurons, the glial cells, and the blood vessels show positive activity.

SE. The neuropil shows stronger positive activity in SE than in AC preparations. Layers I and VI are mild, Layer II moderate, and Layers III, IV, and V moderately strong. The large neurons in Layer V are strongly positive, whereas the rest of the neurons and glial cells are moderate.

AChE. The Stripe of Gennari shows many more positive fibers than other layers. The various layers, particularly the first one, show a number of closely packed horizontal AChE-positive fibers. The large pyramidal cells in Layer V show mild positive activity. It appears that the positive activity is more concentrated in the cell membrane and its processes. The neuropil is negative, and the capillaries show moderate activity (Fig. 3).

MAO. The neuropil in the uppermost layer shows a strong reaction, gradually decreasing down to Layer VI. The neurons are mildly positive, the positive activity being concentrated in the cytoplasm.

SDH. All the layers show enzyme activity similar to LDH, except that the intensity of positive activity is milder in SDH preparations. The large neurons in Layer V show strong positive activity.

LDH. The neuropil LDH activity in Layers I, II, and VI is moderate, moderately strong in Layers III and IV, and strong in Layer V. The same is true of neurons in these layers, although the large pyramidal cells show more positive activity than other neurons. The enzyme reaction extends into the cell processes. The glial cells are moderately strong in LDH.

HIPPOCAMPUS

Anatomically the hippocampus has been described in detail by Lorente de No (1934), who found no difference in structure between the lower mammals and primates (including the rhesus monkey). Following the classification of Lorente de No, in order to facilitate adequate description, the hippocampus has been further subdivided into *Fascia dentata; Ammon's horn; Subicular area* (*Prosubiculum, Subiculum, Presubiculum,* and *Parasubiculum*); and *Gyrus hippocampus.*

Fascia dentata

This area is composed of three layers (Figs. 5–10): (a) *stratum moleculare,* having superiorly placed small round cells and deeply

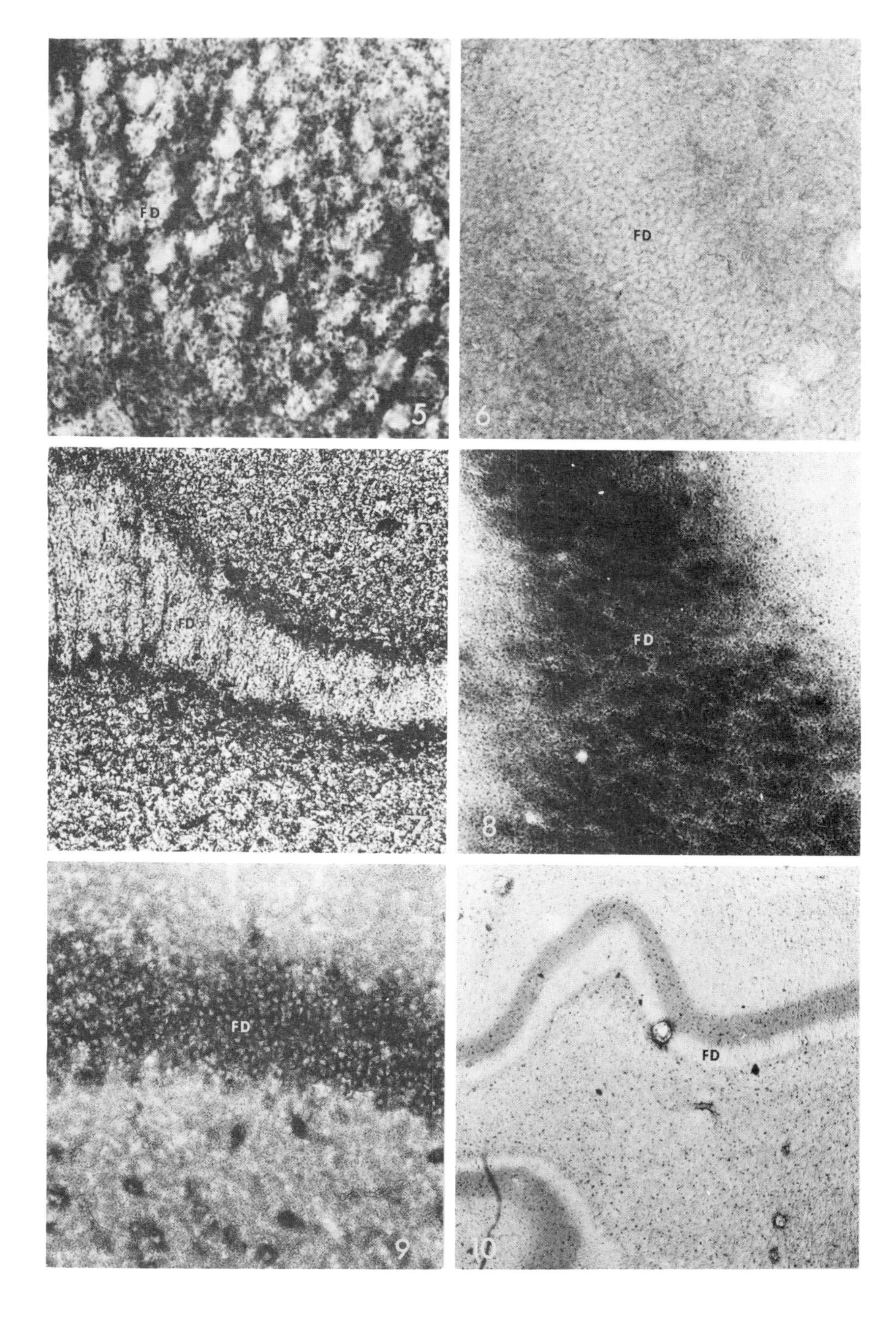
FD
5
FD
6
FD
7
FD
8
FD
9
FD
10

FIGURE 5. Fascia dentata, LDH: the granule cells show strong cytoplasmic activity. The large nuclei of these cells show a negative reaction. The molecular and polymorph layers show moderate enzyme activity. ×148.

FIGURE 6. Fascia dentata, MAO: showing moderate activity in the molecular layer, slightly less in the polymorph layer, and mild reaction in the granular cell cytoplasm. ×148.

FIGURE 7. Fascia dentata, ATPase: showing moderately strong activity in the molecular and polymorph layers and mild activity in the cytoplasm of the granule cells. Note a strong reaction adjoining the granule cell layer. ×148.

FIGURE 8. Fascia dentata, AC: showing moderately strong activity in the cytoplasm of the granule cells compared to a negligible reaction in other layers. ×370.

FIGURE 9. Fascia dentata, SE: showing strong activity in the cytoplasm of the granule cells and in the neurons of the polymorph layer as compared to other layers. The neuropil shows mild enzyme activity. ×148.

FIGURE 10. Fascia dentata and part of the Ammon's horn showing negligible AChE reaction in the upper part of the molecular layer compared to moderately strong AChE activity in the inner part adjoining the granule cell layer. The latter shows negligible activity in the cells, whereas the polymorph layer and those of the Ammon's horn show mild reaction in the cells and the neuropil. ×37.

placed, somewhat larger, oval cells. This layer is composed mainly of nerve fibers with very few neurons; (b) *stratum granulosa,* having large granule cells with a large nucleus and a thin rim of cytoplasm and a few fibers; and (c) *stratum polymorphum,* having pyramidal basket cells as well as polygonal and horizontal cells, and also some modified pyramidal cells of Ammon's horn.

The *fascia dentata* shows variable enzyme reactions in the different layers for each enzyme tested. The stratum moleculare shows moderately strong LDH; mild to moderate MAO; moderate SDH; mild SE, AC, ATPase, and AChE; and negative AK activity in the neurons. The neuropil shows moderately strong LDH; mild to moderate SE, SDH, and ATPase; moderate AChE and MAO; mild AC; and negative AK activity. In AChE preparations, the superficial half of this layer shows negligible activity with numerous positive fibers running through it, whereas the inner half shows moderately strong AChE activity (Fig. 10). The nerve fibers passing through the deeper part of the molecular layer show much stronger positive activity than the superficial part. Blood vessels in this layer are moderately strong AChE-positive.

The *stratum granulosa* shows strong to very strong LDH, moderately strong SDH, strong SE, moderate AC and ATPase, mild AChE and MAO, and negative AK activity. There is very little neuropil found between the cells of the stratum granulosa. The polymorph layer shows moderately strong LDH and SDH; moderate AC, SE, ATPase, and MAO; mild AChE; and negative AK activity. The neuropil has moderate LDH, SDH, and MAO; mild AC, SE, and AChE; moderately strong ATPase; and negative AK activity. AK is exclusively localized in the blood vessels. The stratum moleculare and granulosa contain more blood vessels than the polymorph layer.

Ammon's horn

Ammon's horn contains the following layers (Crosby *et al.*, 1962):

Layer I. *Superficial white substance* containing dendrites from cornua ammonis, terminal axons from alveus intermingled with polymorph cells.

Layer II. *Stratum lacunosum* consisting of richly arranged plexuses formed by axons.

Layer III. *Stratum radiation* situated at the border of the pyramidal cell layer formed by the interlacing dendritic processes from these cells.

Layer IV. *Layer of double pyramids*—medium-sized, having dendritic processes on both sides and axons given off from the ventricular surface of the cell body and the dendrite joining alveus and fimbria. As one traces the Ammon's horn toward the fascia dentata, the compact arrangement of the stellate cells breaks up, giving rise to randomly distributed multipolar cells. This part has been named by Lorente de No (1934) the Cornua Ammonis No. 4.

Layer V. *Stratum oriens,* consisting of layers of irregularly arranged cells. They are mostly triangular and fusiform in shape, small and medium in size.

Layer VI. *Alveus or deep white substance* is found on the ventricular surface of Ammon's horn and consists of both afferent and efferent fibers.

Layer VII. *Ependymal layer* of the lateral ventricle.

The neurons in Ammon's horn are arranged as (a) a compact layer which is further divided into cornua ammonalis (CA 1, 2, 3) and (b) randomly arranged cells situated at the bend of the fascia dentata (CA4). The neurons in CA1 show less enzymatic reaction than those in CA2, CA3, and CA4. In general the cells show strong LDH; moderately strong SE, AC, and SDH; moderate AChE; and mild ATPase and MAO activity. The positive activity is seen extending into dendritic processes on both sides of these cell layers (i.e., into the stratum radiatum and stratum oriens). The alveus shows very little enzyme activity: the superficial white substance and stratum radiatum show much more LDH, SDH, AChE, AC, and SE activity than the alveus.

In AChE preparations, one can distinguish various layers of Ammon's horn much more easily than in the other enzyme preparations. Layers I and II give much more positive activity than Layer III. Layer V shows the strongest positive reaction. Layer IV shows only a few positive fibers running through it.

In ATPase preparations, Layers I, II, III, and V give much more positive activity than Layers IV and VI. In AC preparations, Layer IV is the strongest; next to it are Layers II and V (moderate), whereas the rest of the layers, including Layer VI, are mild and diffusely stained. In SE preparations, Layers I, II, III, and VI are all diffusely moderately stained. Layers IV and V show much more (moderately strong) positive activity than the other layers.

In MAO preparations, the alveus (Layer VI) gives the strongest (moderately strong) positive activity compared to other layers, whereas Layers I, II, III, and V are moderately diffusely stained. Layer IV shows negligible activity. The blood vessels show AK-positive activity. They are found uniformly distributed in Layer IV, but there are fewer blood vessels in the other layers.

The ependymal layer lining here appears to be in no way different from ependyma in other parts of the lateral ventricle, and hence has been discussed in Chapter XII.

Subicular area

This area extends between the Cornua Ammonis and hippocampal gyrus. It is divisible into *prosubiculum, subiculum, presubiculum,* and *parasubiculum.* The prosubiculum and subiculum areas are five-layered, whereas the presubiculum and parasubicular areas are six-layered (Lorente de No, 1934). In histochemical studies it is quite difficult to differentiate clearly between the layers.

The subiculum is made up of (1) *plexiform layer,* (2) *the layer of islands of small pyramids,* (3) *the layer of large pyramids,* (4) *the layer of small, deep pyramids,* and (5) *the layer of polymorph cells.*

Histochemically, the subicular area can be described under three layers: (a) superficial layers made up of a plexiform layer, (b) middle layers made up of large and small pyramidal cells, and (c) deeper layers made up of polymorph cells.

The superficial layer contains numerous AChE-positive fibers, which are less common in other layers. The neurons in the middle layer show mild activity, whereas those in the deeper layer show very little activity. All the layers show a number of positive fibers running through them.

In MAO preparations, the neuropil in these layers is moderately and diffusely stained. The neuropil of the inner part of the deeper layer shows considerably less activity. The neurons show mildly positive activity.

In SDH and LDH preparations, the superficial layer shows less activity than the middle and deeper ones. The neurons in the middle layer show moderately strong LDH activity and moderate SDH activity. The polymorph cells in the deeper layer also show moderate LDH-positive and mild SDH-positive activity. The neuropil in general shows moderately positive activity.

In AC preparations, the superficial layer and the inner part of the deeper layer show mildly positive activity. The neuropil in the middle layer and the upper part of the deeper layer are moderately positive. The neurons in the middle layer are moderately strong. Small neurons are mildly positive, and the enzyme activity does not extend into cell processes, or at best continues for only a short distance.

Preparations of SE show much more positive activity in both neuropil and neurons. Neurons in the middle layer are strongly positive, and those in the deeper one are moderate. The superficial layer and the neuropil in the rest of the layers show moderately positive activity.

In ATPase preparations, the uppermost part of the superficial layer shows the strongest positive activity. The neuropil in the middle and deep layers is moderately positive, whereas the neurons are mild.

The blood vessels are positive in AK, and all the layers seem to have almost similar vascularity.

Gyrus hippocampus

This area extends between the parasubiculum and rhinal fissure, and is also called the piriform cortex. In the rhesus monkey, it is well developed and is differentiated into six layers: (1) *plexiform layer* containing mostly nerve fibers with a few small neurons, (2) *layer of stellate cells,* (3) *layer of superficial pyramids* made up of superficial pyramidal cells of deeper plexiform layer, (4) *layer of horizontal spindle cells* or deep pyramidal cell layer, (5) *layer of small pyramids,* and (6) *layer of polymorph and spindle-shaped cells.* The histochemical reactions are summarized below:

AC. The neuropil is mildly positive in all the layers, with less activity shown in Layer I than in the others. The larger neurons are strongly positive compared to the mild to moderate activity observed in small- and medium-sized cells.

AK. Small blood vessels show AK-positive activity. The distribution of blood vessels in this area is similar to other areas of the cortex.

ATPase. The superficial part of Layer I shows moderately strong positive activity, whereas the other layers are moderate. The neurons show mild to moderate positive activity in the cytoplasm as well as in the nucleoli. A number of blood vessels, especially the large ones, are also ATPase-positive.

SE. The neuropil in Layer I is mildly positive, whereas in other layers it is moderate. The neurons show variable activity depending on their size, the larger neurons showing stronger activity than the smaller ones (moderate). The neurons in Layers I and VI, however, are considerably less active than those of the other layers.

AChE. It is difficult to discern the different layers in the AChE preparations, because all the layers contain moderately positive nerve fibers running in various directions. The layer of polymorph and spindle cells has the least number of positive fibers compared to the other five layers. The pyramidal cells show mild AChE activity compared to less activity in the cells of Layers I and VI.

MAO. The neuropil in all the layers except Layer VI is moderately positive, whereas Layer VI shows less positive activity. The nerve cells show mild to moderate activity in their cytoplasm.

SDH. The neuropil of all six layers shows moderate activity. The neurons in pyramidal cell layers show stronger reaction than other cells, particularly in the case of larger neurons. The positive activity also extends into the cell processes. The deepest part of Layer VI shows considerably less SDH activity.

LDH. The neuropil in Layers I and II seems to show considerably more enzyme activity (very strong) than in other layers, which are moderately strong. The perikarya show moderate to moderately strong positive activity in all the layers except Layer VI, which shows mild to moderate activity. In the pyramidal cell layers, very strongly positive neurons are also observed occasionally, in which case the enzymatic activity extends for a considerable distance in the cell processes.

TUBERCULUM OLFACTORIUM

The tuberculum olfactorium (TO) of the rhesus monkey has been studied by Lauer (1945). It is bounded laterally by the piriform cortex and dorsally by the caudate nucleus. On its medial side, it is related to the nucleus medialis septi and the fasciculus diagonalis band of Broca. In some places, its boundary is partially marked by the small islands of Calleja. The latter are also observed in the innermost layer of the TO, and the largest island of Calleja lies medial to the anterior end of the TO in the rhesus monkey. In most of the histochemical preparations, the three layers of the tuberculum olfactorium—corticalis, pyramidalis, and multiformis—are easily recognizable due to their varying enzyme reactions, and correspond to the plexiform, pyramidal, and polymorph layers of Lauer (1945). The cortical layer is the outermost layer and consists mostly of small cells. The pyramidal layer is composed predominantly of small granule cells and some small- and medium-sized pyramidal cells. In some places, particularly on the medial side, clusters of these cells project into the inner layer (Lauer, 1945). The multiformis layer consists of pyramidal cells as well as the characteristic medium- and large-sized neurons.

Histochemically, the neurons of the cortical layer show moderate to moderately strong LDH and SE; moderate SDH, ATPase, AChE, and MAO; mild AC; and negative AK activity. The neuropil is moderate in LDH, ATPase, and SE; moderate to moderately strong in MAO; and mild in SDH, AC, and AChE preparations. The pyramidal layer shows strong LDH, SDH, AC, and SE; moderately strong ATPase and AChE; and mild to moderate MAO activity in the neurons, whereas the neuropil reaction is moderately strong for LDH and SE; moderate for SDH, ATPase, and AChE; and mild for AC and MAO. The multiformis layer shows the same reaction as the pyramidal layer in LDH and SDH, and the enzyme-positive material is observed in the cell processes as well. The cells are moderately strong in ATPase, AC, and SE; moderate in AChE; and mild in MAO preparations. The neuropil is also moderately strong in LDH; moderate in SDH, ATPase, and SE; and mild in AC preparations. In the AK preparations, only blood vessels are positive, and the cortical layer is the least vascular of the three layers.

AMYGDALOID COMPLEX

Detailed study of this area has been made in the macaque by Lauer (1945). Sterotaxic studies on the brain of squirrel, java, and cebus monkeys and the tree shrew indicate that all these animals, including the macaque and human, have similar nuclear subdivisions in the amygdaloid complex (Emmers and Akert, 1963; Shantha *et al.*, 1968; Manocha *et al.*, 1968; Tigges and Shantha, 1969). The amygdaloid complex has been divided into the following structures.

Anterior amygdaloid area (AAA)

This area is situated between the claustrum laterally and the cortex lateroanteriorly. It continues to increase in size as it is traced posteriorly, where it is situated between the piriform cortex and the substriatal gray. This nucleus is composed mostly of small polymorph cells, with some medium-sized neurons in the dorsal region of the cephalic end. The neurons show strong LDH; moderately strong AC, SE, and SDH; mild to moderate ATPase; mild AChE and MAO; and negative AK activity. The neuropil shows moderate LDH, SDH, ATPase, and MAO; mild to moderate SE; mild AC and AChE; and negative AK activity.

Nucleus centralis amygdalae (CEA)

This nucleus is caudal to the middle of the amygdaloid complex and extends in the caudal direction. The cells are somewhat larger than those of the nucleus medialis amygdalae described below. Histochemically, this area shows slightly stronger enzyme activity than the AAA.

Nucleus intercalatus amygdalae (ITA)

This nucleus is situated between the AAA and area claustralis amygdalae anteriorly and between the nucleus centralis amygdalae (CEA) and the commissura anterior posteriorly. It is made up of small neurons condensed into small masses. Histochemically, the ITA is similar to the AAA except that the ITA shows somewhat lesser degree of enzyme reactions.

Area claustralis amygdalae (ACA)

The ACA is situated between the ITA and claustrum in the anterior part of the amygdaloid complex. It is made up of medium-sized neurons, which are fairly compactly arranged. Histochemically, the ACA shows stronger positive activity than the AAA, and is somewhat similar to the nuclei basalis accessorius and medialis amygdalae.

Nucleus medialis amygdalae (MA)

This is a polyhedral mass of cells medial to, and extending much farther caudally than, the AAA. Posteriorly this nucleus becomes thin and lies along the optic tract. It consists predominantly of small- and medium-sized neurons with a central core of denser, darkly staining neurons. The neurons of this nucleus give strong LDH, AC, and SE; moderately strong SDH; mild to moderate ATPase, MAO, and AChE; and negative AK activity. The neuropil shows moderately strong LDH; moderate SE, AC, SDH, ATPase, AChE, and MAO; and negative AK activity.

Nucleus corticalis amygdalae (CTA)

This nucleus is situated superficially on the most medial part of the amygdaloid complex. It extends to the same level as the accessory basal nucleus. It is likewise made up of small- and medium-sized neurons and gives almost the same type of histochemical activity as the MA.

Nucleus basalis accessorius (BAA)

The BAA is divided into a medial and lateral part [nucleus basalis accessorius medialis (BAM) and nucleus basalis accessorius lateralis (BAL)] towards its caudal end. The cells of BAL are larger than those of BAM. This area gives almost the same type of histochemical activity as the MA. The neurons in the BAL group show slightly more enzymatic activity than the medial group; on the other hand, the neuropil in the BAM shows more enzyme activity than the neuropil in the BAL.

Nucleus basalis amygdalae (BA)

The BA is situated in the basal part of the amygdaloid complex and extends almost throughout the corpus amygdaloidum. Like the BAA, it is divided into nuclei basalis lateralis (BLA) and medialis amygdalae (BMA). The BLA is composed of large neurons, while the medial part is composed of small cells. The cellular aggregation seems to be fairly uniform. The BLA has the largest neurons of any nucleus in the amygdaloid complex, and the cells are multipolar in nature. Histochemically, the BLA stands out distinctly compared to the BMA and shows more enzyme activity than any other nucleus. The neurons of the BLA show strong LDH, AC, and SE; moderately strong SDH; moderate ATPase; mild to moderate AChE and MAO; and negative AK activity. The neuropil shows strong LDH; moderately strong SE and SDH; moderate AC, MAO, AChE, and ATPase; and negative AK activity. The glial cells show moderately strong SE and SDH, strong LDH, moderate ATPase and MAO, and mild AC activity. The blood vessels are rich in AK and AChE, and this area is more vascular than other parts of the amygdaloid complex.

Nucleus lateralis amygdalae (LA)

This is the most lateral nucleus in the amygdaloid complex and is situated in line with the claustrum. Somewhat sickle-shaped, it is present throughout the length of the amygdaloid complex. The LA is separated laterally from the surrounding cortical areas by nerve fiber bundles, while medially it is in contact with the BLA. It is mainly composed of small neurons with a few medium-sized cells, rather densely arranged. Histochemically the LA is similar to the CEA.

CLAUSTRUM

The physiological functions of the claustrum are unknown, but anatomically and phylogenetically, it is said to be related to the cortex of the sulcus rhinalis. The CL is located lateral to the putamen and between the external and extreme capsules. It consists of small- and medium-sized cells with a predominance of the latter. The neurons are mostly triangular, fusiform, or pyramidal. Berke (1960) observed that the fusiform cells in the claustrum of the rhesus monkey are arranged with their maximum diameter in a vertical plane. The neurons are rich in LDH, giving a strongly positive reaction; they show moderate to moderately strong SDH, SE, ATPase, and AC, mild to moderate AChE, mild MAO, and a negative AK reaction. In the latter, only the blood vessels are positive, and their number is much less than in the adjoining putamen. The neuropil shows moderately strong LDH; moderate ATPase; mild SDH, SE, AC, and MAO; and negligible AChE activity. The glial cells also show moderate LDH, SE, and ATPase; mild AC activity; and negligible to negative activity in the other enzyme preparations. Similar observations have been made in the squirrel monkey by Manocha and Bourne (1967) and Manocha *et al.* (1967).

EXTERNAL AND EXTREME CAPSULE AND CAPSULA INTERNA

The fibers of the external (CE) and extreme capsule (CET) are found on either side of the claustrum (CL). The former separates the CL from the putamen, and the latter separates it from the insular cortex. In the rhesus monkey, the CE is composed primarily of corticotegmental fibers and a few cortical association fibers, whereas the capsula extrema contains largely the fibers of cortical association which interconnect the frontal, insular, and temporal cortices (Berke, 1960). The capsula interna (CI) is located between the nuclei caudatus and lentiformis and the thalamus. It is continuous caudally with the pes pedunculi and cephalically with the corona radiata (Riley, 1960). Histochemically, the CE and CET give mild and occasionally moderate LDH reactions, whereas

TABLE I[a]

HISTOCHEMICAL REACTIONS OF THE VARIOUS FIBER SYSTEMS IN THE DIENCEPHALON IN THE RHESUS MONKEY BRAIN

	LDH[b]		SDH		AC		ATPase[c]		AK[c]		SE		AChE		MAO	
	F[d]	G[d]	F	G	F	G	F	G	F	G	F	G	F	G	F	G
Capsula Externa (CE)	+ → ++	++±	±	±	±	±	++	++	−	−	±	++	±	−	+	−
Capsula Extrema (CET)	+ → ++	++	±	±	±	±	++	++	−	−	±	++	±	−	+	−
Commissura Anterior (CA)	+	++	±	±	±	±	++	++	−	−	±	++	±	−	±	−
Capsula Interna (CI)	+	++±	+	+	+	±	++	++	−	−	±	+	±	−	+	−
Lamina Medullaris Extrema Thalami (LME)	+	++±	±	±	± → +	±	+ → ++	+++	−	−	+	++	++±	±	+	−
Lamina Medullaris Interna Thalami (LMI)	± → +	++±	±	±	+	+	++	++	−	−	+	+ → ++	++	±	++	−
Stria Medullaris Thalami (SM)	+	++ → ++±	±	±	±	±	+++	++ → ++±	−	−	+	+ → ++	±	−	+	±
Corpus Callosum (CC)	+	++ → ++±	+	±	+	±	++	++	−	−	++	++	+	−	+	+

Fornix (F) and Fimbria Hippocampi (FH)	++	++± → +++	+	±	+	±	++±	++±	−	−	++	+	±	−	++	+
Optic Tract (TO) and Optic Chiasma (CHO)	++	+++	±	±	+	+	++	++±	−	−	+	++ → ++±	±	−	+ → ++	± → +
Tractus Retroflexus Meynert (TMT)	++	++±	±	±	+	±	++	++	−	−	+	++	±	±	++±	+ → ++
Commissura Posterior (CP)	++	+++	± → +	+	+	±	++	++	−	−	+	++	±	±	++±	++
Commissura Habenularis (CH)	++	+++	+	+	+	±	++	++ → ++±	−	−	+	++	±	±	+++	++

[a] − = negative reaction; ± = negligible activity; + = mild activity; ++ = moderate activity; ++± = moderately strong activity; +++ = strong activity; → = to.

[b] There is a larger number of positive glial elements in the LDH preparation than is seen in other enzyme preparations. This indicates that probably all glial cells react with LDH, whereas in others such as SE, only astrocytes seem to be positive.

[c] ATPase and AK preparations show moderately strong to strongly positive blood vessels in the various fiber systems.

[d] F = Fibers; G = Glial Cells.

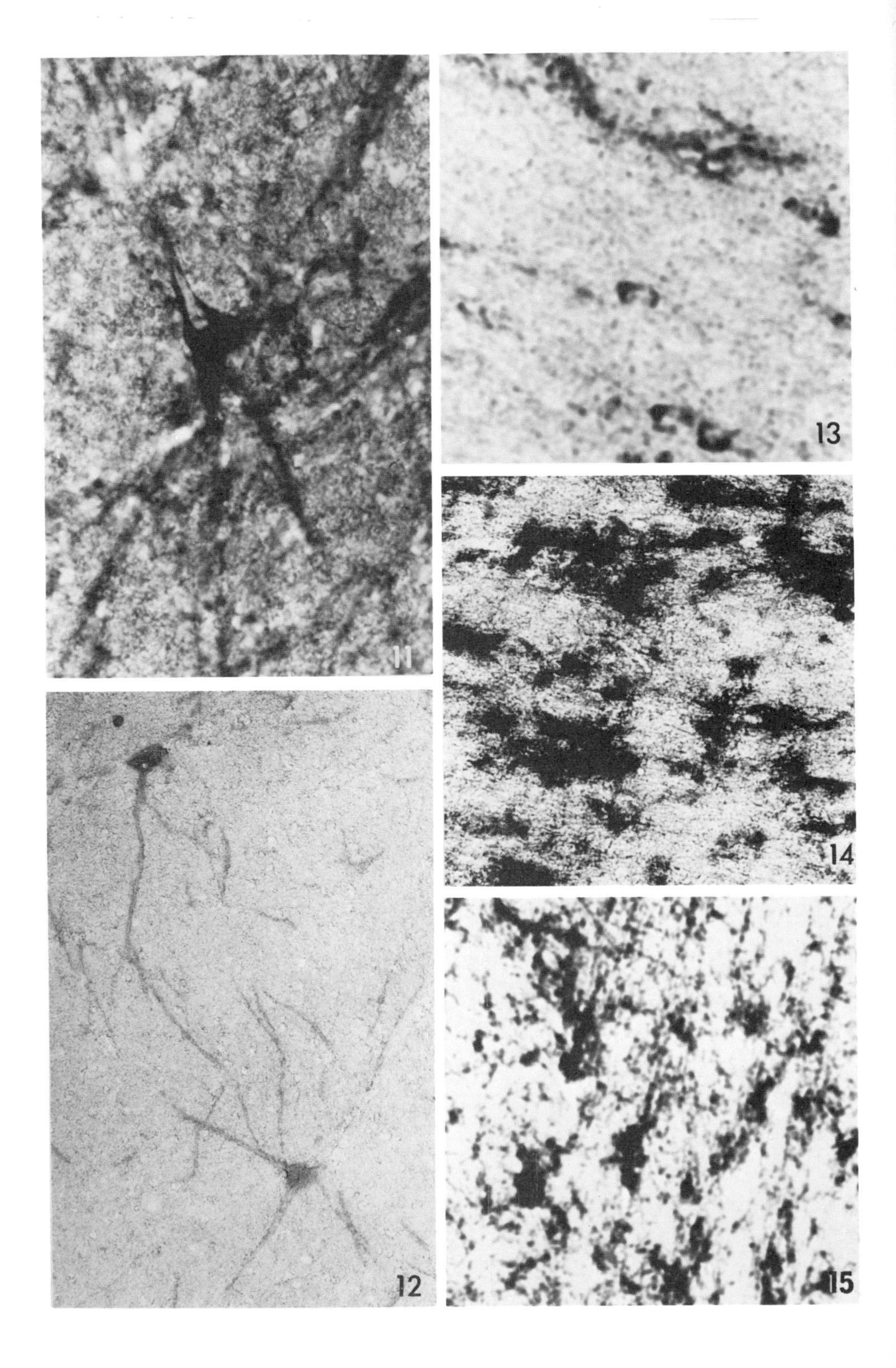
11
12
13
14
15

FIGURES 11, 12. Anterior part of the capsula interna in the vicinity of the globus pallidus showing LDH and AChE activity, respectively, in some of the scattered neurons found in this area. Note the enzyme activity extends into the cell processes for a long distance. ×592, ×148.

FIGURES 13, 15. LDH and ATPase activity in the optic chiasma showing strongly positive glial cells. The ATPase preparation shows, in addition, strong activity in the blood vessels. ×592, ×370.

FIGURE 14. ATPase reaction in the corpus callosum showing strongly positive blood vessels and glial cells. ×370.

the CI fibers are negligible. All these fibers, however, show negligible activity in SDH, AC, and SE preparations and are negative for AK. The CE, CET, and CI show a large number of fibers positive for AChE. In SE and SDH, the small neurons of the CE and CET are moderately and mildly stained. In LDH the glial cells are moderately strong. In MAO, the CE, CET, and CI fibers are mild and the glial cells negligible. In ATPase the fibers and the glial cells are moderate and the blood vessels strong. The latter are AK-positive as well. The above observations are similar to those in the squirrel monkey except that in the latter the MAO preparation seems to show a stronger enzyme reaction in the CE and CET than in the rhesus monkey (Manocha and Bourne, 1967; Manocha *et al.*, 1967). Table I outlines the histochemical reactions of some prominent fiber systems in the rhesus monkey.

GENERAL DISCUSSION

OXIDATIVE ENZYMES AND MONOAMINE OXIDASE

The histochemical reactions in the cerebral cortex of the rhesus monkey are essentially similar to those described in other animals. A detailed study has been made on the distribution of SDH in the cerebral cortex of the guinea pig (Friede, 1960), rat, and squirrel monkey (Shantha *et al.*, 1968). The isocortex showed SDH in the upper three layers of the neuropil, particularly the second and third, and in the cell bodies of the fifth and sixth layers. The cingular cortex failed to show histochemically distinguishable laminae (Friede, 1960). In the human cerebral cortex, similar gradations of SDH activity have been observed. The temporal cortex showed weak enzyme activity, having less than the frontal and parietal cortices, while highest activity was observed in the occipital and postcentral cortices. The frontopolar region showed strong enzyme activity in the human being, but weak activity in the guinea pig (Friede and Fleming, 1962). A quantitative study by Robins *et al.* (1956a, 1956b, 1956c) showed similar results. Robins *et al.* showed that the distribution of malic, lactic, and glutamic dehydrogenase in the cerebral cortex in the human is similar to that found in the monkey. Highest activity for MDH, LDH, and AD was in the upper cortical layers with a progressive decline in the enzyme activity in the deeper layers. Other studies on the isocortex of various animals showed similar results regarding the distribution of SDH, CYO, LDH, NAD, NADP diaphorase, and G6PD (Campbell, 1939; Meath and Pope, 1950; Neumann and Koch, 1953; Kuhlman and Lowry, 1956; Shimizu and Morikawa, 1957; Grimmer, 1960; Schiffer and Vesco, 1962; Felgenhauer and Stammler, 1962; Busnuk, 1963; Friede and Fleming, 1963). In the rhesus monkey and some insectivores, rodents, and carnivores, a certain level of SDH activity is typical of each cortical layer, the highest activity being in Layer III of the motor area and Layer IV of the optic area. A mildly diffuse SDH reaction has been observed in the neuropil of the plexiform layer. The underlying layers, particularly the second, third, and fourth, contain granular and small pyramidal cells and show a mild reaction for SDH in the neuropil, except for the fascia dentata, where the granule cells react strongly. Generally the neurons are distinguishable from the neuropil by their milder enzyme reaction. A few cells, however, show stronger SDH activity than the neuropil and come in the category of hyperchromic cells. The ganglionic layer shows a contrast to the upper layers because of the stronger SDH reaction in the perikarya than in the neuropil. When compared to other cell types, the giant pyramidal cells show the strongest SDH reaction (Shantha *et al.*, 1968). The DPN-diaphorase preparations show reactions similar to SDH. The synapses, nerve processes, and perikarya of the pyramidal and granule cells give strong DPN-diaphorase-positive activity.

In Ammon's horn strong reactions for SDH, CYO, and LDH have also been observed in the molecular layer and the pyramidal cells (Shimizu and Morikawa, 1957; Friede, 1960; Friede and Fleming, 1963), compared to an intense G6PD reaction in the pyramidal cells and a weak reaction in the molecular layer (Abe *et al.*, 1963). In the fascia dentata, the molecular layer can be differentiated into two laminae; the SDH and CYO show a strong reaction in the outer and a moderately strong reaction in the inner lamina, in contrast to a weak G6PD reaction in these layers. Kuhlman and Lowry (1956) showed that Ammon's horn is also rich in glutamic dehydrogenase. The pyramidal cells, on the other hand, show negligible SDH and CYO activity and moderate G6PD activity (Felgenhauer and Stammler, 1962; Friede *et al.*, 1963). Friede and Fleming (1963) observed that the pyramidal cells of the fascia dentata showed no SDH reaction. On the other hand, our studies indicate that they are strongly LDH- and moderately SDH-positive.

The higher activity of the glycolytic enzymes compared to the enzymes of the citric acid cycle only confirms the well-known fact that the brain uses carbohydrates as its chief substance for metabolic activities. The activity of succinic and malic dehydrogenase and cytochrome oxidase is located more in the axons and dendrites, whereas lactic dehydrogenase and glucose-6-phosphate dehydrogenase (Friede *et al.*, 1963) are more active in the perikarya. This is in agreement with the view of Friede, Adams, and other workers that aerobic respiration is carried out by neuronal processes and the perikarya are mainly responsible for glycolytic activity. Robins *et al.* (1956a, 1956b, 1956c) demonstrated an equal ratio between the activity of enzymes intermediating aerobic and anaerobic metabolism in all the layers of the cortex.

MAO is localized mainly in the form of strongly positive areas on the peripheries of cell processes, particularly in the synaptic region. The plexiform layer shows strong activity for MAO and SDH, including the spherules of the collateral branches of the Cajal cells. The granule cells and the pyramidal cells show activity of AChE on these sites (Shanthaveerappa *et al.*, 1963). These observations together with those of De Robertis *et al.* (1961) suggest that there are both cholinergic and adrenergic types of synapses in the cerebral cortex. Hydén (1967) and his co-workers have done extensive study on the neuron-glial relationship. In a biochemical sense, the nerve cells and the glia are interlocked in an energetic system, the two units of which can swing between two positions (Hydén, 1967). Hydén and Lange (1964) showed that SDH activity is high in neurons and low in the glial cells during sleep, whereas in wakefulness the enzyme picture is reversed. In the present study, most of the oligodendroglia give a mild to moderate SDH and strong LDH reaction, thus the oligodendroglia are involved in anaerobic glycolysis in addition to the pentose monophosphate shunt metabolism. Friede *et al.* (1963) explained that since oligodendroglia have a role in myelination, ribose is produced for synthesis of nucleic acids, while $NADPH_2$, required for the synthesis of fatty acids, is formed by the reduction of coenzyme by NADP-dependent enzymes of the shunt.

The biochemistry and histochemistry of the brain in relation to myelination have been reviewed by Adams and Davison (1965). Wender and Kozik (1967) demonstrated that the myelination gliosis parallels the increase of activity of AC, ATPase, TPPase, and cholinesterase. Glial cells showing respective enzyme activity are aggregated in clusters, which, according to Wender and Kozik (1967), is a general characteristic for the glial proliferation in the early period of myelination gliosis. Friede (1962) showed earlier that the proliferation of glial cells connected with myelination is under rigid control and the axons represent the essential organizers, controlling glial proliferation. The astrocytes contain much smaller amounts of active oxidative enzymes than do the oligodendroglia (Wallace *et al.*, 1963; Adams, 1965; Friede, 1966). Friede observed moderate and diffuse activity of LDH, NAD-linked α-glycerophosphate dehydrogenase, $NADPH_2$-tetrazolium reductase, and NADP-linked dehydrogenase in the protoplasmic astrocytes throughout the gray matter of human and rat brain. The above results are obvious because normal astrocytes are believed to contain very few mitochondria (Luse, 1956), and most of the DPN-linked diaphorase activity in the neuropil can, therefore, be attributed to the processes of oligodendroglia (Friede, 1962; Adams, 1965). Ishii and Friede (1967) concluded that the glial cells which show marked activity of NAD diaphorase are the oligodendroglial cells.

The activity of oxidative enzymes is much lower in the white matter than in the gray matter. Friede (1961) showed an inverse relationship in the distribution of DPN diaphorase in the axons and glial cells in the white matter. Tracts with marked activity in the axons did not show glial cells with any significant activity, and *vice versa.* Oligodendroglia in the white matter of the rhesus monkey brain show a marked degree of LDH but very little SDH activity. These glial cells are characterized by the presence of elongate processes, and a diffuse enzyme activity is observed for long distances into these processes. Ishii and Friede (1967) stated that these glial cells also show marked activity of G-6-P dehydrogenase but little of cytochrome oxidase. Potanos *et al.* (1959) described the oligodendrogliomas as showing a high rate of oxygen consumption in normal and neoplastic tissues. The astrocytes do not seem to show any marked activity of oxidative enzymes.

CHOLINESTERASES

The majority of the neurons of the cerebral cortex show negligible AChE activity. The pyramidal cells and Betz cells show a mildly positive reaction, except for a few scattered neurons which are moderately positive. A large number of fine fibers also react positively, especially in the fiber layer, which shows the strongest activity of all the layers. In general, AChE activity varies from layer to layer and gradually decreases from the first to the seventh layer. The studies of Koelle (1954,

1963), Gerebtzoff (1959), Goldberger (1961), Shanthaveerappa *et al.* (1963), and Shantha *et al.* (1968) on AChE activity in the cerebral cortex of various other animals showed that the enzyme was localized in the plexiform layer, pyramidal cell layer, granule cell layer, and lamina multiformis in a decreasing order. The surface of cell membranes and their proximal processes were the main sites of AChE activity. The cytoplasm of most neurons was generally negative, although Goldberger (1961) described an AChE reaction in the cytoplasm of all the pyramidal cells. The Betz cells in the precentral cortex also showed a strong AChE reaction (Okinaka *et al.*, 1961; Foldes *et al.*, 1962). Krnjevic and Silver (1963a, 1963b, 1965), in the study of cat and rabbit cerebral cortex, showed that in the deep pyramidal cells of sensory and motor cortex, the AChE activity was present in the cytoplasm of the cells or in the form of small granules at the cell membrane. Pope *et al.* (1952) and Okinaka *et al.* (1961) described the distribution of AChE in the individual layers and concluded that the strong activity in the first layer is primarily due to the reaction of the neuropil.

Krnjevic and Silver (1965, 1966) described a tangential system of fine fibers whose terminal network was closely related to the deep pyramidal cells. This network provided a cholinergic innervation for these cells. Feldberg and Vogt (1948) showed the presence of cholinergic as well as noncholinergic neurons in the individual areas of the cortex. According to these workers, the pyramidal tract (upper motor neurons) was noncholinergic in nature, and the cholinacetylase content of this area has been derived from the cholinergic neurons converging on the pyramidal cells. Burgen and Chipman (1951) earlier showed clearly a parallelism between acetylcholine, choline acetylase, and cholinesterase in most regions of the dog cerebral cortex.

The rhinencephalon (paleocortex and archicortex) shows very high activity of AChE compared to the neocortical areas (Girgis, 1967). Gerebtzoff (1959) observed that the area retrosubicularis shows high AChE activity and constitutes a transition zone between the neocortex of the limbic lobe and the paleocortex of the hippocampal gyrus.

In a detailed study of the AChE content in the developing forebrain, Krnjevic and Silver (1966) suggested, on the basis of the complete absence of AChE in the embryonic cortical layers, that the cerebral cortex is developed from tissue which has no AChE activity and which may not give rise to cholinergic elements. The AChE-containing fibers of the adult cortex, therefore, have their origin in other areas of the brain, from which they travel into the pallium during development. Also interesting is a study by Bennett *et al.* (1966) on the weight measurements and the AChE activity of the rat brain. These workers showed that there is no difference in the AChE content of the brain of 100- and 150-day-old rats in spite of a significant gain in weight.

In the fascia dentata, fine fibers and the inner half of the stratum moleculare show moderately strong reactions. The activity in the stratum granulosa and stratum polymorpha is generally mild. In Ammon's horn the activity varies in the seven layers described in the text, with strongest activity in the fifth layer. In the subicular area, the middle layer of large and small cells shows the greatest activity. In Ammon's horn of the rabbit, Lowry *et al.* (1954) described maximum enzyme activity in the molecular layer and in the layer of pyramidal cells, indicating that AChE was present in both the fiber plexus of the neuropil and the perikarya. The distribution of AChE in the hippocampal region of the rat has been studied in detail by Mathisen and Blackstad (1964). These authors have shown that the layers containing terminal commissural fibers are poor in AChE, whereas enzymes are abundant in zones receiving cingulum fibers, or where axonal arborization of basket cells form axosomatic synapses. The studies of Shute and Lewis (1961, 1963, 1967) show that the hippocampus of the rat received large numbers of cholinergic fibers via the fimbria. They believed that these fibers formed a part of a massive AChE-containing component of the ascending reticular activating system and may be cholinergic in nature.

The amygdaloid nuclei show variations in the enzyme content varying from negligible to mild and moderate. The nuclei basalis and basalis accessorius show the maximum activity. De Giacomo (1962) and Ishii and Friede (1967) showed very high AChE activity in the medial and lateral part of the basal amygdaloid nucleus including all the extensions and very little activity in the lateral amygdaloid nucleus. The central amygdaloid nucleus is mild in our preparations. It also gives a positive reaction in the rabbit (Gerebtzoff, 1959), cat (Krnjevic and Silver, 1965), and coypu (Girgis, 1967). Ishii (1957) and Shute and Lewis (1961), however, observed a negative reaction in the central amygdaloid nucleus in the rat brain.

BChE activity in the brain of man, rhesus monkey, cat, and rat has been observed in the oligodendroglial cells of white matter, in axons, and in nerve cells in the gray matter (Friede, 1967). The regional variations in the distribution of BChE activity are not as marked as in the AChE content (Ord and Thompson, 1952; Cavanagh *et al.*, 1954). There is considerable agreement among the various workers that comparatively high levels of AChE occur in the white matter and that the glial cells are the major repositories. Okinaka *et al.* (1961) showed that those regions which show strong AChE activity also contain fair quantities of BChE.

BChE has been correlated with metabolic processes of the myelin sheaths of neuroglia (Ord and Thompson, 1952).

The simple esterases are found in the perikarya of all the neurons, whereas the neuropil shows mild activity. The positive activity in the pyramidal cells also extends into the dendritic processes for some distance. The activity is diffuse as well as granular in nature.

PHOSPHATASES

While the neurons and the axonal and dendritic processes are negative in the AK preparations, the synaptic regions are positive both on the cell surface and on the cell processes in the form of granular deposits in the cerebral cortex. Nandy and Bourne (1963) and Shantha *et al.* (1968) showed AK activity in the nucleolus, nucleus, and cytoplasm of the neurons. The plexiform layer gives a more prominent reaction than the other layers of the cortex. Bourne (1957) showed that the ground substance of the cerebral cortex exhibited consistent AK activity. It is stronger with glycerophosphate and carboxy-phenyl-phosphate and weaker with pyridoxal phosphate; there is no diffusion into the strikingly negative pyramidal cells and their processes. The blood vessels are strongly reactive. The distribution of AK in the vessels varies among different species. Enzyme activity is consistent in the intima, whereas the adventitia showed variable enzyme activity in various species (Shimizu, 1950). A detailed study by Bannister and Romanul (1963) showed an intense AK activity in the endothelium of small arteries at their origin from larger ones. The AK activity is continuous from the origin of a small artery to a capillary. These authors also observed that more intense AK activity existed in the walls of the unperfused small cerebral arteries than in the perfused arteries, which may either be due to greater diffusion of phosphatase from small arteries during perfusion or some of the activity formed here may have been due to retained blood (Bannister and Romanul, 1963). Naidoo (1963) believed that the AK activity in the lipid phagocytes adjacent to the area of an injury is derived from the capillaries of the area, and it has been observed that the AK activity increases greatly at the site of damage unrelated to structural elements.

The neurons of the cerebral cortex give a strong AC reaction with pyramidal cells showing particularly strong enzyme activity. Acid phosphatase is randomly distributed in the cytoplasm, but with a concentration of activity in the beadlike structures which are probably the synapses. The strong reaction in the cytoplasm of the neurons is very prominent against a very mild background and passes into the cell processes only for a short distance. It is hard to pinpoint a correlation between the AC activity in the cytoplasm and the distribution of lysosomes, although a number of workers have indicated a direct relationship. In a histochemical study of the heavy metals in the rat brain at various ages, using the sulfide silver method, Brun and Brunk (1968) showed that the heavy metals were located in different types of lysosomes, and a diffuse background staining may be due to free metals in the cytoplasm. A marked increase in the AC activity in the neurons and macrophages in Tay–Sachs disease (Wallace *et al.*, 1964) and various types of demyelinating and lipid storage diseases have been noted and may be due to the release of enzymes from the lysosomes all over the cytoplasm. This is interesting in the context that Wallace *et al.* (1964) also showed that the activity of oxidative enzymes and ATPase in Tay–Sachs disease was displaced to peripheral and paranuclear positions whereas the AC activity was always localized throughout the perikarya along with the lipid bodies. In Ammon's horn the AC activity was extremely strong in the H2 segment (Fleischhauer, 1959) and in the granule cell layer of the fascia dentata (Friede, 1966).

Schiffer *et al.* (1967) showed the presence of AC and SE in the glial cells of human nervous tissue. In the cortical glia, particularly in the marginal astrocytes, AC activity predominates, whereas in the oligodendrocytes of the white matter, SE activity is prominent. This probably suggests a functional relationship of the oligodendroglia in the myelin metabolism. According to Schiffer *et al.*, the AC activity in the cortical glia may be related to different functional conditions. Firstly, the cortical glia, which are supposed to maintain a constant environment for the nerve cells and in the processes, may be engaged in the removal of waste products like lipofuscin. Secondly, superficial glial cells are sensitive to changes occurring in the subarachnoidal spaces and act as a barrier to the subarachnoidal fluid (Glees, 1955).

Adenosine triphosphatase activity is prominent in the capillaries and the neuropil, whereas in the neurons the activity is significant only in the cell membrane, and may be related to molecular transport across the membrane. Similar observations were reported by Becker *et al.* (1960), Novikoff *et al.* (1961), and Fahn and Côté (1968). Becker *et al.* (1960) showed that the ATPase activity of the cell membrane is specific because the membranes are negative if ADP is substituted for ATP. Novikoff (1967), however, was not clear as to whether the nucleotide phosphatase activity shown by *in situ* staining techniques is indeed a "transport ATPase" identical with or similar to that studied by biochemists. Becker *et al.* (1960) described the distribution pattern of calcium-activated ATPase and AChE as being closely parallel and implied that both enzymes are localized in the cortical plexuses of axons and dendrites (Hess and Pope, 1959) and cytologically at the surface membranes of the cell bodies and processes (Pope and Hess, 1957).

All types of pyramidal cells, including the multiform cells and also some granule cells of the cerebral cortex, showed thiamine pyrophosphatase (TPPase)-positive Golgi material in the form of a network (Shanthaveerappa and Bourne, 1965). They showed that the giant pyramidal cells exhibit a highly complex, darkly staining Golgi network which becomes progressively smaller in amount and stainability in large, medium, and small pyramidal cells, respectively. All these cells show an extension of the network into the dendritic processes. These extensions start from a darkly staining TPPase-positive "basal body" located at the junction of the dendritic processes and the neuron. The granule cells have a very simple network, and only a few of them have separate TPPase-positive vesicles and granules instead of a network. Granule cells do not show any extension of TPPase-positive material into the dendritic processes.

The neurons in the hippocampal gyrus show a TPPase-positive Golgi network (Shantha and Bourne, 1966). Similar observations are made in the neurons of Ammon's horn, but the amount, stainability, complexity of the network, and the extension of TPPase-positive strands into the cell processes vary from region to region. Neurons in the CA1 region (classification after Lorente de No, 1934) contain a thin TPPase-positive Golgi network, with a small thin strand of Golgi material extending into the apical dendrites only. Some of the cells of this region have TPPase-positive vesicles with the absence of a network. The neurons of Region CA2 contain a thick, darkly stained TPPase-positive network with a medium-length strand of Golgi material found mostly in the apical dendrites. Free vesicles and granules are also found in these cells. The neurons of Regions CA3 and CA4 have a very complex, darkly stained TPPase-positive network made of thin strands having vesicular enlargements. It is also not uncommon to see more than two thin strands of TPPase-positive Golgi material in a cell process. In some cells of this region, two to three cell processes containing TPPase-positive Golgi material are observed. The cells of the fascia dentata show discrete vesicular, granular, and comma-shaped masses of TPPase-positive material in their cell cytoplasm.

In a study on the histochemical method for cyclic 3′,5′-nucleotide phosphodiesterase (3′,5′-AMPase), Shantha *et al.* (1966) showed the presence of this enzyme in the plexiform layer, inner and outer granular cell layers, and lamina multiformis of the rabbit cerebal cortex. The enzyme was mainly localized in the neuropil and on the cell membranes, whereas the cell cytoplasm showed negligible activity. The glial cells also showed strong activity, indicating that the glial cells and the neuropil play an important role in the breakdown of 3′,5′-AMP in the cerebral cortex.

REFERENCES

Abe, T., Yamada, Y., and Hashimoto, P. H. Histochemical study of glucose-6-phosphate dehydrogenase in the brain of normal adult rat. *Med. J. Osaka Univ.* **14**:67–98 (1963).

Adams, C. W. M. (ed.) Histochemistry of the cells in the nervous system. In: *Neurohistochemistry,* pp. 253–331. Elsevier, Amsterdam, 1965.

Adams, C., and Davison, A. The myelin sheath. In: *Neurohistochemistry* (C. W. M. Adams, ed.), pp. 332–400. Elsevier, Amsterdam, 1965.

Bannister, R. G., and Romanul, F. C. A. The localization of alkaline phosphatase activity in cerebral blood vessels. *J. Neurol. Neurosurg. Psychiatry.* **26**:333 (1963).

Becker, N. H., Goldfischer, S., Shin, Woo-Yung, and Novikoff, A. B. The localization of enzyme activities in the rat brain. *J. Biophys. Biochem. Cytol.* **8**:649–663 (1960).

Bennett, E. L., Diamond, M. C., Morimoto, H., and Herbert, M. Acetylcholinesterase activity and weight measures in fifteen brain areas from six lines of rats. *J. Neurochem.* **13**:563–572 (1966).

Berke, J. J. The claustrum, the external capsule, and the extreme capsule of *Macaca mulatta. J. Comp. Neurol.* **115**:297–321 (1960).

Bourne, G. H. Phosphatase in the central nervous system. *Nature* **179**:1247–1248 (1957).

Brun, A., and Brunk, U. Histochemical study of heavy metals in the rat brain at various ages. *Acta Pathol. Microbiol. Scand.* **72**:451–452 (1968).

Burgen, A. S. V., and Chipman, L. M. Cholinesterase and succinic dehydrogenase in the central nervous system of the dog. *J. Physiol.* (*London*) **114**:296–305 (1951).

Busnuk, M. M. Ob osobennostiakh raspredeleniia suklsindgedrogenazy v kore. (On the peculiarities of succindehydrogenase distribution in the cerebral cortex of mammals.) *Zh. Vysshei. Nerv. Deyatel' im. I. P. Pavova* **13**:731–740 (1963).

Campbell, A. C. P. Variation in vascular and oxidase content in different regions of the brain of the cat. *Arch. Neurol. Psychiat.* **41**:223–242 (1939).

Cavanagh, J. B., Thompson, R. H. S., and Webster, G. R. The localization of pseudocholinesterase activity in the nervous tissue. *Quart. J. Exptl. Physiol.* **39**:185–197 (1954).

Crosby, E. C., Humphrey, T., and Lauer, E. W. *Correlative Anatomy of the Nervous System.* McMillan, New York, 1962.

de Giacomo, P. Distribution of cholinesterase activity in the human central nervous system. *Proc. Fourth Intern. Cong. Neuropath, Munich, 1961.* Vol. I, pp. 198–205. Thieme, Stuttgart, 1962.

De Robertis, E., Bellegrino de Iraldi, A., Rodriquez, L. A., and Salganicoff, F. Cholinergic and non-cholinergic nerve endings in rat brain. I. Isolation and subcellular distribution of acetylcholine and acetylcholinesterase. *J. Neurochem.* **9**:23–35 (1961).

Emmers, R., and Akert, K. *A Stereotaxic Atlas on the Brain of the Squirrel Monkey.* University of Wisconsin Press, Madison, 1963.

Fahn, S., and Côté, L. J. Regional distribution of sodium-potassium activated adenosine triphosphatase in the brain of the rhesus monkey. *J. Neurochem.* **15**:433–436 (1968).

Feldberg, W., and Vogt, M. Acetylcholine synthesis in different regions of the central nervous system. *J. Physiol.* (*London*) **107**:372–381 (1948).

Felgenhauer, K., and Stammler, A. Das Verteilungsmuster der Dehydrogenasen und Diaphorasen im Zentralnervensystem des Meerschweinchens. *Z. Zellforsch.* **58**:219–233 (1962).

Fleischhauer, K. Zur Chemoarchitektonik der Ammonsformation. *Nervenarzt.* **30**:305–309 (1959).

Foldes, F. F., Zzigmond, E. K., Foldes, V. M., and Erdos, E. G. The distribution of acetylcholinesterase and butyrylcholinesterase in the human brain. *J. Neurochem.* **9**:559–572 (1962).

Friede, R. L. Histochemical investigation on succinic dehydrogenase in the central nervous system. IV. A histochemical mapping of the cerebral cortex of the guinea pig. *J. Neurochem.* **5**:156–171 (1960).

Friede, R. L. A histochemical study of DPN-diaphorase in human white matter; with some notes on myelination. *J. Neurochem.* **8**:17–30 (1961).

Friede, R. L. A quantitative study of myelination in hydrocephalus. *J. Neuropathol. Exptl. Neurol.* **21**:645–648 (1962).

Friede, R. L., *Topographic Brain Chemistry.* Academic Press, New York and London, 1966.

Friede, R. L. A comparative histochemical mapping of the distribution of butyrylcholinesterase in the brains of four species of mammals, including man. *Acta Anat.* **66**:161–177 (1967).

Friede, R. L., and Fleming, L. M. A mapping of oxidative enzymes in the human brain. *J. Neurochem.* **9**:179–198 (1962).

Friede, R. L., and Fleming, L. M. A mapping of the distribution of lactic dehydrogenase in the brain of the rhesus monkey. *Am. J. Anat.* **113**:215–234 (1963).

Friede, R. L., Fleming, L. M., and Knoller, M. A comparative mapping of enzymes involved in hexosemonophosphate shunt and citric acid cycle in the brain. *J. Neurochem.* **10**:263–277 (1963).

Gerebtzoff, M. A. *Cholinesterases.* International Series of Monographs on Pure and Applied Biology (Division: Modern Trends in Physiological Sciences), (P. Alexander and Z. M. Bacq, eds.). Pergamon Press, New York, 1959.

Girgis, M. Distribution of cholinesterase in the basal rhinencephalic structures of the coypu (*Myocastor coypus*). *J. Comp. Neurol.* **129**:85–95 (1967).

Glees, P. *Neuroglia: Morphology and Function.* Blackwell, Oxford, 1955.

Goldberger, M. The effects of lysergic acid diethylamide (LSD-25) upon the histochemical reactions for cholinesterase in the central nervous system. *Acta Anat.* **46**:185–191 (1961).

Grimmer, W. Die Verteilung der Succinodehydrogenase-aktivität im Gehirn von *Macaca mulatta. Z. Anat. Entwicklungsgeschichte* **122**:414–440 (1960).

Hess, H. H., and Pope, A. Intralaminar distribution of adenosine-triphosphatase activity in rat cerebral cortex. *J. Neurochem.* **3**:287–299 (1959).

Hydén, H. (ed.) Dynamic aspects of the neuron-glia relationship. A study with microchemical methods. In: *The Neuron.* Elsevier, Amsterdam, 1967.

Hydén, H., and Lange, P. W. Rhythmic enzyme changes in neurons and glia during sleep and wakefulness. *Life Sci.* **3**:1215–1221 (1964).

Ishii, T., and Friede, R. L. A comparative histochemical mapping of the distribution of acetylcholinesterase and nicotinamide adenine-dinucleotide-diaphorase activities in the human brain. *Intern. Rev. Neurobiol.* **10**:231–275 (1967).

Ishii, Y. The histochemical studies of cholinesterase in the central nervous system. I. Normal distribution in rodents (in Japanese). *Arch. Histol. Jap.* **12**:587–611 (1957).

Koelle, G. B. The histochemical localization of cholinesterases in the central nervous system of the rat. *J. Comp. Neurol.* **100**:211–228 (1954).

Koelle, G. B. (ed.) Cytological distribution and physiological functions of cholinesterases. In: *Handbuch der Experimentellen Pharmakologie,* pp. 187–298. Springer-Verlag, Berlin, 1963.

Krnjevic, K., and Silver, A. The distribution of "cholinergic" fibres in the cerebral cortex. *J. Physiol.* (*London*) **168**:39P–40P (1963a).

Krnjevic, K., and Silver, A. Cholinesterase staining in the cerebral cortex. *J. Physiol.* (*London*) **165**:3P–4P (1963b).

Krnjevic, K., and Silver, A. A histochemical study of cholinergic fibers in the cerebral cortex. *J. Anat.* (*London*) **99**:711–759 (1965).

Krnjevic, K., and Silver, A. Acetylcholinesterase in the developing forebrain. *J. Anat.* (*London*) **100**:63–89 (1966).

Kuhlman, R. E., and Lowry, O. H. Quantitative histochemical changes during the development of the rat cerebral cortex. *J. Neurochem.* **1**:173–180 (1956).

Lauer, E. W. The nuclear pattern and fiber connections of certain basal telencephalic centers in the macaque. *J. Comp. Neurol.* **82**:215–254 (1945).

Lorente de No, R. Studies on the structure of the cerebral cortex. Continuation of the study of the Ammonic system. *J. Psychol. Neurol.* **46**:113–177 (1934).

Lowry, O. H., Roberts, N. R., Leiner, K. Y., Wu, M. Y., Farr, A. L., and Albers, R. W. The quantitative histochemistry of brain. III. Ammon's horn. *J. Biol. Chem.* **207**:39–49 (1954).

Luse, S. A. Electron microscope observations of the central nervous system. *J. Biophys. Biochem. Cytol.* **2**:531–541 (1956).

Manocha, S. L., and Bourne, G. H. Histochemical mapping of succinic dehydrogenase and cytochrome oxidase in the diencephalon and basal telencephalic centers of the brain of squirrel monkey (*Saimiri sciureus*). *Histochemie* **9**:300–319 (1967).

Manocha, S. L., Shantha, T. R., and Bourne, G. H. Histochemical mapping of the distribution of monoamine oxidase in the diencephalon and basal telencephalic centers of the brain of squirrel monkey (*Saimiri sciureus*). *Brain Res.* **6**:570–586 (1967).

Manocha, S. L., Shantha, T. R., and Bourne, G. H. *A Stereotaxic Atlas of the Brain of the Cebus Monkey* (*Cebus apella*). Oxford University Press, London, 1968.

Mathisen, J. S., and Blackstad, T. W. Cholinesterase in the hippocampal region. *Acta Anat.* **56**:216 (1964).

Meath, J. A., and Pope, A. Histochemical distribution of indophenol oxidase and acid phosphatase in rat cortex. *Federation. Proc.* **9**:204 (1950).

Naidoo, D. Alkaline phosphatase at the site of cerebral injury. *Acta Histochem.* **15**:182–184 (1963).

Nandy, K., and Bourne, G. H. Alkaline phosphatases in brain and spinal cord. *Nature* **200**:1216–1217 (1963).

Neumann, K. H., and Koch, G. Übersicht über die feinere Verteilung der Succinodehydrogenase in Organen und Geweben verschiedener Säugetiere, besonders des Hundes. *Z. Physiol. Chem. Hoppe-Seylers* **295**:35–61 (1953).

Novikoff, A. B. Enzyme localizations with Wachstein-Meisel procedures: real or artifact. *J. Histochem. Cytochem.* **15**:353–354 (1967).

Novikoff, A. B., Drucker, J., Shin, W., and Goldfischer, S. Further studies of the apparent adenosine triphosphatase activity of cell-membranes in formolcalcium fixed tissues. *J. Histochem. Cytochem.* **9**:434 (1961).

Okinaka, S., Yoshikawa, M., Uono, M., Muro, T., Mozai, T., Igata, T., Tanabe, H., Ueda, S., and Tomonaga, M. Distribution of cholinesterase activity in the human cerebral cortex. *Am. J. Physi. Med.* **40**:135–145 (1961).

Ord, M. G., and Thompson, R. H. S. Pseudocholinesterase activity in the central nervous system. *J. Biochem.* **51**:245–251 (1952).

Pope, A., and Hess, H. H. Cytochemistry of neurones and neuroglia. In: *Metabolism of the Nervous System* (D. Richter, ed.). Pergamon Press, London, 1957.

Pope, A., Caveness, W., and Livingston, K. E. Architectonic distribution of acetylcholinesterase in the frontal isocortex of psychotic and nonpsychotic patients. *Arch. Neurol. Psychiat.* **68**:425–443 (1952).

Potanos, J. N., Wolf, A., and Cowen, D. Cytochemical localization of oxidative enzymes in human nerve cells and neuroglia. *J. Neuropathol. Exptl. Neurol.* **18**:627–635 (1959).

Riley, H. L. *An Atlas of the Basal Ganglia, Brain Stem, and Spinal Cord.* Hafner, New York, 1960.

Robins, E., Smith, D. E., and Eydt, K. M. The quantitative histochemistry of the cerebral cortex. I. Architectonic distribution of the chemical constituents in the motor and visual cortices. *J. Neurochem.* **1**:54–67 (1956a).

Robins, E., Smith, D. E., Eydt, K. M., and McCaman, F. E. The quantitative histochemistry of the cerebral cortex. II. Architectonic distribution of nine enzymes in the motor and visual cortices. *J. Neurochem.* **1**:68–76 (1956b).

Robins, E., Smith, D. E., and Eydt, K. M. The quantitative histochemistry of the cerebral cortex. III. Analyses at 50 μ intervals compared with analyses by architectonic layers in the motor and visual cortices. *J. Neurochem.* **1**:77–83 (1956c).

Schiffer, D., and Vesco, C. Contribution to the histochemical demonstration of some dehydrogenase activities in the human nervous tissues. *Acta Neuropathol.* **2**:103–112 (1962).

Schiffer, D., Fabiani, A., and Monticone, G. F. Acid phosphatase and non-specific esterase in normal and reactive glia of human nervous tissue. A histochemical study. *Acta Neuropathol.* **9**:316–327 (1967).

Shantha, T. R., and Bourne, G. H. The thiamine pyrophosphatase technique as an indicator of morphology of the Golgi apparatus in the neurons. IV. Studies on the spinal cord, hippocampus, and trigeminal ganglion. *Acta Histochem.* **11**:337–351 (1966).

Shantha, T. R., Woods, W. D., Waitzman, M. B., and Bourne, G. H. Histochemical method for localization of cyclic 3′, 5′-nucleotide phosphodiesterase. *Histochemie* **7**:177–190 (1966).

Shantha, T. R., Manocha, S. L., and Bourne, G. H. Enzyme histochemistry of the cerebral cortex of squirrel monkey and rat. *Acta Histochem.* **30**:218–233 (1968).

Shanthaveerappa, T. R., and Bourne, G. H. The thiamine pyrophosphatase technique as an indicator of the morphology of the Golgi apparatus in the neurons. II. Studies on the cerebral cortex. *La Cellule.* **65**:201–209 (1965).

Shanthaveerappa, T. R., Nandy, K., and Bourne, G. H. Histochemical studies on the mechanism of action of the hallucinogens *d*-lysergic acid diethylamide tartrate (LSD-25) and *d*-2-bromo-lysergic acid tartrate (BOL-148) in rat brain. *Acta Neuropath.* 3:29–39 (1963).

Shimizu, N. Histochemical studies on the phosphatase of the nervous system. *J. Comp. Neurol.* **93**:201–218 (1950).

Shimizu, N., and Morikawa, N. Histochemical studies of succinic dehydrogenase of the brain of mice, rats, guinea pigs, and rabbits. *J. Histochem. Cytochem.* **5**:334–345 (1957).

Shute, C. C. D., and Lewis, P. R. The use of cholinesterase techniques combined with operative procedures to follow nervous pathways in the brain. *Bibliotheca. Anat.* **2**:34–49 (1961).

Shute, C. C. D., and Lewis, P. R. Cholinesterase-containing systems of the brain of the rat. *Nature* **199**:1160–1164 (1963).

Shute, C. C. D., and Lewis, P. R. The ascending cholinergic reticular system: neocortical, olfactory, and subcortical projections. *Brain* **90**:497–520 (1967).

Tigges, J., and Shantha, T. R. *A Stereotaxic Brain Atlas of the Tree Shrew (Tupaia glis)*. Williams & Wilkins, Baltimore, 1969.

Wallace, B. J., Volk, B. W., and Lazarus, S. S. Glial cell enzyme alterations in infantile amaurotic family idiocy (Tay-Sachs disease). *J. Neurochem.* **10**:439 (1963).

Wallace, B. J., Volk, B. W., and Lazurus, S. S. Fine structure localization of acid phosphatase activity in neurons of Tay-Sachs diseases. *J. Neuropathol. Exptl. Neurol.* **23**:676–691 (1964).

Wender, M., and Kozik, M. Histochemistry of cerebral white matter in relation to myelination of mouse brain. *J. Hirnforsch.* **10**:79–88 (1967).

IV

BASAL GANGLIA and SEPTAL AREA

BASAL GANGLIA

Nucleus caudatus and putamen

The nucleus caudatus (CD) and putamen (PUT) show similar histochemical reactions, hence a common description will suffice (Figs. 16–18, 24–27, 32, 37, 43, 46). The CD and PUT consist predominantly of small and medium-sized neurons with a few large cells. The small neurons, however, dominate the cell population. The neurons show strong LDH and SDH activity, and the reaction is somewhat variable in different neurons. A mild reaction is seen in MAO; moderate in AC; and moderately strong to strong in the ATPase, SE, and AChE preparations. The neuropil shows strong AChE; moderately strong LDH; moderate SDH, SE, and ATPase; and mild MAO and AC activity. AK reacts negatively except in the blood vessels, which show strong AK, AChE, and ATPase activity. The glial cells, which are numerous in both the CD and PUT, give a moderate LDH, mild to moderate SE, and a mild SDH reaction. The nerve fibers passing through these structures show a mild to moderate MAO and a mild to negligible AChE activity. It may be of interest that both the CD and PUT give a little stronger reaction in the area bordering the capsula interna, the significance of which is not understood at this time. The bridges of gray matter between the CD and PUT give the same histochemical reaction as the main body of either.

A high level of oxidative enzyme activity in the CD and PUT in the neurons as well as neuropil has been described in various animals (Shimizu and Morikawa, 1957; Shimizu *et al.*, 1957; Tyler, 1960; Friede and Fleming, 1963; Duckett and Pearse, 1964; Manocha and Bourne, 1967). This is in agreement with the views of Himwich and Fazekas (1941) that the nucleus caudatus has one of the highest oxygen consumptions found in brain tissue. This high level compares to a much weaker reaction in the neuropil of the globus pallidus discussed below and is also in agreement with the observation of Shimizu and Morikawa (1957), Manocha and Bourne (1967), and Manocha *et al.* (1967). The CD and PUT also show strong reaction for the phosphorylases (Adams, 1965).

The AChE activity in CD and PUT is strong, and with an increase in the incubation time, the staining becomes so dense that demarcation of its localization in the perikarya or neuropil becomes extremely difficult. In this connection Shute and Lewis (1963) suggested that short incubation time ensures high selectivity for the sites of enzyme concentration. A close examination reveals that the AChE activity is mainly localized in the neuropil of the CD and PUT. Okinanka *et al.* (1961) showed that strong cholinesterase activity is found not only in the nerve cells and glia but also in the ground substance as well as in the myelin sheaths of the nerve fibers.

High AChE activity also characterizes the nucleus ansa lenticularis and the lenticular fasciculus (Ishii and Friede, 1967). The AChE reaction in the rhesus monkey is also identical to that in the human striatum (high activity), cerebral cortex (low activity), and thalamic nuclei (intermediate reaction) [de Giacomo, 1962]. Koelle (1954), in a detailed study of AChE in the CNS of the rat, observed that the intensity of AChE activity in the correlating centers showed a marked difference in the various components. For example, the globus pallidus showed weak activity in contrast to the CD and PUT, which showed the strongest activity. Ishii (1957) showed a species difference as far as the distribution of AChE is concerned and found that the intensity of AChE activity is the highest in mice and rats, intermediate in guinea pigs, and lowest in rabbits. Marked AChE activity in the CD and PUT has also been shown by Gerebtzoff (1959), Okinaka *et al.* (1960), de Giacomo (1962), and Friede (1966). The latter estimated that the AChE activity in the CD and PUT is 29.9 and 38.9 times that in the cerebral cortex (Nachmansohn, 1939; Okinaka *et al.*, 1951, 1961; Foldes *et al.*, 1962). Ashby *et al.* (1952) estimated that the AChE activity in the caudate nucleus is 2500% greater than in the cerebrum. Gerebtzoff (1959) showed that the lenticular nucleus is one of the richest sources of AChE.

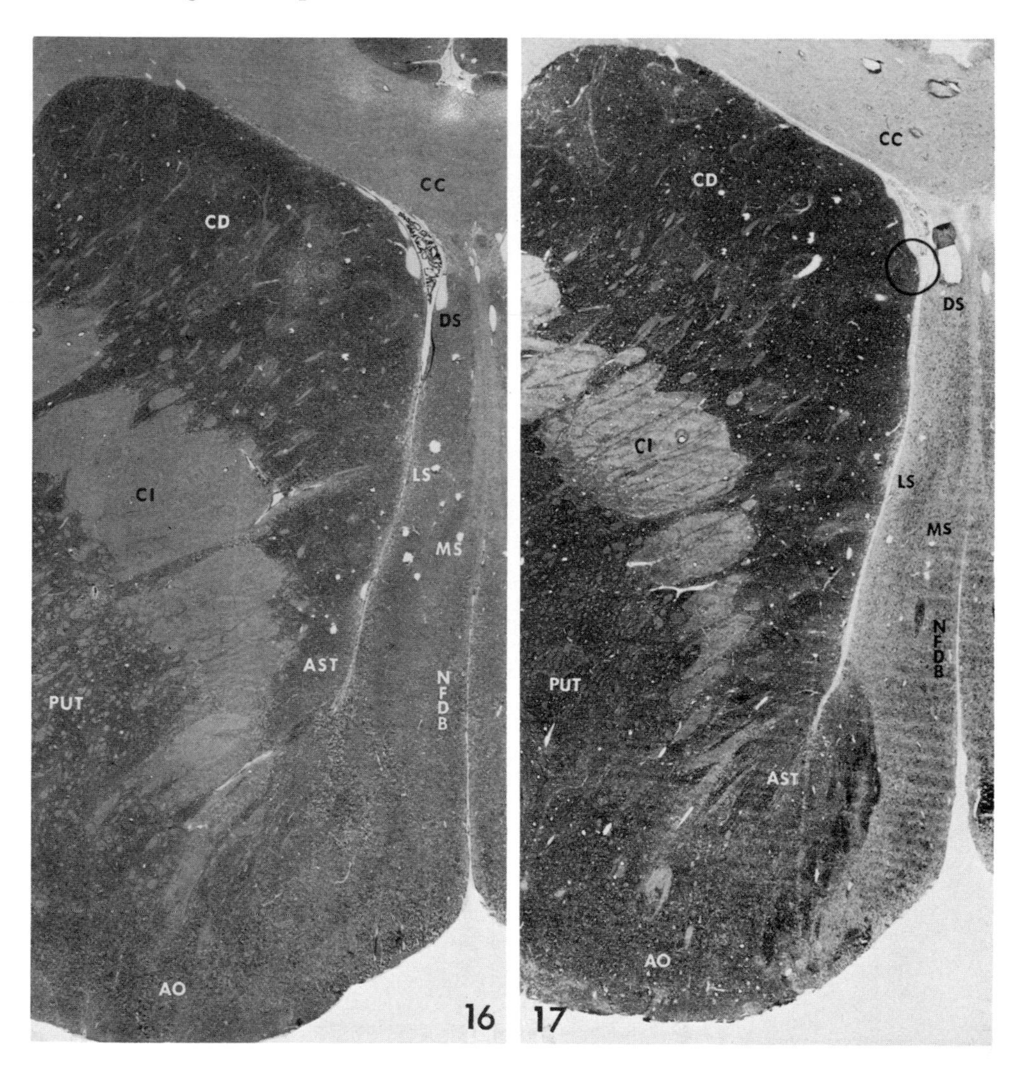

FIGURES 16, 17. LDH and SE reactions, respectively, showing variations in the activity of these enzymes in the septal and olfactory areas as well as the basal ganglia at this level. ×7, ×7.

In the developing brain, Duckett and Pearse (1967) observed that oxidative and hydrolytic enzymes are present in the analogue of human basal ganglia as early as the second month of embryonic life. This is in contrast to the activity of AChE which appears in the sixth month of prenatal life. Earlier, Bargeton–Farkas and Pearse (1965) also showed that in the growing basal ganglia of the rat, the nonspecific esterase distribution is similar to that of SDH as shown by Friede (1959).

Globus pallidus

The globus pallidus is almost triangular and is divisible into two parts, external and internal, separated from each other by the fibers of the lamina medullaris externa and interna (LGE and LGI) [Figs. 11, 12, 19, 20, 26, 27, 32, 37, 43, 46, 47]. The externa (LGE) separates the globus pallidus from the putamen. The internal part, which is located ventromedially to the fibers of the LGI, is further divided into medial and lateral parts by a few transverse fibers. Histochemically, the cells give a strong LDH and moderately strong and strong SDH reactions, and the enzyme activity passes on to the cell processes for a considerable distance. A strong cytoplasmic reaction is observed in AC and SE, moderately strong in ATPase, and moderate in AChE preparations. The AK preparations are negative, with only the blood vessels showing moderately strong activity, and the number of blood vessels is much smaller than in the nucleus caudatus and the putamen. The MAO reaction is mild in the cells, neuropil, and the cell processes of the GP. The neuropil is moderate in LDH, mild in SDH, mild to moderate in AC, SE, and AChE, and moderate in ATPase preparations. The glial cells are strong in LDH; negligible in SDH, AC, and AChE; mild in SE, and moderate in ATPase preparations. The large number of fibers passing through the globus pallidus gives a mild LDH and AC, a negligible reaction in the SDH, and negative in AK preparations. These results are similar in some respects and different in others from a number of previous studies on the globus pallidus which suggest that the GP is rich in ATPase, 5′-nucleotidase, and AChE; but poor in acid and alkaline phosphatases, glucose-6-phosphatase, glucose-6-phosphate dehydrogenase, amylophosphorylase, and monoamine oxidase (Campbell, 1939; Ishii, 1957; Shimizu and Morikawa, 1957; Shimizu and Okada, 1957; Shimizu *et al.,* 1957; Amakawa, 1959; Matunami, 1959; Yamada, 1961; Thomas and Pearse, 1961; Felgenhauer and Stammler, 1962; Abe *et al.,* 1963). In the squirrel monkey, a stronger MAO reaction has been observed in the nerve fibers than in those of the rhesus monkey. This may be due to a species difference. In lower animals such as the rodents, MAO activity in the nerve fibers seems to be stronger than observed in monkeys (Wawrzyniak, 1965). In the developing brain of the guinea pig, Wawrzyniak observed that the activity of MAO increases in proportion to the number of fibers as well as to the progress of myelination, and that there is a definite correlation between the appearance of nerve fibers and the MAO activity.

Shimizu and Kumamoto (1952) described the medium-sized neurons of the globus pallidus of rodents as showing glycogen. Sato (1959) observed in the rat that glycogen was seen only during a certain postnatal period, and that the GP did not show any glycogen four weeks after birth. Shimizu and Okada (1957) and Amakawa (1959) failed to demonstrate amylophosphorylase in the globus pallidus of the rodent.

There has been a controversy in the literature with regard to the embryonic origin of the GP and whether it is derived from the hypothalamus or from the telencephalon (Spatz, 1922; Mettler and Marburg as quoted in von Bonin, 1959). An origin from the telencephalon indicates a cortical nature, whereas a hypothalamic origin would put the GP on an entirely different basis from the striatum. The present histochemical study does not resolve this controversy.

Nucleus basalis (Meynert)

The nucleus basalis of Meynert (BM) is considered to be a functional part of the globus pallidus (GP) [Kodama, 1929; Gorry, 1963], but since it is different from the GP histochemically and anatomically, it may be described as a separate nucleus (Figs. 21, 22). Iijima *et al.* (1968) showed in the squirrel monkey that the cells of the BM revealed peripheral distribution of Nissl substance as shown in the neurosecretory neurons of the supraoptic and paraventricular nuclei (Scharrer and Scharrer, 1945; Bargmann, 1948; Palay, 1953; Iijima *et al.*, 1967a).

Histochemically, the neurons of the nucleus basalis give a very strong LDH, AC, SE, and ATPase reaction; moderate SDH; mild to moderate AChE and MAO; and a negative AK reaction. The activity of LDH, SE, and ATPase extends into the cell processes for a considerable distance. In the other enzyme preparations the activity is negligible to mild (SDH, AC, AChE, MAO, AK) in the cell processes. Similar histochemical results have been observed for nucleus ansa peduncularis (NAP). Figure 23 shows the AChE activity in this nucleus. The glial cells close to neurons give a moderate to strong reaction in the LDH, SDH, and ATPase preparations. In the squirrel monkey, the satellite cells have shown large amounts of amylophosphorylase, aldolase, and glucose-6-phosphate dehydrogenase, in addition to LDH and SDH (Iijima *et al.*, 1968). This may indicate that the satellite cells are well equipped with the enzymes of the glycolytic pathway and the TCA cycle. Iijima *et al.* (1968) believed that because of an intimate relationship between the satellite cells and the neurons, they should be considered as a functional unit (as was earlier suggested by Hydén and Pigon, 1960; Shantha *et al.*, 1967; and Iijima *et al.*, 1967a, 1967b).

It has been observed in some areas of the brain that there is an inverse relationship between the activity of MAO and AChE, except for certain

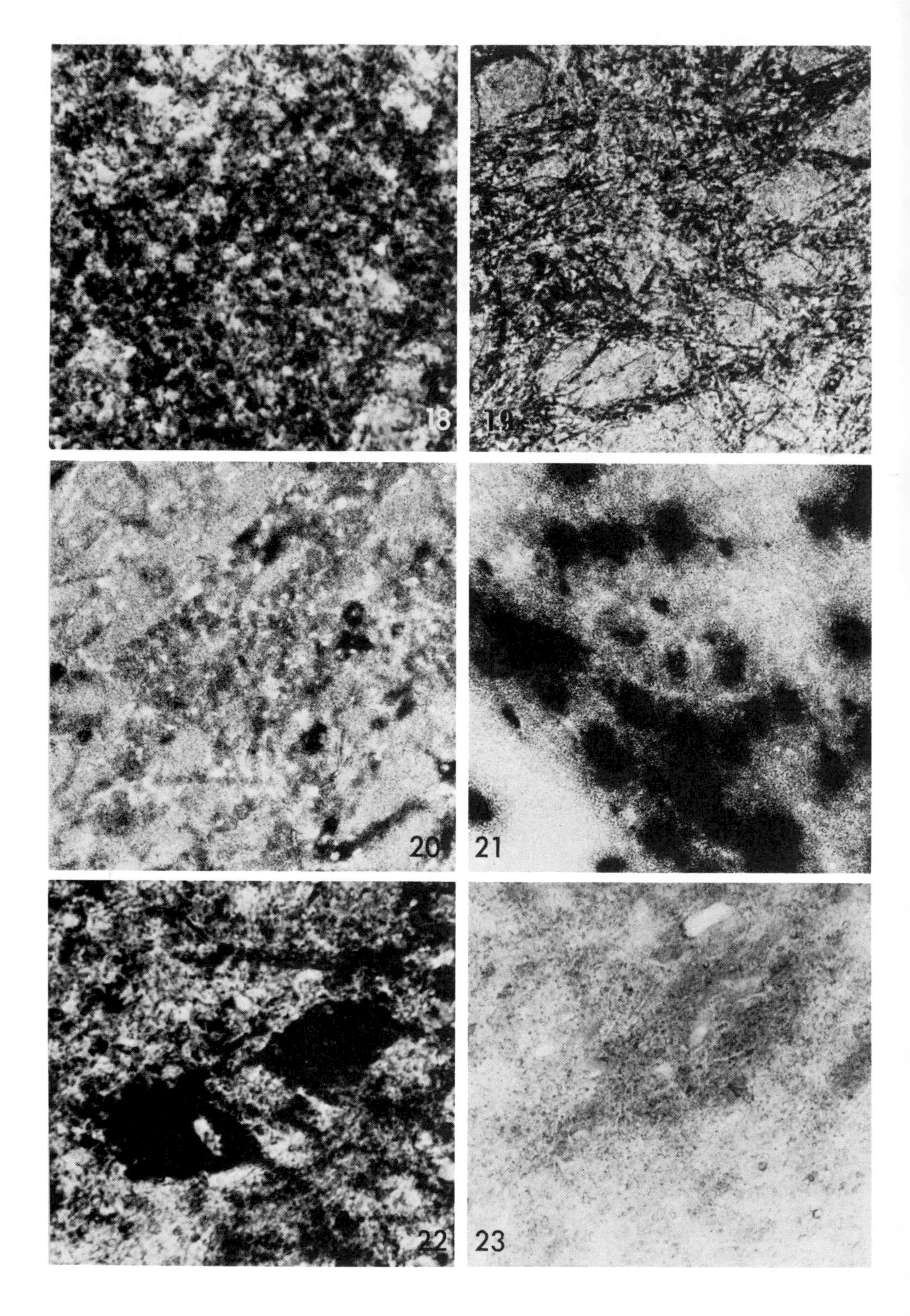
18
19
20
21
22
23

FIGURE 18. LDH reaction in the putamen showing strong activity in the neurons and moderate in the neuropil. ×148.

FIGURE 19. LDH activity in the globus pallidus showing strong enzyme activity in the cells extending into the cell processes for a considerable distance. The fiber bundles crossing through the area show negligible reaction. ×148.

FIGURE 20. SE activity in globus pallidus showing strong activity in the neurons and moderate in the glial cells and the neuropil. ×148.

FIGURE 21. AC reaction in the cells of the nucleus basalis (Meynert) showing very strong activity for this enzyme. ×148.

FIGURE 22. LDH preparation of the cells of the nucleus basalis (Meynert). Note a strong reaction in the perikarya as well as in the cell processes. ×592.

FIGURE 23. AChE activity as seen in the nucleus ansa peduncularis. Note negligible to mild activity in the cells and moderate in the neuropil. ×148.

areas such as the nucleus locus coeruleus, interpenduncularis, and dorsalis vagi (Shimizu *et al.*, 1959; Manocha and Bourne, 1966a, 1966b, 1966c). The histochemical reactions show that the neurons of the BM also come in the same category, giving a mild to moderate reaction in AChE and MAO. It may then be assumed that the MAO activity might be involved in the metabolism of the visceral region of the brain rather than exclusively participating in the adrenergic function of the neurons (Shimizu *et al.*, 1959). Some authors feel that regions of the brain rich in MAO may be poor in SDH and vice versa (Shimizu and Morikawa, 1957; Shimizu *et al.*, 1959). In that case, the BM may be an exception to this rule and show vigorous aerobic metabolism in spite of its poor vascularity.

However, it cannot be denied that histochemically the BM is different from the globus pallidus, of which it is generally considered a part. Being rich in LDH, SDH, AC, SE, MAO, AChE, and ATPase and thus showing a high metabolic activity in its neurons, the BM may also be functionally distinct from the globus pallidus.

SEPTAL AREA

The septal area in this study includes the portion of the medial hemisphere wall which lies below the corpus callosum and between the frontal cortex rostrally and the anterior commissure caudally. It has been grouped into dorsal, ventral, medial, and caudal nuclei by the description of Stephen and Andy (1964) [Figs. 16, 17, 24–27]. The dorsal group contains the nucleus dorsalis septi (DS), the ventral group the nucleus lateralis septi (LS), and the medial group the nucleus medialis septi (MS) and nucleus fasciculus diagonalis band of Broca (NFDB). The nucleus accumbens septi (AST) has also been included in this group because of its proximity. The caudal group consists of nuclei commissura anterior (NCA), stria terminalis (NST), triangularis septi (TRS), and fimbrialis septi (NSF). The subfornical organ is also described in this section under a separate heading.

Dorsal and ventral group

Since the nuclei dorsalis and lateralis sepi are similar histochemically, a joint description will suffice. The DS is represented by an oval mass located in the dorsal part of the septum, separated laterally from the nucleus caudatus by the lateral ventricle and consisting mostly of small and medium-sized cells. The LS is located in the ventral part of the lateral aspect of the septum bounded by the lateral ventricle; the MS consists of small and medium-sized cells with a predominance of the former type. The neurons of the DS and LS are strong in LDH, moderate in SDH, moderate (DS) to moderately strong (LS) in AC, mild (DS) to moderate (LS) in MAO, moderately strong in SE and ATPase, mild in AChE, and negative in the AK preparations. In the latter, only the small capillaries show positive enzyme activity. The enzyme activity passes on into the cell processes in somewhat weaker form in the LDH, SDH, AChE, and ATPase preparations. Glial cells are moderate to moderately strong in LDH, mild in ATPase, and moderate in SE preparations. The neuropil shows moderate activity in LDH, SDH, SE, MAO, and ATPase, whereas the enzyme reaction is mild in AChE and AC.

Medial group

The MS is located ventromedial to the LS. In its rostral part, it is bounded by the NFDB ventromedially. The latter forms a vertical band of cells located along the medial region of the septum. The AST is that part of the head of the nucleus caudatus which is related to the septum pellucidum. It is generally included in the septal area because it extends around the ventricle into this area, and its connections resemble those of the septal region (Lauer, 1945). The MS contains small and medium-sized cells, a contrast to the mostly large and medium-sized multipolar neuronal composition of the NFDB.

Histochemically, the MS shows strong LDH (somewhat stronger than the LS); moderately strong SDH, SE and ATPase; moderate AC; mild to moderate AChE; mild MAO; and negative AK in the neurons, whereas

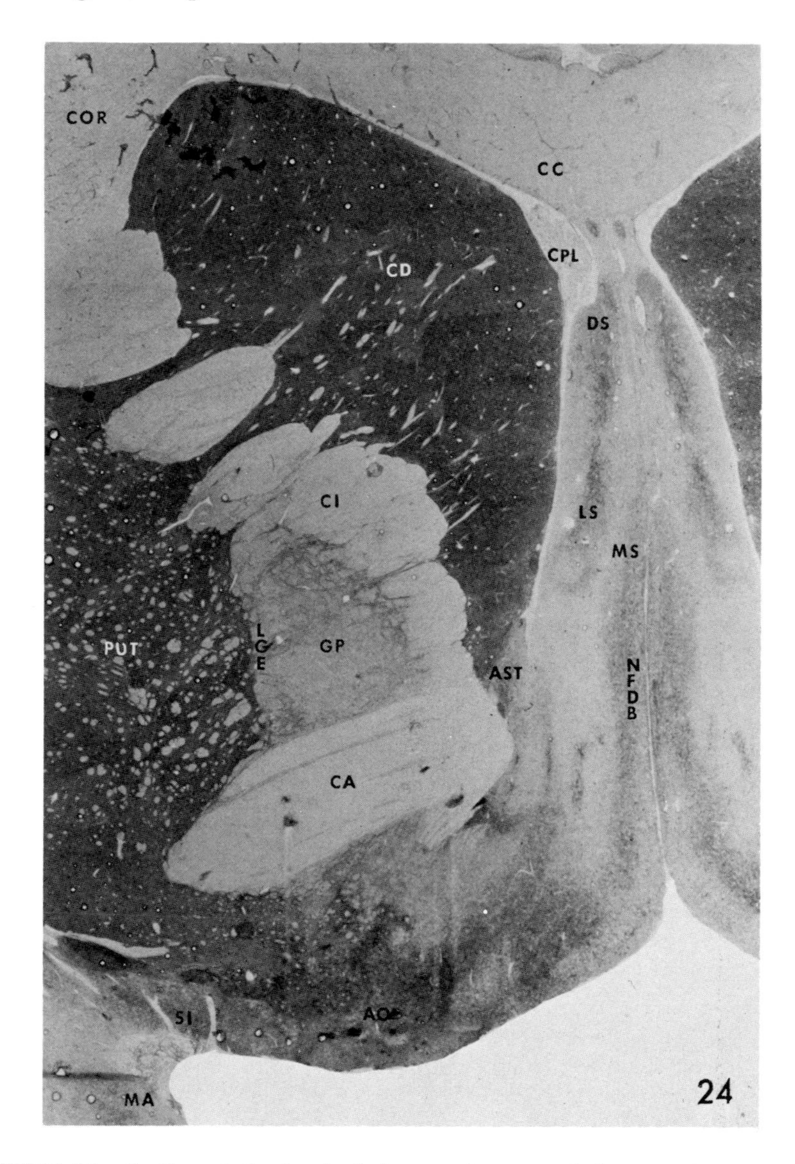

FIGURE 24. Section at the level of the anterior part of the commissura anterior showing AChE activity in the various nuclei. Note a particularly strong AChE reaction in the nucleus caudatus and putamen. ×7.

the neuropil is slightly weaker enzymatically except in MAO where the cells show less activity than the neuropil. In the coypu rat, Girgis (1967) observed strong AChE reactions in the perikarya and cell processes of the nucleus medialis septi. The rhesus monkey, however, has only mild to moderate AChE activity.

The NFDB cells give a very prominent reaction in a number of enzyme preparations. The activity is very strong in LDH and ATPase and extends into the cell processes. It is moderate to moderately strong in SDH, and only a mild reaction extends into the cell processes. In AChE preparations, the NFDB shows much stronger fine granular enzyme activity than the other septal nuclei. The neurons are moderately strong in AChE, strong in SE and mild in MAO. In the latter the neuropil is stronger than the neurons; however, the neuropil reaction is weaker in most preparations than the cells. The glial cells give a moderately strong LDH and a moderate ATPase reaction. Koelle (1954), Shute and Lewis (1963), and Girgis (1967) in the rat observed a strong AChE reaction in NFDB, whereas Gerebtzoff (1959) did not show any appreciable reaction in this animal.

Histochemically the AST shows the same reactions as the nucleus caudatus and may be an integral part of the latter. Morphologically, also, its cell population greatly resembles that of the nucleus caudatus (Manocha and Bourne, 1967; Manocha *et al.*, 1967; Shantha and Manocha, 1969). Girgis (1967) observed in the coypu that the AST stains as intensely in AChE as do the other parts of the corpus striatum.

Caudal group

The nucleus commissura anterior (NCA) is represented by groups of cells rostral and ventral to the commissura anterior (CA) where the latter approaches the midline. It consists of small, medium-sized, and a few large neurons. Dorsolaterally, its neurons are not distinguishable from those of the NST (Lauer, 1945). The nucleus stria terminalis (NST) is also located above the CA, medial to the capsula interna and globus pallidus, and ventromedial to the head of the nucleus caudatus. The nucleus triangularis septi (TRS) is located dorsal to the more rostral part of the anterior commissure and appears as a small dense cell mass projecting upward between the two descending columns of fornix. Most of the neurons are small and oval, ovoid or fusiform. The nucleus fimbrialis septi (NSF) consists of loosely arranged cells found between the fibers of the fimbria-fornix complex. The cell size is small but slightly larger than those of the TRS.

Histochemically, the neurons show moderate (NCA and NST) and moderately strong to strong (TRS and NSF) LDH activity and a mild (TRS and NSF) to moderate (NCA) or moderately strong (NST) SDH reaction. The ATPase reaction in the neurons is moderate (NST and NSF) and strong (NCA and TRS) and resembles to some extent the AC reaction, which is moderately strong (NST and NSF) to strong (NCA and TRS). The nerve cells show moderate (NSF) to moderately strong (NCA, NST, and TRS) SE activity. AChE and MAO do not react strongly and mild AChE (NST, TRS) and MAO (NST, TRS, NSF) to moderate AChE (NCA) and MAO (NCA) activity is shown. Small variations in the enzyme activity in the individual neurons of the same nucleus have been observed and may represent the functional state of a particular cell at a particular time. The reactions described above are based on a majority of the cells. There did not seem to be any definite correlation between the cell size and the enzyme reaction, and reaction may be observed which is mild in a smaller cell and moderate in a larger one, and vice versa. The alkaline phosphatase preparations do not show any activity in the neurons or the neuropil; only the blood vessels react positively. The latter also show a moderately strong ATPase reaction. The glial cells, particularly in the nucleus stria terminalis, give moderate to moderately strong LDH, moderate ATPase, and mild SE reactions.

The neuropil is generally weaker in enzyme activity than the neurons, except in the MAO preparations where the neuropil is moderate compared to a mild reaction in the neurons. In ATPase, the intensity of reaction in the neurons is less than in the neuropil.

Subfornical organ

The subfornical organ (SFO) is located in the middle of the anterior wall of the third ventricle at the level of the interventricular foramen and consists of a main body and stalk. In the squirrel monkey, Akert *et al.* (1961) measured the dimensions of the SFO and found it to be 0.95 mm dorsoventrally, 0.45 mm rostrocaudally, and 0.55 mm mediolaterally. It appears that three fluid systems meet at the SFO: (1) the cerebrospinal fluid of the ventricle, which penetrates into the depths of the SFO through the ependymal canaliculi found on its ventricular surface (Andres, 1965); (2) the fluid of the subarachnoidal space, which reaches the SFO through the transverse fissure (Sprankel, 1960); and (3) the blood supply of the SFO, which enters through a branch of the anterior cerebral artery and also through a vertebral artery by way of the choroid arteries (Spoerri, 1963).

The subfornical organ contains small cells which are generally referred to as "parenchymal cells." The histological and electron microscope

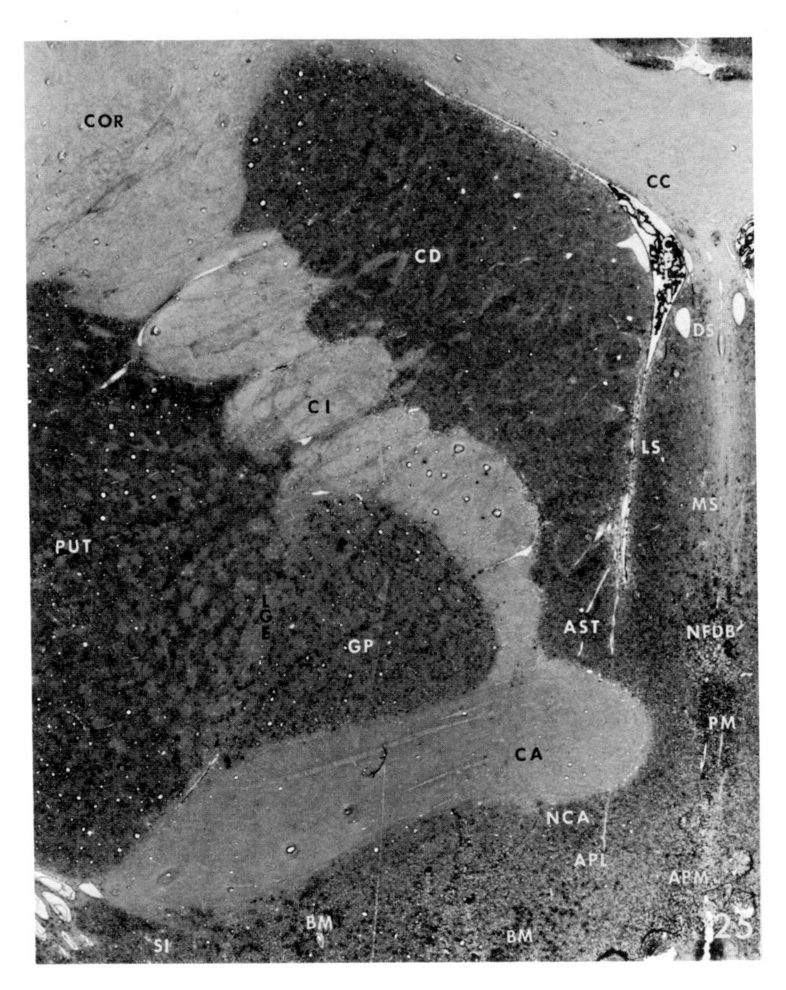

FIGURE 25. Acid phosphatase: section at the level of the anterior part of the commissura anterior showing acid phosphatase activity. Note very strong reactions given by the Nucleus fasciculi diagonalis Brocae, Nucleus praeopticus medianus, and the neurons of the basalis Meynert. ×7.

studies of Andres (1965) and histochemical studies of Nakajima *et al.* (1968) have shown that there are neurons in the SFO in addition to other glial or ependymal-like cells. Asida (1943) believed that they were undifferentiated cells and may be "genetically" regarded as nerve cells because of their origin from cells very similar to neuroblasts. Sprankel (1957) reported that they were related to "typical" nerve cells and are in transitional form. Wislocki and Leduc (1952) believed that they were derived from ependymal cells as medullary epithelium.

The neurons of the main body of the SFO give strong LDH and AC, moderately strong SDH and ATPase, mild MAO, moderate SE and mild to moderate AChE reactions. This is in contrast to the findings of Atherton (1963) who described a negative AChE reaction in the SFO of the chick. The AK preparations show a negative reaction except for moderate to moderately strong activity in the blood vessels. In LDH, the enzyme activity is restricted to the thin periphery of the neurons, and some small round glial cells give a very strong reaction. The neuropil of the SFO is loose due to the presence of wide intercellular spaces, containing a large number of blood vessels. The neuropil gives strong LDH and ATPase reactions, moderate SDH, negligible MAO and AChE, and a mild AC and SE reaction. The glial cells are strong in LDH and ATPase, mild to moderate in SDH, negligible in MAO, negligible to mild in SE and AC, and negative in AChE preparations. The stalk of the SFO is slightly weaker in enzyme activity than the main body, except in the ATPase preparations which show strongly positive cells and neuropil. Mild AChE activity is observed in some nerve processes. It could not be determined whether or not they correspond to the nerve fibers of unknown origin and destination described by Akert *et al.*, 1961. The Bodian preparations of Nakajima *et al.* (1968) also showed a number of fibers originating from the body of the SFO. It appears from all the above studies that the SFO contains both cholinergic and adrenergic neurons and fibers.

The present study on the rhesus monkey and that on the squirrel monkey (Nakajima *et al.*, 1968) have shown that the nerve cells of the SFO show in varying degrees glycolytic and aerobic pathway enzymes. This differs from the observations of Shimizu *et al.* (1957) and Shimizu and Morikawa (1957) in the rat, rabbit, and guinea pig. They believe that the anaerobic glycolysis is the predominant feature of this organ, whereas Nakajima (1964, 1966), having worked on the enzymes of the tricarboxylic acid cycle and the Warburg–Dickens pathway, concluded that aerobic, not anaerobic, glycolysis is predominant in the SFO. In our studies and in the squirrel monkey (Nakajima *et al.*, 1968), a number of nerve cells and oligodendroglia in the SFO are rich in enzymes of the TCA cycle and pentose cycle; thus, they presumably have the capacity of producing ATP and reduced nicotinamide adenine dinucleotide phosphate ($NADPH_2$). The ATP in the neurons is probably used as energy for synaptic transmission, active transport, secretion, and various other metabolic processes; whereas $NADPH_2$ is used for synthetic processes such as the production of fatty acids and some amino acid conversion—for example, conversion of phenylalanine to tyrosine (Nakajima *et al.*, 1968).

The function performed by the SFO is still a matter of speculation and little can be said with certainty. A number of functions such as neurosecretion, chemoreception for cerebrospinal fluid (CSF), or the regulation of quantity or quality of CSF have been attributed to the SFO. Weindl (1965) believed that the SFO has a metabolic as well as a receptive function. The present histochemical study, and those referred to above, have shown that the SFO is enzymatically active and is capable of performing any one of all the functions mentioned.

REFERENCES

Abe, T., Yamada, Y., and Hashimoto, P. H. Histochemical study of glucose-6-phosphate dehydrogenase in the brain of the normal adult rat. *Med. J. Osaka Univ.* **14**:67–98 (1963).

Adams, C. W. M. (ed.) Histochemistry of the cells in the nervous system. In: *Neurohistochemistry*, pp. 253–331. Elsevier, Amsterdam, 1965.

Akert, K., Potter, H. D., and Anderson, J. W. The subfornical organ in mammals. I. Comparative and topographical anatomy. *J. Comp. Neurol.* **116**:1–14 (1961).

Amakawa, F. Studies on the polysaccharide synthesis of the central nervous tissue with special reference to the phosphorylase reaction and several enzymes (in Japanese). *Kumamoto Igaku Zassi* 33: Suppl. 8, 1–37 (1959).

Andres, K. H. Der Feinbaus des Subfornikalorganes von Hund. *Z. Zellforsch.* **68**:445–473 (1965).

Ashby, W., Garzoli, R. F., and Schuster, E. M. Relative distribution patterns of three brain enzymes: carbonic anhydrase, choline esterase and acetyl phosphatase. *Am. J. Physiol.* **170**:116–120 (1952).

Asida, K. Beiträge zur Kenntnis der Morphologie und Entwicklungsgeschichte des subfornikalen Organs. *Keijo J. Med.* **12**:45–94 (1943).

Atherton, G. W. An investigation of the specificity of cholinesterase in the developing brain of the chick. *Histochemie* 3:214–221 (1963).

Bargeton-Farkas, E., and Pearse, A. G. E. Aspects histoenzymologiques de la maturation du systeme neureux. *J. Neurol. Sci.* **2**:213–240 (1965).

Bargmann, W. Über die neurosekretorische Verknüpfung von Hypothalamus und Neurohypophyse. *Z. Zellforsch.* **34**:610–634 (1948).

Campbell, A. C. P. Variation in vascular and oxidase content in different regions of the brain of the cat. *Arch. Neurol. Psychiat.* **41**:223–242 (1939).

de Giacomo, P. Distribution of cholinesterase activity in the human central nervous system. *Proc. Fourth Intern. Cong. Neuropath., Munich, 1961.* Vol. I, pp. 198–205. Thieme, Stuttgart, 1962.

Duckett, S., and Pearse, A. G. E. The nature of the solitary active cells of the central nervous system. *Experientia* **20**:259–260 (1964).

Duckett, S., and Pearse, A. G. E. Histoenzymology of the developing human basal ganglia. *Histochemie* **8**:334–341 (1967).

Felgenhauer, K., and Stammler, A. Das Verteilungmuster der Dehydrogenasen und Diaphorasen im Zentralnervensystem des Meerschweinchens. *Z. Zellforsch.* **58**:219–233 (1962).

Foldes, F. F., Zzigmond, E. K., Foldes, V. M., and Erdos, E. G. The distribution of acetylcholinesterase and butyrylcholinesterase in the human brain. *J. Neurochem.* **9**:559–572 (1962).

Friede, R. L. Histochemical investigations on succinic dehydrogenase in the central nervous system. I. Distribution in the developing rat's brain. *J. Neurochem.* **4**: 101–111 (1959).

Friede, R. L. A quantitative mapping of alkaline phosphatase in the brain of the rhesus monkey. *J. Neurochem.* **13**:197–203 (1966).

Friede, R. L., and Fleming, L. M. A mapping of the distribution of lactic dehydrogenase in the brain of the rhesus monkey. *Am. J. Anat.* **113**:215–234 (1963).

Gerebtzoff, M. A. *Cholinesterases.* International Series in Monographs on Pure and Applied Biology (Division: Modern Trends in Physiological Sciences), (P. Alexander and Z. M. Bacq, eds.). Pergamon Press, New York, 1959.

Girgis, M. Distribution of cholinesterase in the basal rhinencephalic structures of the coypu (*Myocastor coypus*). *J. Comp. Neurol.* **129**:85–95 (1967).

Gorry, J. D. Studies on the comparative anatomy of the ganglion basale of Meynert. *Acta Anat.* **55**:51–104 (1963).

Himwich, H. E., and Fazekas, B. J. F. Comparative studies of the metabolism of the brain of infant and adult dogs. *Am. J. Physiol.* **132**:454–459 (1941).

Hydén, H., and Pigon, A. A cytophysiological study of the functional relationship between oligodendroglial cells and nerve cells of Deiter's nucleus. *J. Neurochem.* **6**:57–72 (1960).

Iijima, K., Shantha, T. R., and Bourne, G. H. Enzyme-histochemical studies on the hypothalamus with special reference to the supraoptic and paraventricular nuclei of squirrel monkey (*Saimiri sciureus*). *Z. Zellforsch.* **79**:76–91 (1967a).

Iijima, K., Bourne, G. H., and Shantha, T. R. Histochemical studies on the distribution of some enzymes of the glycolytic pathway in the olfactory bulb of the squirrel monkey (*Saimiri sciureus*). *Acta Histochem.* **27**:1–9 (1967b).

Iijima, K., Shantha, T. R., and Bourne, G. H. Histochemical studies on the nucleus basalis of Meynert of the squirrel monkey (*Saimiri sciureus*). *Acta Histochem.* **30**:96–108 (1968).

Ishii, T., and Friede, R. L. A comparative histochemical mapping of the distribution of acetylcholinesterase nicotinamide adenine-dinucleotide-diaphorase activities in the human brain. *Intern. Rev. Neurobiol.* **10**:231–275 (1967).

Ishii, Y. The histochemical studies of cholinesterase in the central nervous system. I. Normal distribution in rodents (in Japanese). *Acta Histol. Jap.* **12**:587–611 (1957).

Kodama, S. Pathologisch-anatomische Untersuchungen mit Bezug auf die sogenannten Basalganglien und ihre Adnexe. *Schweiz. Arch. Neurol. Psychiat.* **8**:1–206 (1929).

Koelle, G. B. The histochemical localization of cholinesterases in the central nervous system of the rat. *J. Comp. Neurol.* **100**:211–228 (1954).

Lauer, E. W. The nuclear pattern and fiber connections of certain basal telencephalic centers in the macaque. *J. Comp. Neurol.* **82**:215–254 (1945).

Manocha, S. L., and Bourne, G. H. Histochemical mapping of monoamine oxidase and lactic dehydrogenase in the pons and mesencephalon of squirrel monkey. *J. Neurochem.* **13**:1047–1056 (1966a).

Manocha, S. L., and Bourne, G. H. Histochemical mapping of succinic dehydrogenase and cytochrome oxidase in the spinal cord, medulla oblongata, and cerebellum of squirrel monkey (*Saimiri sciureus*). *Exptl. Brain Res.* **2**:216–229 (1966b).

Manocha, S. L., and Bourne, G. H. Histochemical mapping of succinic dehydrogenase and cytochrome oxidase in the pons and mesencephalon of squirrel monkey (*Saimiri sciureus*). *Exptl. Brain Res.* **2**:230–246 (1966c).

Manocha, S. L., and Bourne, G. H. Histochemical mapping of succinic dehydrogenase and cytochrome oxidase in the diencephalon and basal telencephalic centers of the brain of the squirrel monkey (*Saimiri sciureus*). *Histochemie* **9**:300–319 (1967).

Manocha, S. L., Shantha, T. R., and Bourne, G. H. Histochemical mapping of the distribution of monoamine oxidase in the diencephalon and basal telencephalic centers of the brain of squirrel monkey (*Saimiri sciureus*). *Brain Res.* **6**:570–586 (1967).

Matunami, T. Histochemical study of lactic dehydrogenase in the brain. I. The histochemical distribution of lactic dehydrogenase in the brain of adult rats (in Japanese). *Med. J. Osaka Univ.* **11**:3617–3631 (1959).

Nachmansohn, D. Cholinesterase dans le systeme nerveux central. *Bull. Soc. Chim. Biol.* (*Paris*) **21**:761–796 (1939).

Nakajima, Y. Histochemical studies on the distribution of amylophosphorylase in the subfornical organ of the rat. *Bull. Tokyo Med. Dental Univ.* **11**:391–402 (1964).

Nakajima, Y. Histochemical studies on the carbohydrate metabolism of the subfornical organ in the rat. *Bull. Tokyo Med. Dental. Univ.* **13**:125–146 (1966).

Nakajima, Y., Shantha, T. R., and Bourne, G. H. Histological and histochemical studies on the subfornical organ of the squirrel monkey. *Histochemie* **13**:331–345 (1968).

Okinaka, S., Yoshikawa, M., and Goto, J. Studies on cholinesterase. I. On the distribution of cholinesterase in the human brain. *Tohoku J. Exptl. Med.* **55**:81–85 (1951).

Okinaka, S., Yoshikawa, M., Uono, M., Mozai, T., Toyota, M., Muro, T., Igata, A., Tanabe, H., and Ueda, T. Histochemical study on cholinesterase of the human hypothalamus. *Acta Neuroveget.* (*Vienna*) **22**:53–62 (1960).

Okinaka, S., Yoshikawa, M., Uono, M., Muro, T., Mozai, T., Igata, A., Tanabe, H., Ueda, T., and Tomonaga, M. Distribution of cholinesterase activity in the human cerebal cortex. *Am. J. Phys. Med.* **40**:135–145 (1961).

Palay, S. L. Neurosecretory phenomena in the hypothalamo-hypophyseal system of man and monkey. *Am. J. Anat.* **93**:107–141 (1953).

Sato, S. Postnatal changes of glycogen in the brain of rats (in Japanese). *Med. J. Osaka Univ.* **11**:869–873 (1959).

Scharrer, E., and Scharrer, B. Neurosecretion. *Physiol. Rev.* **25**:171–181 (1945).

Shantha, T. R., Iijima, K., and Bourne, G. H. Histochemical studies on the cerebellum of squirrel monkey (*Saimiri sciureus*). *Acta Histochem.* **27**:129–162 (1967).

Shantha, T. R., and Manocha, S. L. The brain of chimpanzee (*Pan troglodytes*). II. Basal ganglia, septal area, epithalamus, dorsal thalamus, metathalamus, subthalamus, and hypothalamus. In: *Handbook of Chimpanzee* (G. H. Bourne, ed.), Vol. I, pp. 250–317. S. Karger, Basel and New York, 1969.

Shimizu, N., and Kumamoto, T. Histochemical studies on the glycogen of the mammalian brain. *Anat. Rec.* **114**:479–498 (1952).

Shimizu, N., and Morikawa, N. Histochemical studies of succinic dehydrogenase of the brain of mice, rats, guinea pigs, and rabbits. *J. Histochem. Cytochem.* **5**:334–345 (1957).

Shimizu, N., and Okada, M. Histochemical distribution of phosphorylase in the rodent brain from newborn to adult. *J. Histochem. Cytochem.* **5**:459–471 (1957).

Shimizu, N., Morikawa, N., and Ishii, Y. Histochemical studies of succinic dehydrogenase and cytochrome oxidase of the rabbit brain with special reference to the results in the paraventricular structures. *J. Comp. Neurol.* **108**:1–21 (1957).

Shimizu, N., Morikawa, N., and Okada, M. Histochemical studies of monoamine oxidase of the brain of rodents. *Z. Zellforsch.* **49**:389–400 (1959).

Shute, C. C. D., and Lewis, P. R. Cholinesterase-containing systems of the brain of the rat. *Nature* **199**:1160–1164 (1963).

Spatz, H. Über Bezielungen zwischen der substantia nigra des Mittelhirnfurbes und dem globus pallidus des Linsenkern. *Anat. Anz.* **55**:159–180 (1922).

Spoerri, O. Über die Gefassversorgung des Subfornikalorgans der Ratte. *Acta Anat.* (*Basel*) **54**:333–348 (1963).

Sprankel, H. Zur Zytologie des Subfornikalen Organes bei Affen. *Verh. Deutsch. Zool. Ges.* **4**:44–51 (1957).

Sprankel, H. Über die Beziehung des plexus des dritten Ventrikels zum Subfornikalen Organ bei den Primaten. *Naturwissenschaften* **47**:383–384 (1960).

Stephen, H., and Andy, O. J. Cytoarchitectonics of the septal nuclei in old world monkeys (*Cercopithecus* and *Colobus*). *J. Hirnforsch.* **7**:1–8 (1964).

Thomas, E., and Pearse, A. G. E. The fine localization of dehydrogenases in the nervous system. *Histochemie* **2**:266–282 (1961).

Tyler, H. R. Enzyme distribution in human brain. Lactic and malic dehydrogenases. *Proc. Soc. Exptl. Biol. Med.* **104**:79–83 (1960).

von Bonin, G. The basal ganglia. In: *Introduction to Stereotaxis with an Atlas of the Human Brain* (Schaltenbrand and Bailey, eds.), pp. 317–330. Grune and Stratton, New York, 1959.

Wawrzyniak, M. The histochemical activity of some enzymes in mesencephalon during the ontogenic development of the rabbit and the guinea pig. II. Histochemical development of acetylcholinesterase and monoamine oxidase in the nontectal portion of the midbrain of the guinea pig. *Z. Mikr. Anat. Forsch.* **73**: 261–305 (1965).

Weindl, A. On the morphology and histochemistry of the subfornical organ, organum vasculosum laminae terminales and area postrema in rabbits and rats. *Z. Zellforsch.* **67**:740–775 (1965).

Wislocki, G. B., and Leduc, E. H. Vital straining of the hematoencephalic barrier by silver nitrate and trypan blue, and cytological comparison of the neurohypophysis, pineal body, area postrema, intercolumnar tubercle and supraoptic crest. *J. Comp. Neurol.* **96**:371–413 (1952).

Yamada, Y. Histochemical observations on alternation of activities of cytochrome oxidase in the brain of rat from late fetal life. *Med. J. Osaka Univ.* **11**:383–400 (1961).

V

DIENCEPHALON

HYPOTHALAMUS

The hypothalamus is separated dorsally from the thalamus by the hypothalamic sulcus, rostrally by the anterior commissure, medially by and Forel's field, basally by the optic chiasma, optic tract and cerebral the third ventricle, laterally by the optic tract, thalamus, zona incerta peduncles and posteriorly by a line drawn behind the corpus mammillaris and the area posterior hypothalamica. The cytoarchitecture and the nuclear pattern of the hypothalamus of the rhesus monkey has been studied by Crouch (1934a), Krieg (1948), and Kesarev (1965). The classification of the nuclei varies from one author to another. We have followed the classification used by us in an earlier study on the chimpanzee brain (Shantha and Manocha, 1969) and have divided the hypothalamus into (1) area preoptica, (2) area anterior, (3) region of supraoptic and paraventricular nuclei, (4) tuberal region, areas (5) lateralis, (6) dorsalis, and (7) posterior, and (8) nucleus corpus mammillaris hypothalami.

Area preoptica

Situated above the optic chiasma, this region expands gradually as one traces it toward the commissura anterior. The area preoptica is the unevaginated telencephalon and does not belong to the hypothalamus. It is further divided into the area preoptica medianus (APM), consisting mainly of small cells sprinkled with a few medium-sized neurons, the area preoptica lateralis (APL), located laterally and containing both small and medium-sized neurons, and the nucleus preopticus medianus (PM) which contains more compactly arranged, darkly Nissl staining cells (Figs. 26, 27). The latter also extends above the commissura anterior.

Area anterior hypothalamica

This region consists of a diffuse cell mass situated between the area preoptica and the nuclei dorso- and ventromedialis hypothalami, and contains mostly small, oval and fusiform neurons which stain lightly in the Nissl stain.

The areas preoptica and anterior hypothalami and the nucleus arcuatus hypothalami (see under tuberal region) give almost identical histochemical reactions, with only a few minor variations. The APL exhibits comparatively less enzyme activity than other areas, whereas the PM shows greater enzyme concentration than the other nuclei, probably due to a large aggregation of cells with fewer nerve fibers passing through it. All these nuclei show strong AC, moderately strong SE and LDH, moderate SDH and ATPase, and mild MAO. The neuropil is positive for most of the above enzymes, but the degree of positive activity is less than that of the neurons, except for AChE and MAO, where the positive activity is much greater in the neuropil. The blood vessels are AK-positive.

Region of supraoptic (SOH) and paraventricular (PH) nuclei

The SOH is a spindle-shaped mass situated dorsolaterally to the optic chiasma. Most of the neurons are large, some are medium-sized, and they are generally oval. These cells have prominent nucleoli with peripherally distributed Nissl material that does not extend into the cell masses (Figs. 32–37, 43, 44). Kesarev (1965) found that one cubic millimeter of this nucleus contains 46,000 cells in rhesus, 25,000 cells in chimpanzee, and 12,000 cells in man, indicating almost four times as many cells in the rhesus as in a comparable cell mass in man, the largest variation that he noted in any nucleus. The PH is a fusiform cell mass located vertically along the third ventricle. It is composed of small, medium-sized, and large neurons which are oval, fusiform, and spindle-shaped. The nucleoli are most prominent in the larger neurons. These cells show Nissl substance throughout the cytoplasm (Figs. 28–30, 32, 37, 43, 44). Both the SOH and the PH, which are neurosecretory in function, show similar enzyme activity. The neurons show very strong AC; strong SE, LDH, and ATPase; moderately strong SDH; mild to moderate AChE and MAO; and negligible AK activity. There are variations in the enzyme content from cell to cell, but the enzyme activity in all these neurons extends for some distance into the cell processes. The nucleoli show positive activity for all the enzymes except AChE

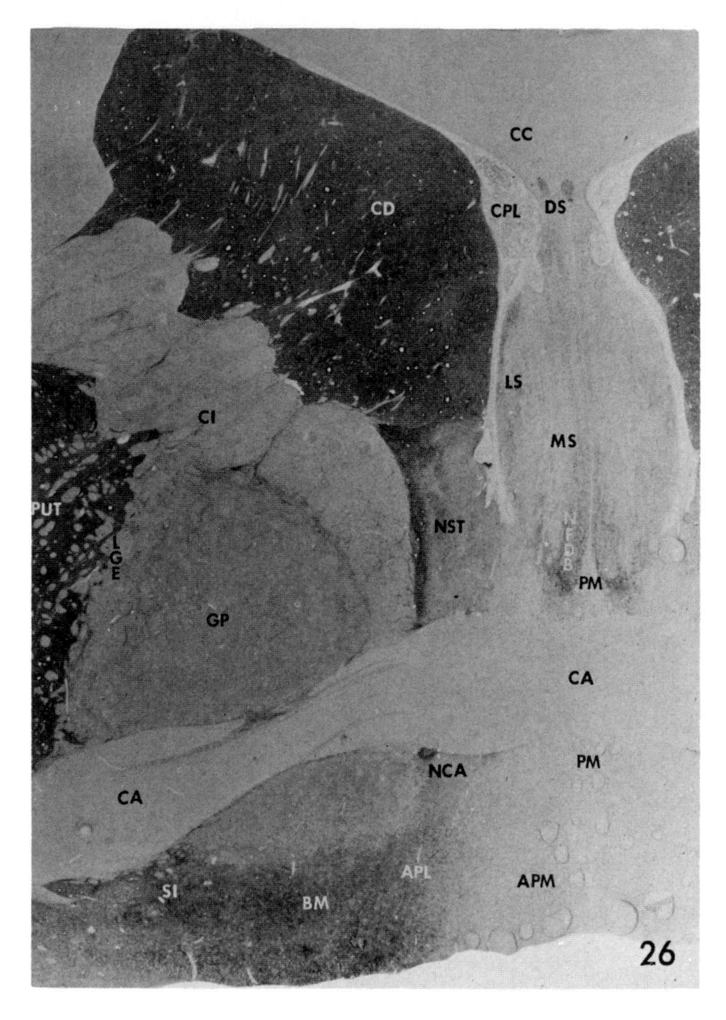

FIGURE 26. Section at the level of the middle of the commissura anterior showing different degrees of AChE distribution in the various nuclei. ×7.

and AK, and the fibers passing through these nuclei give a moderately strong MAO activity. The glial cells show strong LDH; mild to moderate AC, SE, ATPase, and MAO; mild SDH; and negative AChE and AK activity. The blood vessels are very well stained in the AChE, AK, and

ATPase preparations; AK and AChE appear to stain only capillaries whereas the ATPase stains the larger blood vessels, in addition to capillaries. Blood vessels are more numerous in the SOH than in the PH, but in general these nuclei contain many more blood vessels than the surrounding hypothalamic area.

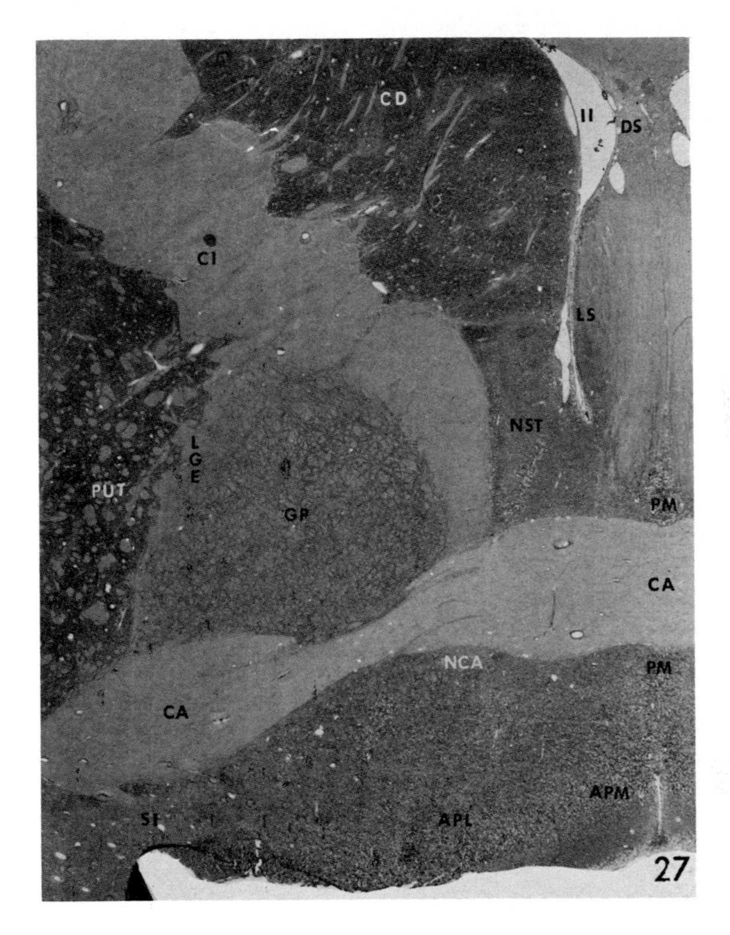

FIGURE 27. Section at the same level as Fig. 26, showing LDH activity. ×7.

Tuberal region

This region is located between the anterior hypothalamic area, the nuclei supraopticus and paraventricularis hypothalami, and the area posterior and corpus mammillaris. This area consists of nuclei dorsomedialis (DMH) [Figs. 31, 45–47], ventromedialis (VMH) [Figs. 31, 45–47],

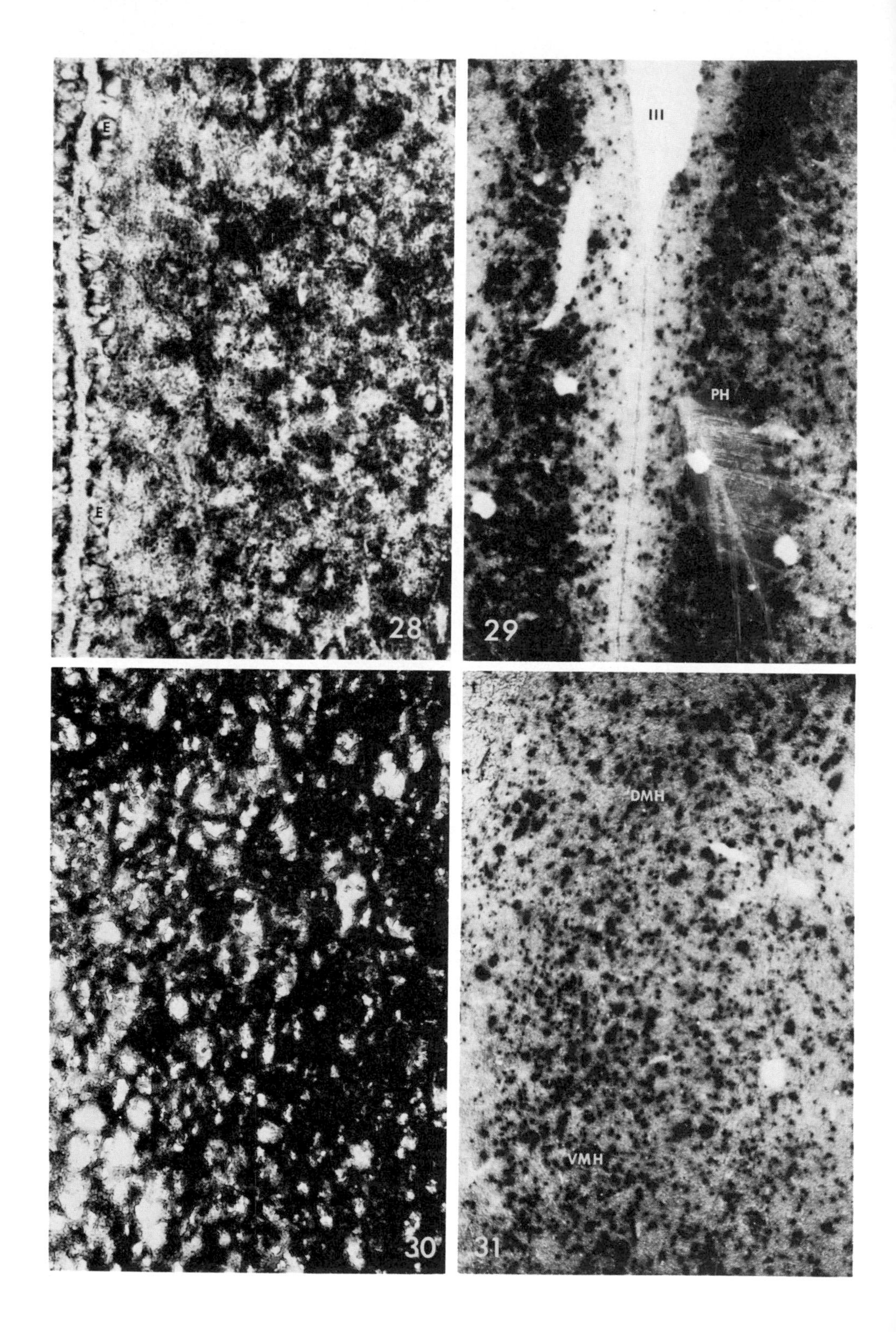

E
E
28
III
PH
29
30
DMH
VMH
31

FIGURES 28–30. Photomicrographs of the nucleus paraventricularis hypothalami showing LDH, AC, and ATPase activity, respectively. The neurons show strong activity in the AC and LDH, whereas the neuropil is stronger in the ATPase preparations. ×370, ×37, ×370.

FIGURE 31. A low power photomicrograph of the nuclei dorsomedialis and ventromedialis hypothalami showing AC activity. Note the strong reaction in the neurons. ×37.

infundibularis (IF), arcuatus (AH), tuberalis (T), and perifornicalis (P) hypothalami.

The DMH is located lateral to the third ventricle, starting from a point caudal to the PH, extending rostrally to the anterior hypothalamic area. The DMH is ill-defined compared to the ventromedially located VMH, and merges to a great extent with the surrounding nuclei. It consists mainly of small and medium-sized neurons. The VMH is a well-defined nucleus compared to the DMH. It is bounded laterally by the subependymal layer and the area lateralis hypothalami and superolaterally by the DMH and the descending fibers of the fornix. It also consists mainly of small and medium-sized neurons, with medium-sized cells found primarily in the lower lateral part and the small neurons in the dorsal part. The IF is an oval nucleus located rostral to the mammillary bodies, and consists of small and medium-sized oval and triangular cells. The AH is situated adjacent to the third ventricle ependyma, rostral to the premammillaris. It contains small, oval and fusiform neurons. The nucleus tuberalis hypothalami is very difficult to identify in the rhesus monkey, although it is well-defined in man and the chimpanzee and has been divided into three distinct masses. The existence of the nucleus perifornicalis (P) in the rhesus monkey is disputed; Papez and Aronson (1934) describe this nucleus in rhesus, whereas Krieg (1948) and Heiner (1960) reported its absence. It is, however, distinct in man and chimpanzee (Krieg, 1953; Heiner, 1960; Shantha and Manocha, 1969), but seems to be absent in java and cebus monkeys (Shantha *et al.*, 1968; Manocha *et al.*, 1968). This nucleus is believed to be located near the descending fibers of the fornix before they enter the posterior hypothalamic area. In our histochemical preparations such a cellular mass was not recognizable, probably due to the intermixing of these cells with the surrounding nuclei with similar histochemical properties. These hypothalamic areas (DMH, VMH, IF, T, ALH, SMH, PRM, ADH, APH), described in the preceding and following pages, show similar histochemical activities, with minor variations in the degree of positive activity. The neurons show strong AC; moderately strong SE and LDH; moderate SDH, ATPase, and MAO; mild AChE; and negative AK activ-

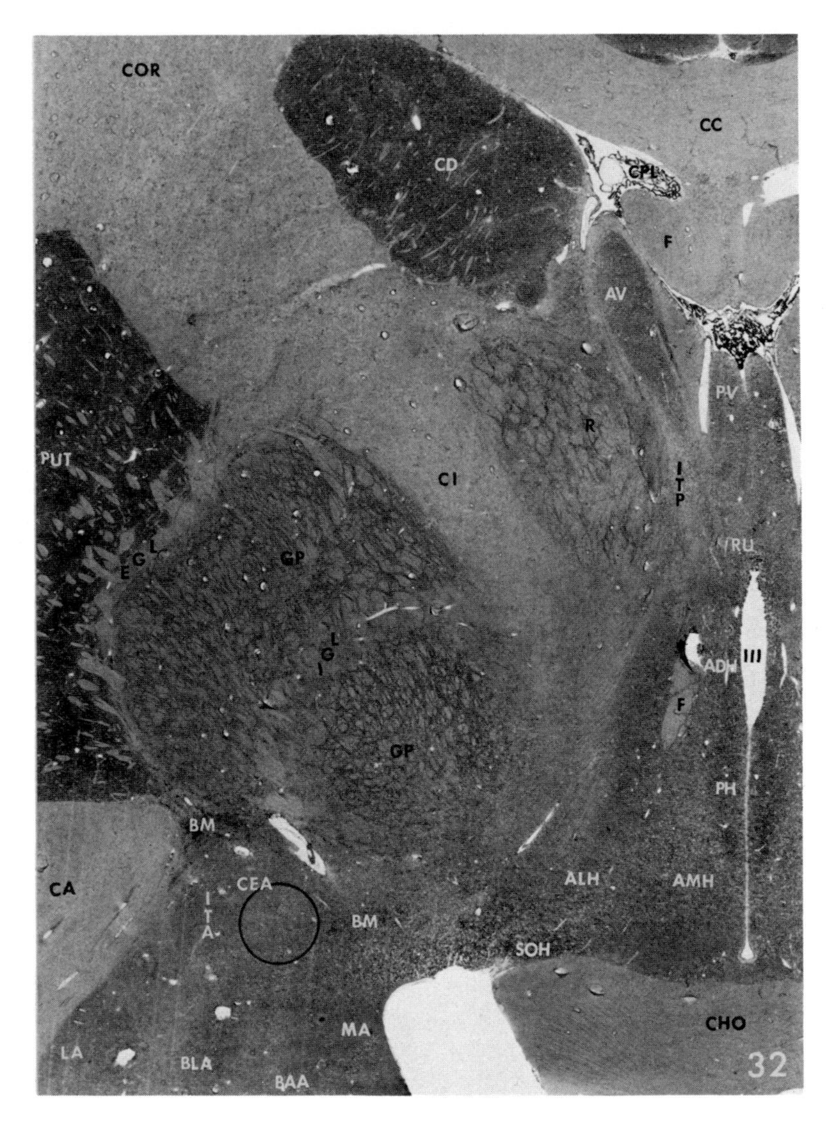

FIGURE 32. Section at the level of the optic chiasma showing different grades of LDH activity in the basal ganglia, amygdaloid complex, thalamus, and hypothalamus. ×7.

ity. The neuropil also shows similar positive activity, but it is not as strongly positive as the neurons for most of the enzymes, with the exception of AChE and MAO. The VMH shows considerably less AChE content than the other nuclei, both in the cells and neuropil. The neurons in this nucleus show AChE activity in the cell cytoplasm, with more enzyme activity at the periphery of the cell than in the central part. In MAO, LDH, SDH, and ATPase preparations, it is not uncommon to find the presence of mild to negligible activity in the nucleolus, but it is quite difficult to say why only certain neurons show a positive nucleolus. As in other areas, variations in the degree of positive activity are found in different neurons and may indicate different metabolic states in different neurons at a particular time. A large number of nerve fibers traversing these nuclei show moderately strong MAO content. Glial cells in all these areas of hypothalamus show moderately strong LDH; moderate ATPase; mild SE, SDH, and MAO; negligible AC; and negative AK and AChE activity. Some glial cells show stronger LDH activity than others. The blood vessels in these regions show positive activity for ATPase, AK, and AChE.

Area lateralis hypothalami (ALH)

This is a very extensive and prominent area of the hypothalamus, extending from the area preoptica to the middle of the mammillary bodies (Figs. 32, 37, 43–47). Its cells diffuse medially with other nuclei extending between its anteroposterior portions. Laterally, the ALH is bounded by the tractus opticus, pedunculus cerebri, internal capsule, and the fornix. The ALH consists of small and medium-sized neurons, with a few large multipolar cells resembling cells of the nucleus basalis (Meynert) and probably belonging to that group.

Area dorsalis hypothalami (ADH)

This region extends between the anterior and posterior hypothalamic area, and is situated dorsal to the PH and DMH (Figs. 43, 46, 47). Caudally, this area seems to represent a zone of transition between the thalamus and hypothalamus. It consists of small neurons with a few medium-sized cells in the dorsal section.

Area posterior hypothalami (APH)

Considerably smaller in the rhesus, as in the java and cebus monkeys, than in the chimpanzee, this region is the posteriormost part of the hypothalamus. It is made up of small and medium-sized neurons intermixed with a few larger cells (Figs. 52–58). Histochemical activity in the ALH, ADH, and APH is described in tuberal region.

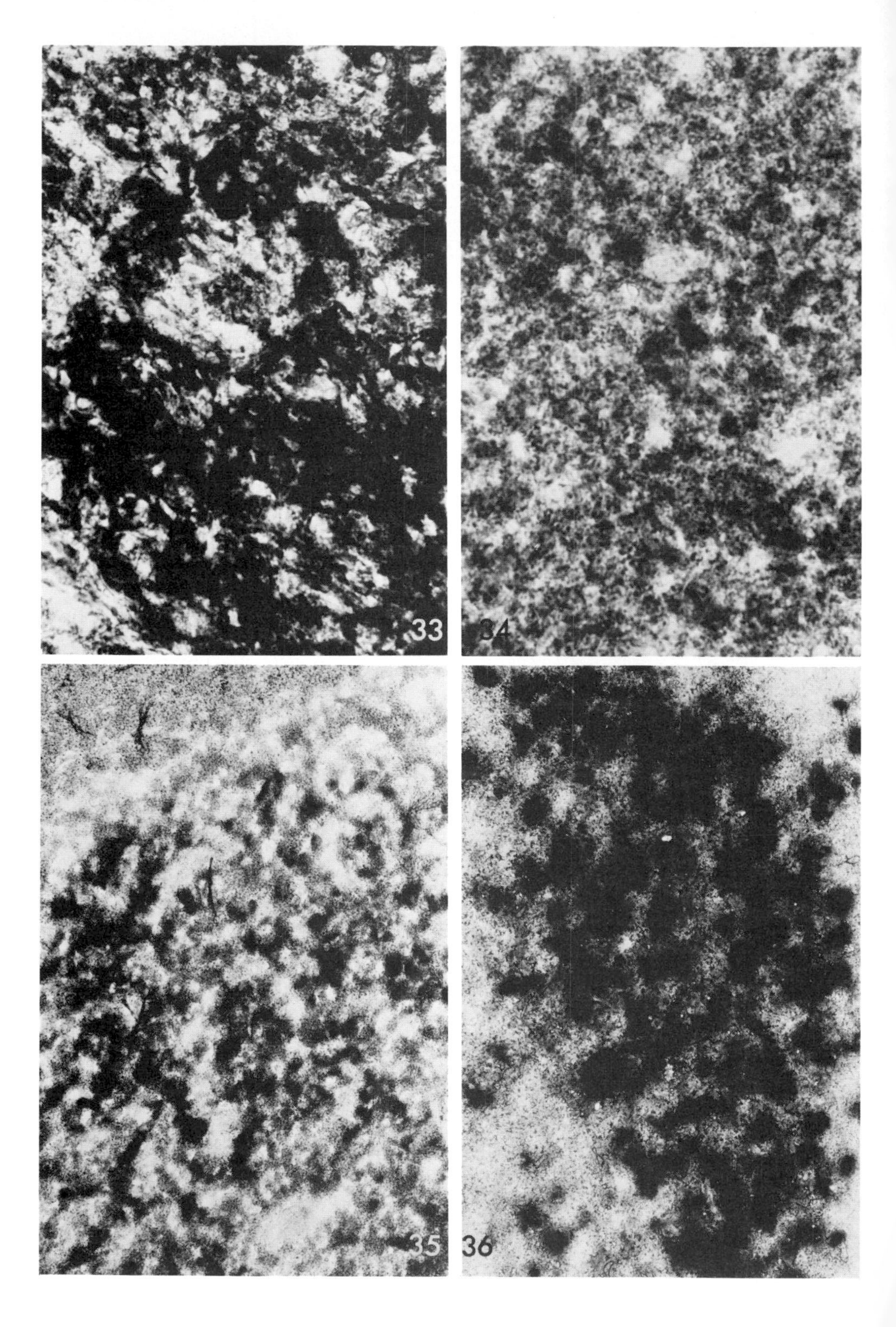
33
34
35
36

FIGURES 33–36. Nucleus supraopticus hypothalami showing the activity of LDH, SDH, SE, and AC, respectively. Note strong LDH and AC activity and moderately strong SDH and SE activity in the neurons. Most of these cells do not show the enzyme reaction extending into the cell processes. ×370, ×148, ×148, ×148.

Nucleus corpus mammillaris

Situated at the base of the diencephalon on both sides of the midline, this body forms the prominent hemispherical elevations. The nuclear complex consists of (a) the nucleus premammillaris (PRM), (b) corpus mammillaris, consisting of medial (MM), intercalate (MI), and lateral (ML) nuclei, and (c) nucleus supramammillaris (SMH) [Figs. 38–42, 50–54]. The griseum periventriculare hypothalami (GPH) has also been described in this section.

The nucleus premammillaris is a small, ill-defined nucleus situated between the posteriormost part of the VMH and the anterior part of the corpus mammillaris. It consists mainly of small, oval and fusiform cells and was quite difficult to locate in our histochemical preparations. The nucleus supramammillaris is situated dorsomedial to the MM between the fibers of the mammillothalamic tract. It has been thought that this nucleus is the caudoventral specialization of the posterior hypothalamic area (Riley, 1960), and histochemical studies indicate this possibility. The SMH consists mainly of small neurons, oval or fusiform, and is traversed by the commissura supramammillaris.

The nucleus mammillaris medialis (MM) constitutes the bulk of the corpus mammillaris projecting at the base of the diencephalon. It is divided into a compactly arranged medial part and a loosely arranged lateral part. This nucleus consists mainly of medium-sized and small neurons, the medial part containing larger neurons than the lateral part. The nucleus intercalatus (intermedius) corpus mammillaris (MI) is not as cellular as the medial part and is placed between the medial and lateral nuclei. The majority of the cells are small with only a few medium-sized neurons. The nucleus lateralis corpus mammillaris (ML) is a well-circumscribed nucleus located between the MI and the nucleus ansa peduncularis. It consists chiefly of medium-sized and large neurons and is divided into small subgroupings. The cells in this group are much larger than those in the other four mammillary nuclei. Kesarev (1965), in Nissl preparations of the rhesus brain, describes the PRM, MM, MI, ML, and SMH as being ill-defined. But in our preparations, except for the PRM, all these nuclei could be easily recognized. This is also true of the java and cebus monkey and the chimpanzee brain (Shantha *et al.*, 1968; Manocha *et al.*, 1968; Shantha and Manocha, 1969).

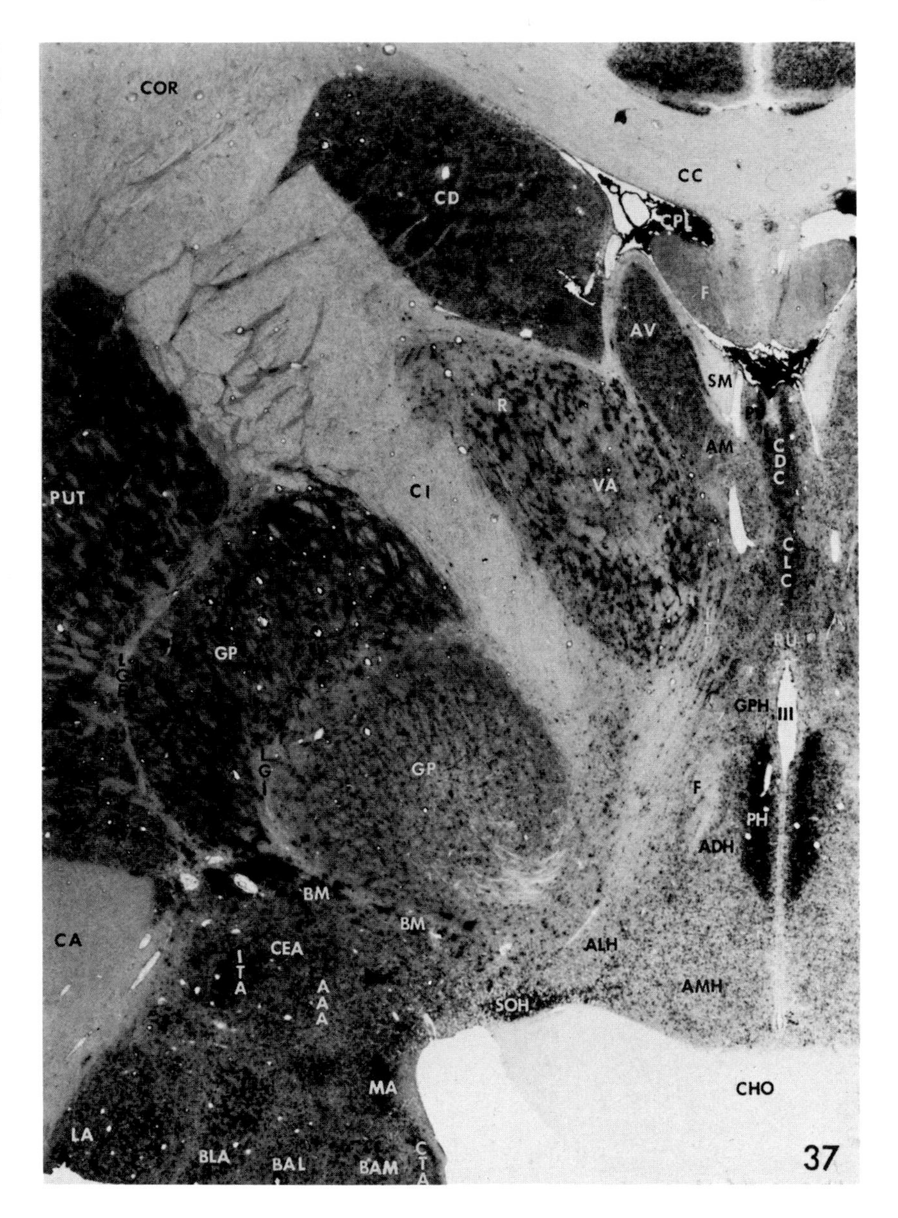

FIGURE 37. Section at the level of the optic chiasma showing varying amounts of AC activity in the basal ganglia, amygdaloid complex, and thalamic and hypothalamic nuclei. Note the strong activity in the paraventricular and supraoptic nuclei of the hypothalamus. ×7.

The degree of enzyme activity in the MM, MI, and ML is variable. Medial parts of the MM show the maximum enzyme content of the three parts; next in order is the ML, while the MI shows the least enzyme content of the three, and is even less active enzymatically than the nucleus supramammillaris. In general, the neurons of the MM and ML show very strong LDH; strong SE, AC, and SDH; moderate ATPase and MAO; mild to moderate AChE; and negative AK activity. The neuropil in these nuclei shows strong LDH; moderate AC, SE, SDH, ATPase, and MAO; mild to moderate AChE; and negligible AK activity. The neuropil in the medial part shows much more enzyme activity than the lateral part. Neurons of the ML show higher cytoplasmic AChE activity than those of the MI and MM. The MI shows moderately strong LDH; moderate AC and SDH; mild to moderate SE, ATPase, and AChE; mild MAO; and negative AK activity. The neuropil shows less positive activity than that of the neuropil in the MM and ML. The fibers in the mammillary body at this level show negative AChE. Blood vessels are positive for AChE, AK, and ATPase. The MM, especially in the medial part, is more vascular than any other part of the corpus mammillaris.

The griseum periventriculare hypothalami (GPH) is made up mainly of small cells which are fusiform and triangular in shape. Ventrally, it is very thin and becomes thickened caudally, ultimately becoming continuous with the periaqueductal gray of the midbrain. The GPH shows strong LDH and ATPase, moderately strong AC, moderate SE and SDH, mild to moderate AChE and MAO, and negative AK activity. This area has very few blood vessels compared to the surrounding hypothalamic areas. In MAO preparations, there are innumerable moderately strong MAO-positive fibers running in various directions. The neuropil of the GPH shows moderately strong LDH and ATPase, moderate MAO, mild to moderate SE and SDH, mild AC and AChE, and negligible AK activity. Glial cells are more distinctly observed in LDH preparations, for which they are strongly positive. They appear to have mild SDH, AC, SE, and MAO; moderate ATPase; and negative AChE and AK activity. The positive activity increases in this area as one traces the area toward the midbrain where it becomes the midbrain central gray.

DORSAL THALAMUS

The dorsal thalamic nuclei of the rhesus monkey have been described in detail by Aronson and Papez (1934), Crouch (1934b), Walker (1937), and Olszewski (1952). The dorsal thalamus is a large oval mass located in an oblique position over the mesencephalon. The thalamic

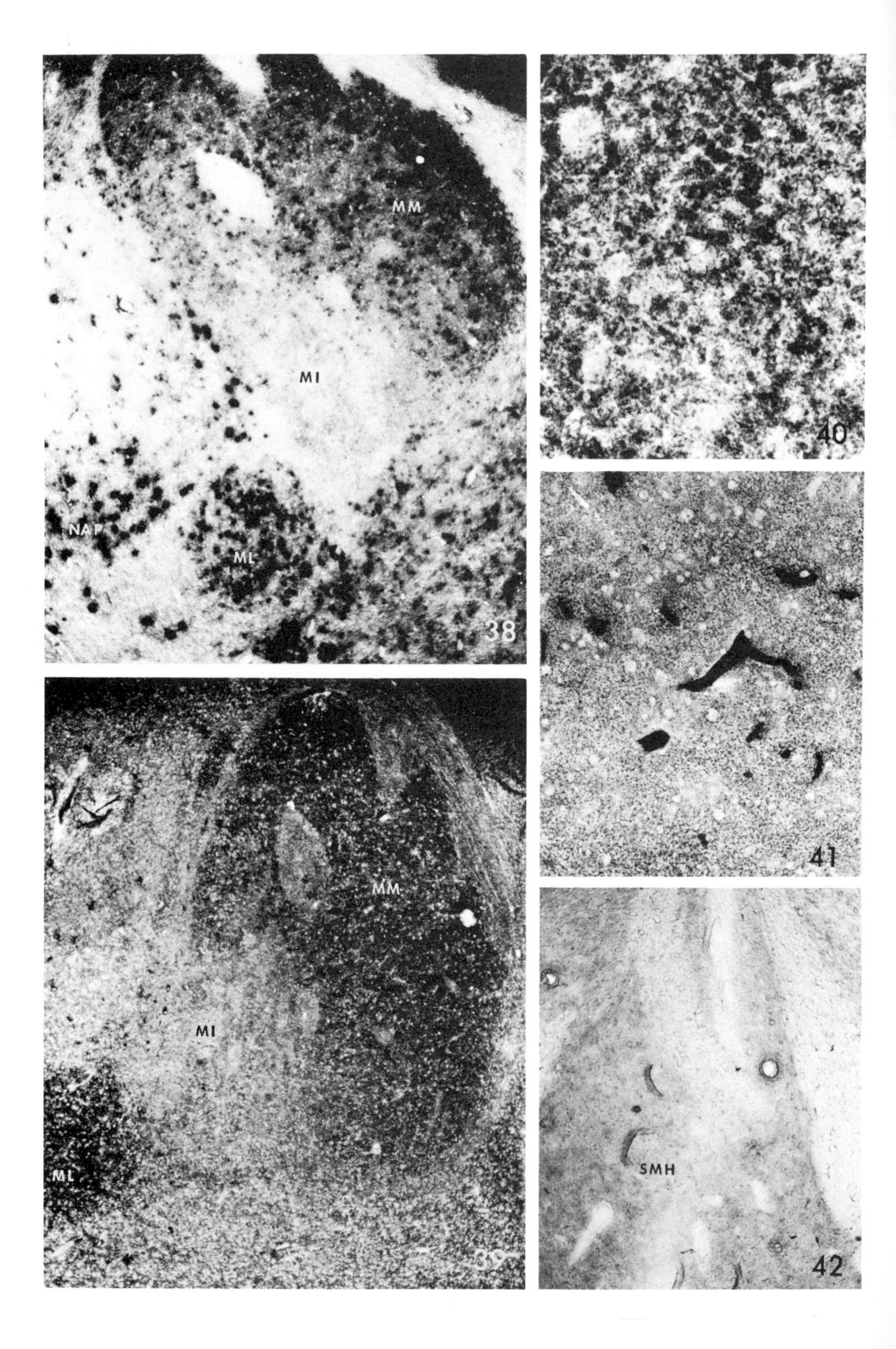
MM
MI
NAP
ML
38
40
41
MM
MI
ML
39
SMH
42

FIGURES 38–41. Photomicrographs of AC, LDH, SDH, and AK preparations showing varying reactions in the components of the mammillary complex. In Figs. 38 and 39, the difference in the positive activity among the medial, intermediate, and lateral nuclei of this body is clearly seen. ×37, ×37, ×148, ×148.

FIGURE 42. AChE preparation showing negligible enzyme activity in the neurons of the nucleus supramammillaris and mild activity in its neuropil. ×37.

nuclei have been grouped under six headings based on the location of the individual nuclei: the anterior, midline, medial, intralaminar, lateral, and posterior nuclei.

Anterior nuclei

This group includes three prominent nuclei (Figs. 32, 37, 43–47): anterior dorsalis (AD), ventralis (AV), and medialis (AM). The AD is a narrow band of compact, medium-sized and slightly elongated cells located dorsomedial to the other anterior nuclei and separated from the third ventricle by a thick layer of white matter. It does not extend as far anteriorly as the other anterior nuclei (Crouch, 1934b). The nucleus anterior ventralis thalami (AV) is less cellular and less compact than the AD and consists of plump, multipolar cells. Ventromedially, it is in contact with the nucleus anterior medialis thalami (AM); therefore, some authors have considered the AV and AM as a single nucleus (Papez and Aronson, 1934; Grünthal, 1934). The AM is V-shaped, and because of less cellularity in the laterodorsal part and a high cell density in the middle part, it has been divided by Olszewski (1952) into pars latocellularis and pars densocellularis, respectively. Since no significant histochemical differences are found, the anterior nuclei are being described here as a single unit.

Histochemically, the neurons of the anterior nuclei give a strong LDH and moderate to moderately strong SDH reaction. The glia are strong in LDH and mild in SDH, whereas the neuropil is moderately strong in LDH and mild to moderate in SDH. In the former, however, the enzyme activity in the neuropil is the strongest in the AD, generally decreasing in the AV and AM. The neuropil is mild in AC and moderate in ATPase. The AC preparations show gradations of enzyme activity with the strongest reaction in the AD, decreasing gradually in the AV and AM. The neurons are mild to moderate in ATPase and strong to very strong in AC. The AV gives the strongest reaction in SE, while the AD gives the weakest. In the latter, the neuropil is mild and the cells are strong; in the AV and AM the neuropil is moderate, and the cells are strong to very strong. In AChE the neuropil of the anterior

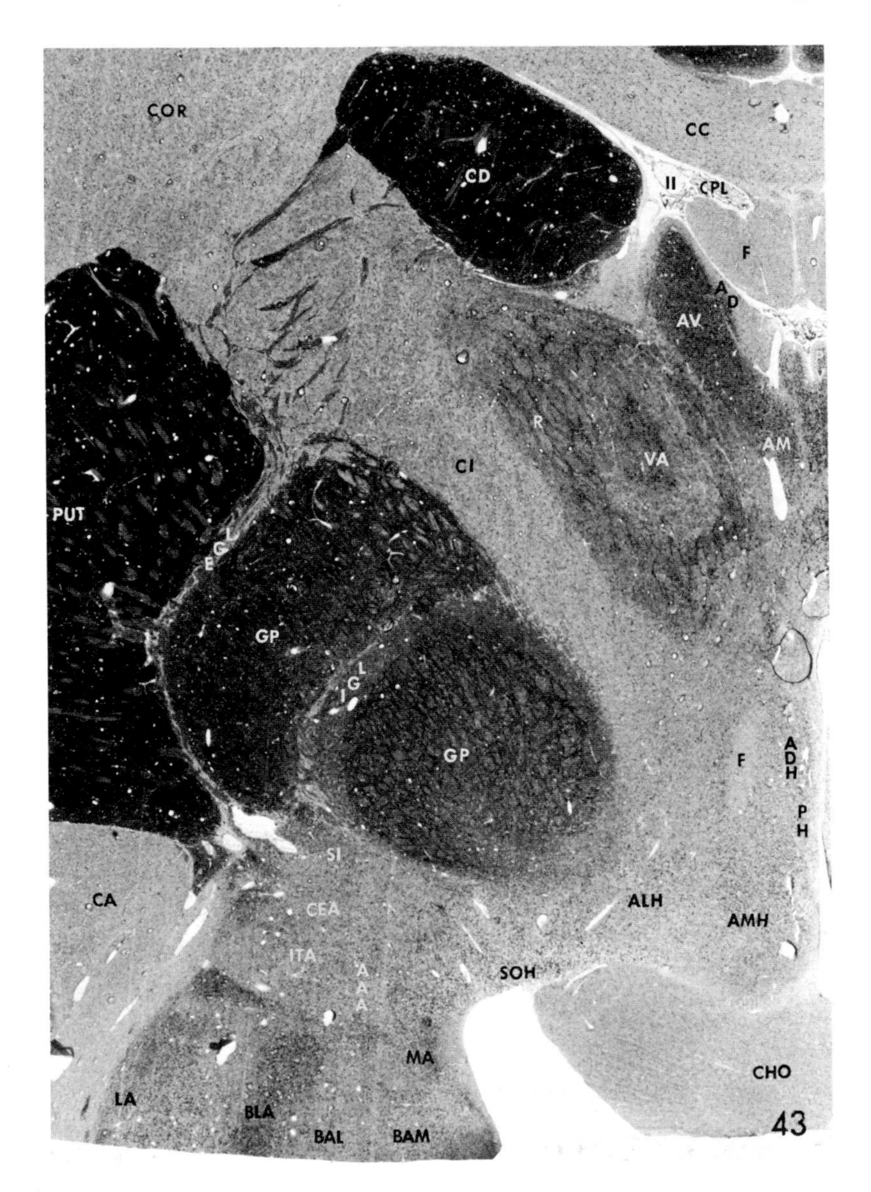

FIGURE 43. SE reaction at the same level as Fig. 37. The lentiform nucleus shows stronger activity than other areas at this level. ×7.

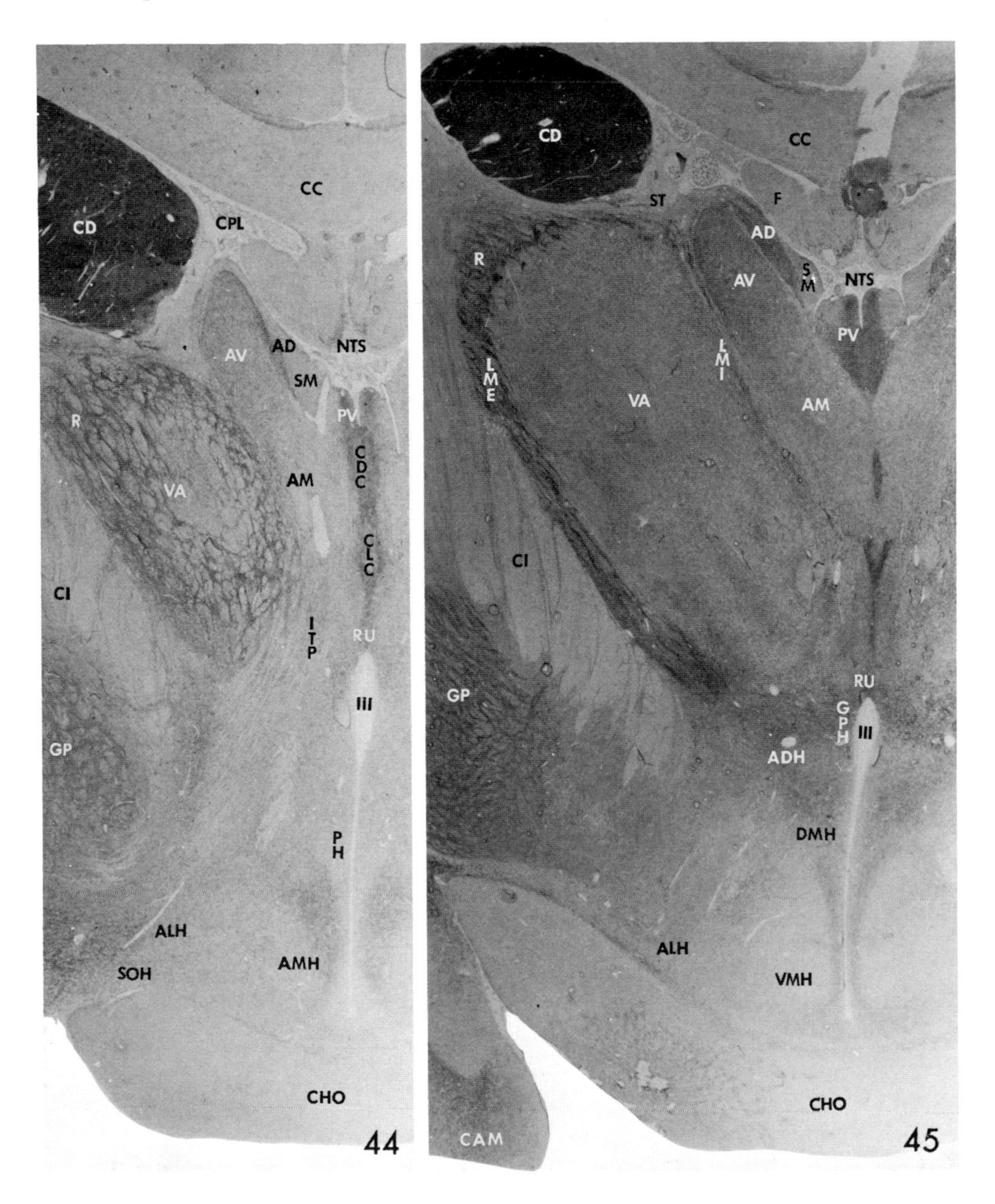

FIGURES 44, 45. Sections at the level of anterior and middle parts of the nucleus ventralis anterior thalami showing AChE activity. The fibers of the lamina medullaris externa and interna are more prominent than the neurons of the thalamic nuclei. ×7, ×7.

nuclei is mild and the cells negligible to mild, whereas in MAO the neuropil and the cells are mild to moderate. In the former, the fibers which form a capsule around the AD and AV give a moderately positive AChE activity. The AK preparations react negatively except for the blood vessels. The AV shows more blood vessels than the AM and AD.

Nuclei of midline

The midline nuclei are represented by groups of cells lying near the midline in the massa intermedia of the thalamus. This group is represented by nuclei parataenialis thalami (PT), paraventricularis thalami (PV) [Figs. 48–50, 52–58], nucleus centralis denso- (CDC) and latocellularis (CLC) thalami (Figs. 37, 44, 48–50), nucleus centralis superior (CS), intermediate (CIM) and inferior (CIF) thalami (Figs. 46, 47, 52–55), and nucleus reuniens thalami (RU) [Figs. 32, 37, 44–47, 52–55].

The PT is represented by an elongated group of medium-sized neurons located ventrolaterally to the stria medullaris thalami (SM). Walker (1938) divided the PT into medial and lateral parts in the macaque brain. The lateral part reaches below the SM almost to the level of the AD. The cells of the lateral part are more loosely arranged than those of the medial part. The neurons of the PT show mild to moderate activity in MAO, moderate in ATPase and AChE, moderate to moderately strong in SDH and SE, and a strong reaction in LDH. The enzyme activity, as a moderate and moderately strong reaction, is observed in the cell processes for a considerable distance in the ATPase and LDH preparations, respectively. The glial cells are also moderate in ATPase and LDH. The neuropil of the PT shows a varying reaction ranging from mild (AC, SE) to moderate (MAO, SDH), to moderately strong ATPase, AChE) and strong (LDH). AK reacts negatively except for the blood vessels, which also show a moderate AChE and a strong ATPase activity.

The PV occupies a position along the thalamic wall of the third ventricle below the subependymal layer of cells and consists mainly of small cells intermingled with a few medium-sized neurons. The PV shows moderate SDH, AChE, and MAO; moderate to moderately strong ATPase; moderately strong to strong SE; and strong to very strong LDH and AC activity in the neurons. The neuropil shows mild AChE and AC; moderate MAO, SDH, and SE; and strong LDH and ATPase activity. The AK preparations show only a few blood vessels which are moderately reactive.

The CDC, CLC, and RU are the midline nuclei located ventrally to the PV. The CDC is bounded dorsolaterally by the AM and laterally

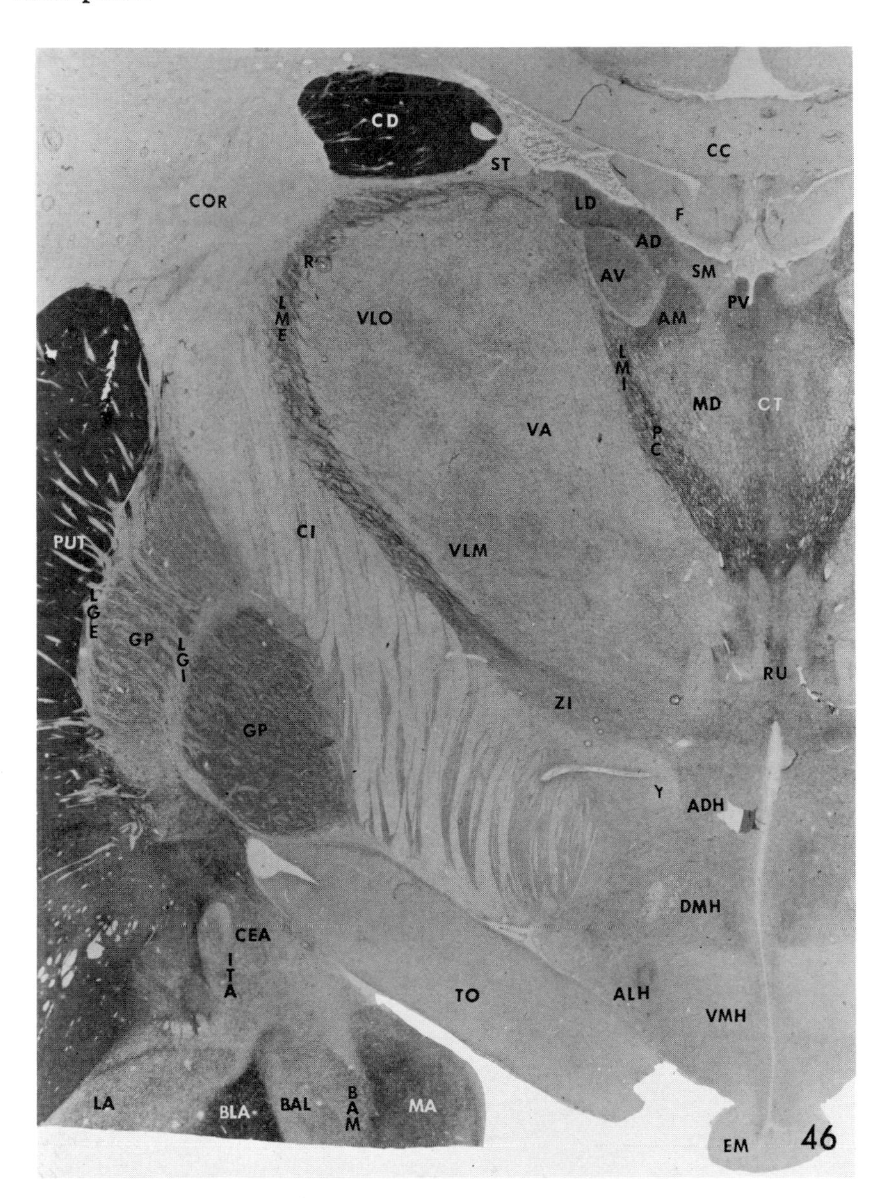

FIGURE 46. Section at the level of the eminentia medialis showing AChE activity in basal ganglia, amygdaloid complex, thalamus, and hypothalamus. ×7.

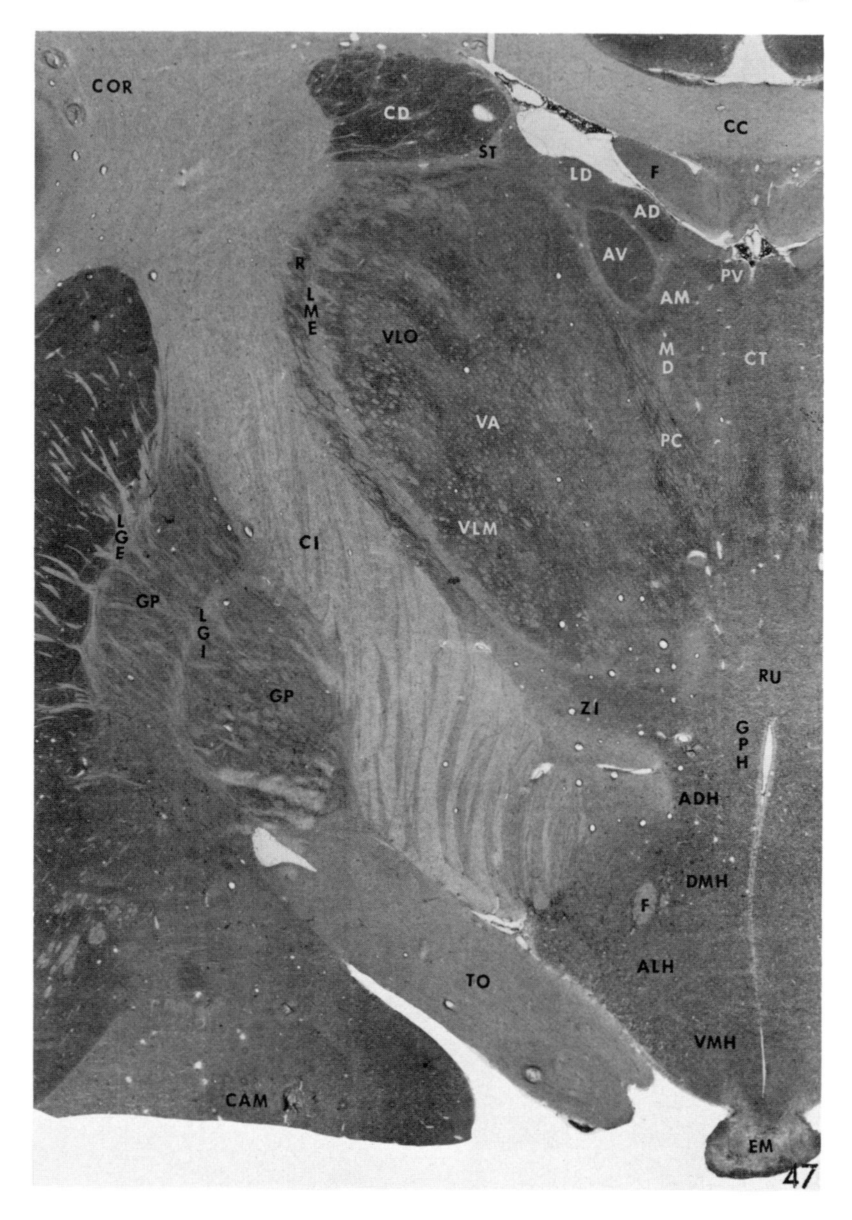

FIGURE 47. LDH preparation at the level of the eminentia medialis showing different grades of enzyme activity in the basal ganglia, and thalamic and hypothalamic nuclei. ×7.

by the MD. The medium-sized neurons are rather closely arranged and are fusiform, triangular, or star-shaped in outline. The CLC has larger cells than the CDC, and these are not as densely arranged. The CLC is located between the CDC and RU. The CDC shows stronger enzyme reaction than the CLC. In the former, LDH gives a strong reaction in the cells and a moderately strong one in the cell processes, glial cells, and neuropil; however, in the latter, the LDH reaction is moderately strong in the cells and moderate in the neuropil, cell processes, and glia. Both SDH and ATPase are one step weaker than the LDH and the blood vessels are strongly positive in ATPase. The AC preparations show strong cells, moderate cell processes, and mild neuropil in the CDC and somewhat weaker cells in the CLC. In SE, the cells in the CDC are strong, and those in the CLC moderate; the neuropil is moderate in the CDC and mild in the CLC. A comparison of AChE and MAO reveals that cells of the CDC are moderate in AChE, whereas those of the CLC are mild. In MAO, the cells of the CDC and CLC are moderately strong and moderate, respectively. The neuropil is moderate in AChE and MAO. The nerve fibers in this region are moderately strong in MAO, and the blood vessels are moderate in AChE. The blood vessels give a moderately strong AK activity.

The nucleus reuniens thalami is found on the ventral margin of the massa intermedia and consists mostly of small and medium-sized neurons with large cell processes and oval, ovoid, or triangular outlines. Toward its oral end, the RU is better demarcated, and on the ventrolateral side it merges with the area dorsalis hypothalami. The RU shows strong activity in AC; a moderately strong reaction in LDH, SDH, and SE; and moderate reaction in ATPase, MAO, and AChE in the neurons. However, in the neuropil the enzyme activity is mild to moderate in AChE and AC, and moderate in LDH, SDH, SE, ATPase, and MAO. The blood vessels show strong ATPase and moderate AChE and AK activity. The number of blood vessels in the RU is slightly less than in the other midline nuclei.

Olszewski (1952) described two additional nuclei in the rhesus monkey: nucleus rotundus (RO), located ventrally to the AM and dorsally to the CLC, and nucleus alaris (AL), located laterally to rotundus between the AM and tractus mammillo-thalamicus. These nuclei have been recognized in our histochemical preparations due to variation in enzyme activity from the neighboring nuclei. The RO consists of small groups of medium-sized plump cells, rather compactly arranged, and is situated in the midline. The cells of the AL are slender and triangular, smaller, and not as closely arranged as those of the RO. The neurons of the latter show strong LDH, ATPase, SE, and AC, moderately strong SDH

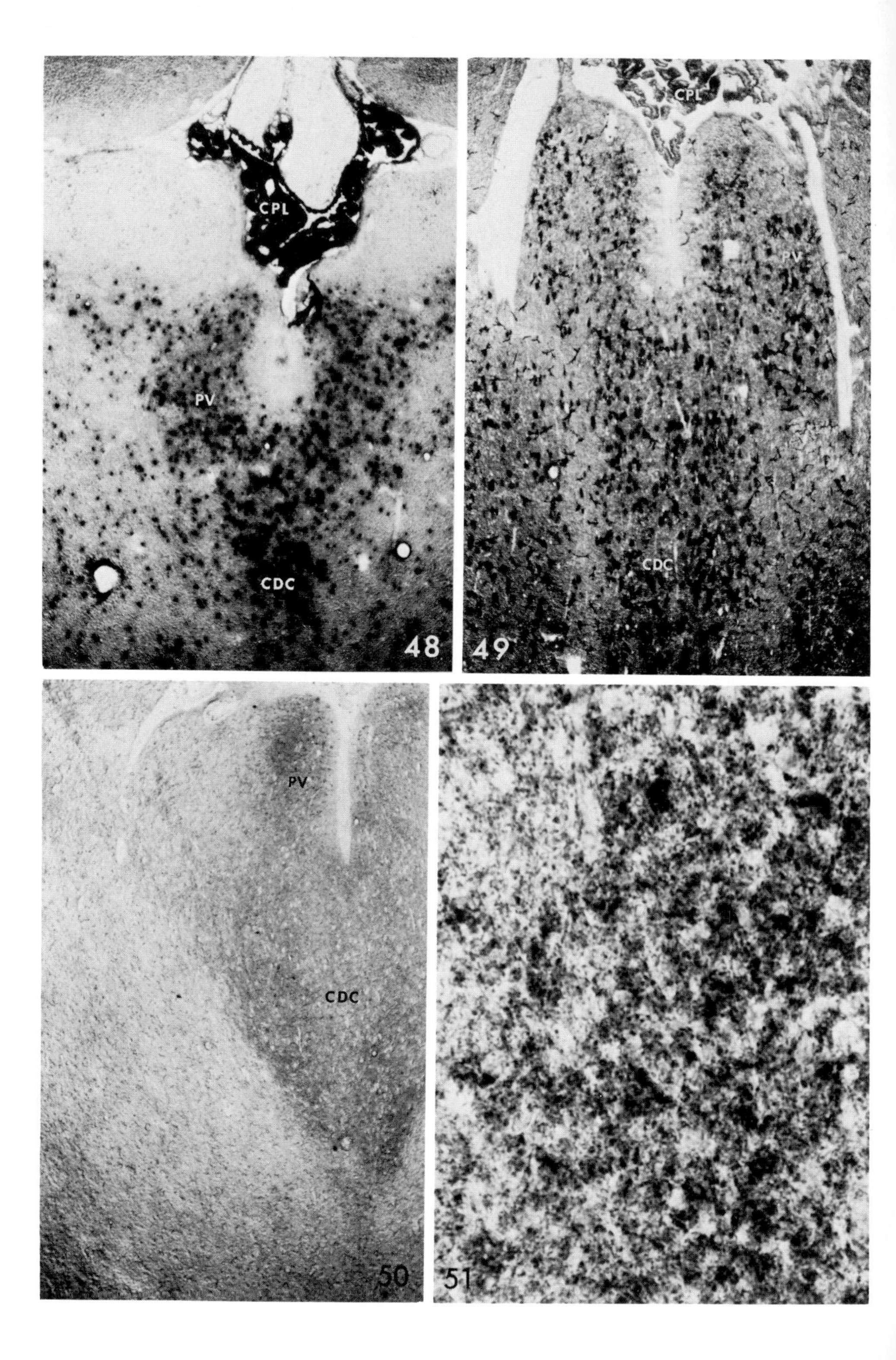
CPL
PV
CDC
48
CPL
PV
CDC
49
PV
CDC
50
51

FIGURES 48–50. Low power photomicrographs of the nucleus paraventricularis thalami showing AC, SE, and AChE activity. Strong activity in the neurons is evident in the AC and SE preparations compared to the mild to negligible reaction in the AChE preparation. ×37, ×37, ×37.

FIGURE 51. LDH preparation showing strong activity in the neurons of the magnocellular part of the nucleus medialis dorsalis thalami. ×370.

and moderate AChE and MAO activity; the neuropil is moderately strong in LDH and moderate in the other enzyme preparations mentioned above. The blood vessels are strong in ATPase, moderately strong in AK, and moderate in AChE in both the RO and the AL. The AL shows mild to moderate AChE and MAO; moderate SDH; and moderately strong LDH, ATPase, AC, and SE in the neurons, with mild SDH and AC, mild to moderate AChE and MAO, moderate ATPase and SE, and a moderately strong LDH reaction in the neuropil. The cell processes in the LDH, ATPase, SDH, and AChE show enzyme activity which is weaker than in the perikarya.

The nuclei centralis superior (CS), intermediate (CIM), and inferior (CIF) are the three divisions of the midline nucleus (CT) bounded dorsally by the PV, laterally by the MD, and ventrally by the RU in their oral extent. According to Olszewski (1952), these nuclei are the caudal extensions of the nucleus centralis densocellularis. Although the distribution of enzymes in the CS, CIF, and CIM is similar, they can be distinguished from each other by the varying size, density, and orientation of cells. The neurons of the CS are densely arranged and are medium-sized, whereas those of the CIM form a narrow layer of vertically arranged medium-sized cells which are not as closely packed as those of the CS. The CIF is dense at the oral end, but caudally the nucleus is sparsely populated. LDH and SDH show a moderately strong to strong reaction in the perikarya and moderate activity in the cell processes and neuropil, whereas in ATPase and MAO, the neurons and the neuropil are moderate. The AC reaction is moderate to moderately strong in the cells and mild in the neuropil, whereas mild activity prevails in both the perikarya and neuropil in the AChE preparations. Variation in enzyme activity has been observed in SE. Most neurons are moderately strong, but some cells of the CIF give a strong reaction. The neuropil is moderate in the CS and comparatively weak in the CIM and CIF.

Medial nuclei

The medial group, located between the midline nuclear group and the intralaminar group, includes the nucleus medialis dorsalis thalami

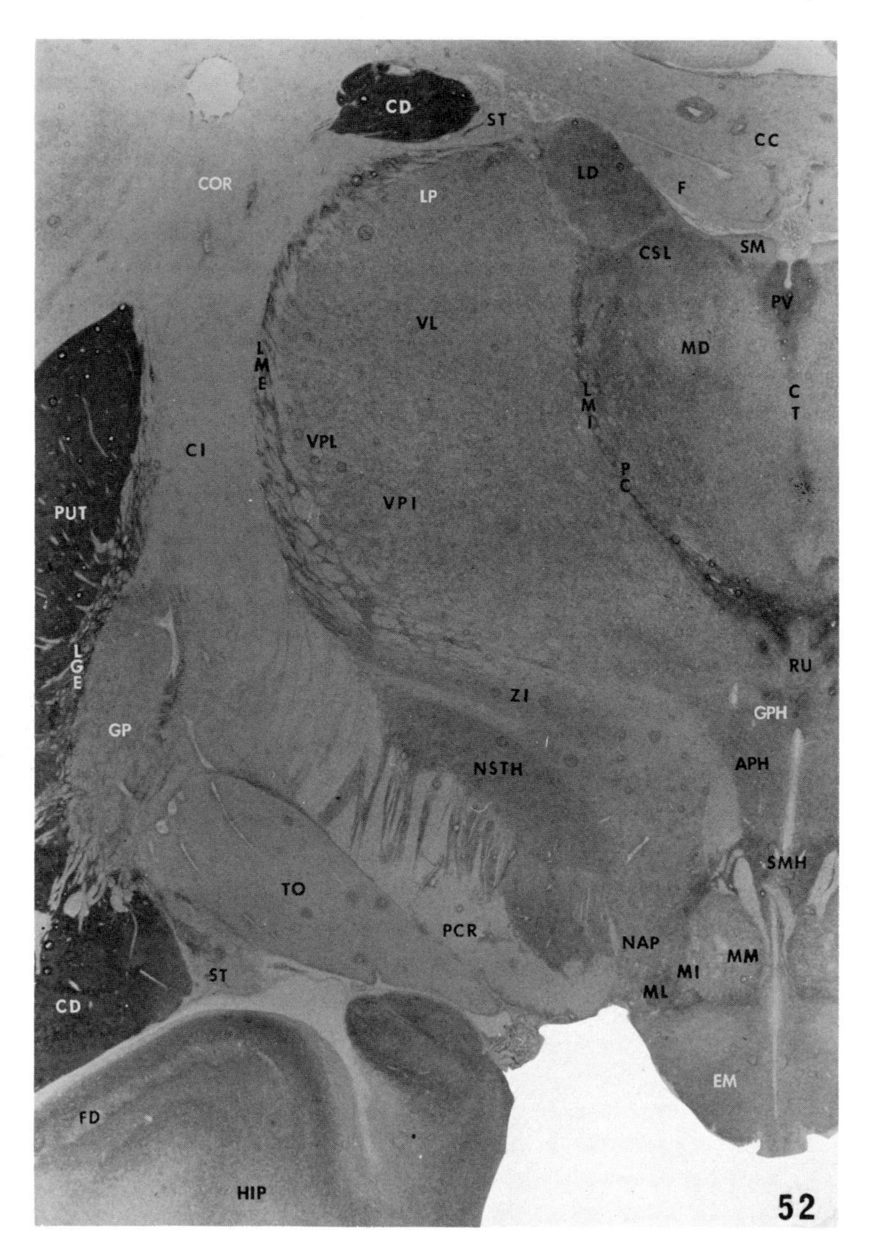

FIGURE 52. Section at the level of the mammillary bodies showing varying grades of AChE activity in the basal ganglia and the thalamic and posterior hypothalamic areas. ×7.

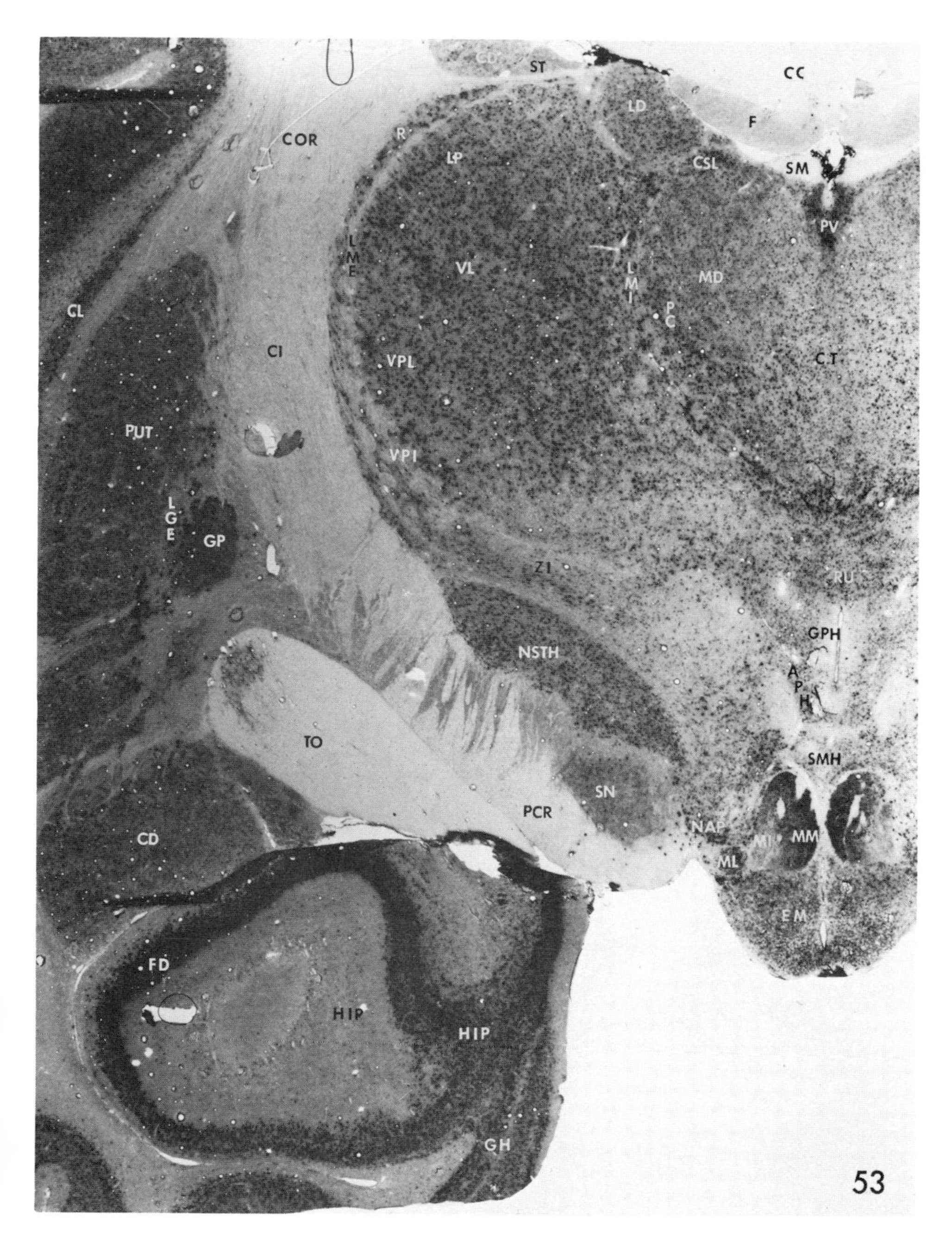

FIGURE 53. Section passing through the mammillary bodies showing acid phosphatase activity at this level. ×7.

(MD) [Figs. 51–59], centrum medianum thalami (CM) [Figs. 54–58], nucleus parafascicularis thalami (PF) [Figs. 54–58], nucleus subfascicularis thalami, and nucleus submedius thalami (SUM) [Figs. 52, 53]. The MD is a very prominent thalamic nucleus and extends rostrally from the level of the AM to the level of the habenular nuclei. Rostrally the MD consists of medial magno- and lateral parvocellular parts. The former consists of scattered large, plump, multipolar cells, whereas the latter consists of more cellular small and medium-sized cells. The peripheral part of the MD, consisting of a few large cells medial to the lamina medullaris interna thalami (LMI), is called the pars multiformis (Olszewski, 1952). At its caudal end, at the level of the habenular nuclei, the above divisions are not easily identifiable, and the MD consists of aggregated masses of medium-sized neurons. The neurons show a strong LDH reaction in the parvo- and magnocellular parts and very strong activity in the multiformis part, whereas the neuropil and the glial cells are moderately strong. The cells in the SDH are moderate in parvocellular, moderately strong in other parts, and mild to moderate in the neuropil. In ATPase the cells are moderate, glial cells mild to moderate, and neuropil moderately strong. In the AC preparations the neuropil is mild, while the cells are moderately strong (parvocellular) to strong (magnocellular and multiformis). The SE reaction is mild to moderate in the neuropil and strong in the neurons. Some neurons in the multiformis area show stronger enzyme activity. The neuropil is moderate in AChE, but the enzyme activity is more prominent in the magnocellular part. The perikarya gives mild AChE activity. The neurons in the magnocellular part give a moderate MAO reaction compared to a mild one in the parvocellular part. The neuropil is mild. The blood vessels are positive in AK and ATPase and are uniformly distributed, although the multiformis area shows a few more vessels than the other parts.

The CM is a well-developed nucleus in the middle third of the thalamic area, demarcated by the LMI, except on the medial and inferior sides where it abuts the PF. The CM consists mostly of medium-sized, slightly elongated neurons, except in its medial part, where larger and more compact cells are observed. The neurons give a mild to moderate AChE, ATPase, and MAO; moderate LDH and SDH, and moderately strong and strong SE and AC reaction. In the neuropil, on the other hand, the enzyme activity is mild in SDH; mild to moderate in AChE, AC, and SE; and moderate in LDH, ATPase, and MAO. The blood vessels are strong in ATPase, moderately strong in AK, and moderate in AChE; they appear to be less numerous in the CM than in the surrounding nuclei.

The PF is located ventromedial to the CM, and toward its rostral

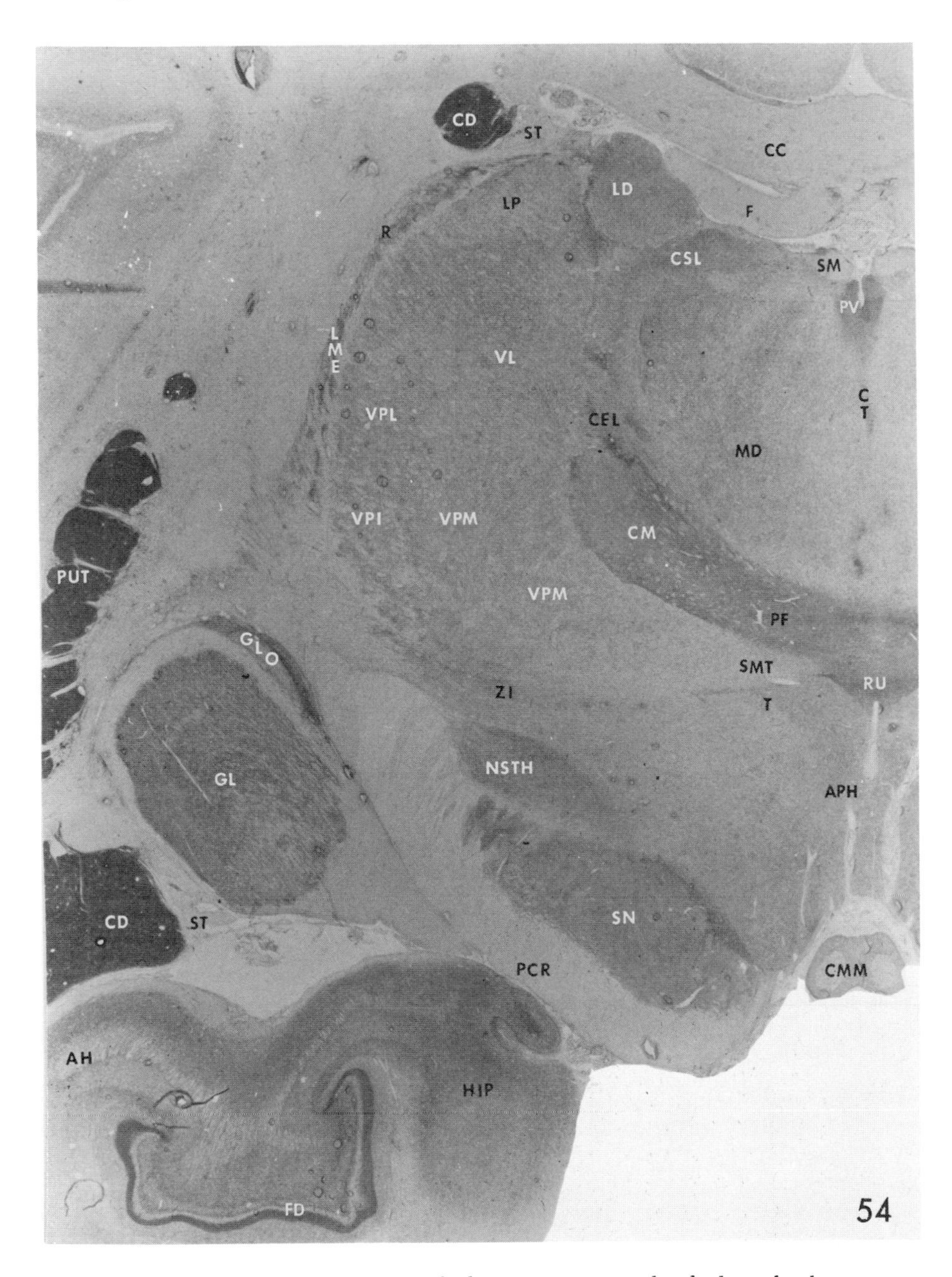

FIGURE 54. AChE preparation of the section at a level through the corpus geniculatum laterale. Note the varying enzyme activity in the different nuclei of the thalamus and other areas. ×7.

end, the neuropil of both nuclei blend smoothly into each other. Its neurons are larger and plumper than those of the CM, a bit more compactly arranged, and elongated, multipolar, or triangular in shape. They give a strong to very strong reaction in SE and AC, moderately strong in LDH, and moderate in SDH, ATPase, AChE, and MAO. The neuropil is moderately strong in LDH and moderate in all other enzymes. The MAO activity in the PF is slightly more than that of the CM although in both nuclei it can be defined as moderate. The blood vessels are ATPase-, AChE-, and AK-positive and are less numerous than the adjoining thalamic nuclei.

The nucleus subfascicularis in the rhesus monkey has been described by Papez and Aronson (1934) as being situated ventral to the medial tip of the CM and consisting of dorsal magno- (medium-sized) and ventral parvocellular (small) neurons. The larger cells show stronger activity than the smaller ones. The latter are moderately strong in SE, LDH, AC, and SDH and moderate in ATPase, AChE, and MAO. The neuropil is slightly stronger than the neurons in ATPase and LDH and weaker in the other preparations. The cell processes also show moderate to moderately strong enzyme activity in the ATPase, SE, and LDH preparations.

The nucleus submedius thalami (SMT), also called the nucleus ventralis medialis (Peele, 1961), has been included in the nucleus paracentralis by Crouch (1934b), but since it is histochemically distinct from the latter and has different fiber connections (Walker, 1937), it may well be described as a separate nucleus. The SMT is located ventromedial to the nucleus paracentralis thalami and consists of medium-sized cells which are fairly closely arranged (Figs. 54–57). The neurons give similar enzyme reactions in the LDH, SE, AC, and SDH preparations compared to the nucleus paracentralis (PC). The MAO reaction in the neuropil is stronger than that of the PC and a bit weaker than that of the PF.

Intralaminar nuclei

The intralaminar group of nuclei is located between the medial and lateral groups in association with the LMI. It includes the nucleus paracentralis thalami (PC) [Figs. 52, 53], nucleus centralis lateralis thalami (CEL) [Figs. 54, 57, 58] and the nucleus centralis superior lateralis thalami (CSL) [Figs. 52–58].

The PC surrounds the rostral part of the MD and consists of medium-sized and large multipolar neurons. At certain places, the fibers of the LMI give this nucleus a patchy appearance, particularly in its oral part. Most of the neurons give a strong to very strong reaction in AC, SE, and LDH; moderately strong in SDH and ATPase; moderate in AChE; and mild in MAO. The glial cells show strong LDH and mild to moderate SE, AC, and SDH activity. The cell processes, with somewhat weaker enzyme activity than the perikarya, have also been observed in the LDH, SDH, SE, and AChE. The neuropil, although interrupted by the fibers of the LMI, is moderately

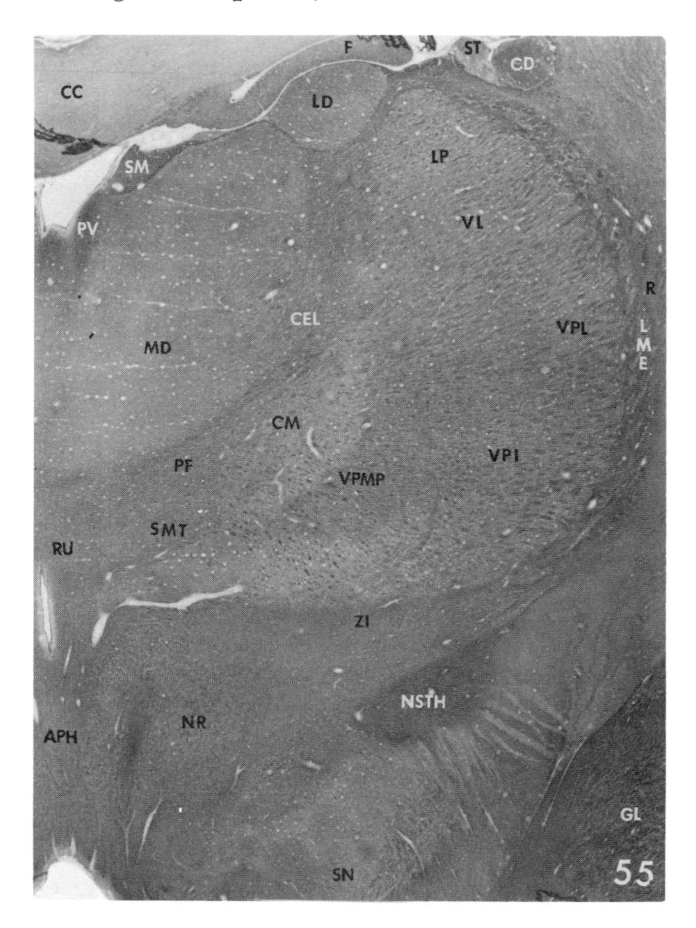

FIGURE 55. Section through the thalamus showing MAO activity. This enzyme reaction is more prominent in the nerve fibers than in the neurons. ×7.

strong in LDH and ATPase; moderate in SE, SDH, AChE, and MAO; and mild in AC.

The CEL is formed as groups of medium-sized cells among the fibers of the LMI, caudal to the PC in a dorsal position. Its maximum dimensions are at the level of the CM. According to Olszewski (1952), the

CEL cells show peripherally arranged Nissl substance and an eccentric nucleus. As in other thalamic nuclei, the neurons are strong in LDH, SDH, AC, and SE; moderate in ATPase and AChE; mild in MAO; and negative in AK. The neuropil is weaker than the perikarya in SE, AC, LDH, and SDH; stronger than the perikarya in ATPase; and the same as the perikarya in AChE and MAO activity. The blood vessels are positive in ATPase, AChE, and AK preparations.

The CSL forms a cap over the MD, and the fibers of the LMI run along its ventral aspect. Rostrally, the CSL is located between the AM, LD, and MD. Most of the neurons are of medium size and are densely arranged, plump, and ovoid or triangular in shape. They are strongly stained in the LDH, SE, and AC; moderately strong in SDH; moderate in ATPase; and mild to moderate in the AChE and MAO preparations. The neuropil is stronger than the perikarya in the ATPase preparations.

Lateral nuclei

This group is represented by those thalamic nuclei which are located between the LMI and LME, anterior to the pulvinar, such as the nucleus reticularis thalami (R), nucleus lateralis dorsalis thalami (LD), lateralis posterior thalami (LP), ventralis anterior (VA), ventralis lateralis (VL), and ventralis posterior (VP) thalami. The R is a semicircular group of neurons which form a sort of cap on the lateral, anterosuperior and anteroinferior aspects of the dorsal thalamus. Ventrolaterally, the R is continuous with the zona incerta; both seem to share a common ventral thalamic origin (Rose and Woolsey, 1943; Rose, 1952) [Figs. 32, 37, 43–47, 52–59, 62, 63]. Crosby *et al.* (1962), however, described it as a constituent of the dorsal thalamus. We have also included it in the lateral nuclei in spite of its location outside the LME. Olszewski (1952) emphasized that the R is cytoarchitecturally identical to the underlying thalamic nuclei and felt that it should be considered a part of them. Scheibel and Scheibel (1966) have shown that the great majority of the neurons of the R send their large axons into the thalamus where they synapse widely with neurons of specific and nonspecific systems. However, distinct histochemical variations are not observed in enzyme reactions in the neurons of the R at various levels, as might be expected if they were parts of different thalamic nuclei at different levels and positions. The perikarya show strong LDH, SE, and AC; moderately strong SDH; moderate ATPase; and mild MAO activity. The AChE activity is moderate in the perikarya as well as in the cell processes. The neuropil gives moderately strong to strong LDH and ATPase; moderate SE; and mild to moderate AC, AChE, MAO, and

SDH activity. The sparsely scattered blood vessels show strong ATPase, moderate AK, and mild AChE activity.

The LD is located caudal to AV in almost the same position; rostrally they are separated from each other at the point of termination by a thick lamella of myelinated fibers (Olszewski, 1952) [Figs. 46, 47, 52–58]. The LD consists mostly of medium-sized cells. The LP is located ventrolateral to the LD and consists of evenly distributed medium-sized, multipolar cells (Figs. 52–59). The LD and LP are similar histochemically. The neurons are strong in AC, moderately strong in LDH and SDH, moderate in ATPase and MAO, and mild in AChE. In the SE preparations, the neurons of LP show stronger activity (strong) than those of the LD (moderately strong), while the neuropil is mild in the latter and moderate in the former. The neuropil in LDH, SDH, and MAO is slightly weaker than the perikarya, whereas in ATPase it is the same (moderate) as the neurons. The neuropil for AChE is moderate and mild in the LD and LP, respectively.

The VA is the anteriormost part of ventral nuclear group and is bounded orally by the LME and LMI (Figs. 37, 43–47). The cells of the VA are predominantly medium-sized, multipolar, and are grouped into clusters due to numerous fibers passing through this nucleus. The VA gives a strong SE, AC, and LDH reaction; moderately strong SDH; and moderate ATPase reaction in the neurons. The neurons show mild AChE and MAO, and negative reaction in AK. The neuropil is weaker than the perikarya except in ATPase.

The VL is the ventrolateral part of the thalamus (Figs. 46, 47, 52–58) and is divisible into pars oralis (VLO), pars medialis (VLM), and pars caudalis (VLC). The medial part consists mostly of small cells compared to the oralis and caudalis. The VLO is very cellular and consists of round, oval, and plump cells arranged in large clusters. The VLC is not as cellular and consists of almost uniformly scattered, large multipolar cells. Histochemically, the VLO and VLC are similar, whereas the VLM shows a one-step weaker enzyme reaction in its smaller neurons and in the neuropil. The reactions described below, therefore, represent the VLO and VLC. The neurons are strong to very strong in SE, AC, and LDH; moderately strong to strong in SDH; moderate in ATPase and AChE; mild in MAO; and negative in AK preparations. A reaction of equal or slightly less intensity is observed in the cell processes in AC, LDH, SDH, and ATPase. The neuropil is strong in LDH; moderate in SE, AC, SDH, ATPase, and AChE; and mild in MAO. The glial cells give a moderately strong LDH, moderate SE, and mild AC and SDH reaction. The blood vessels show very strong ATPase, moderately strong AK, and moderate AChE activity.

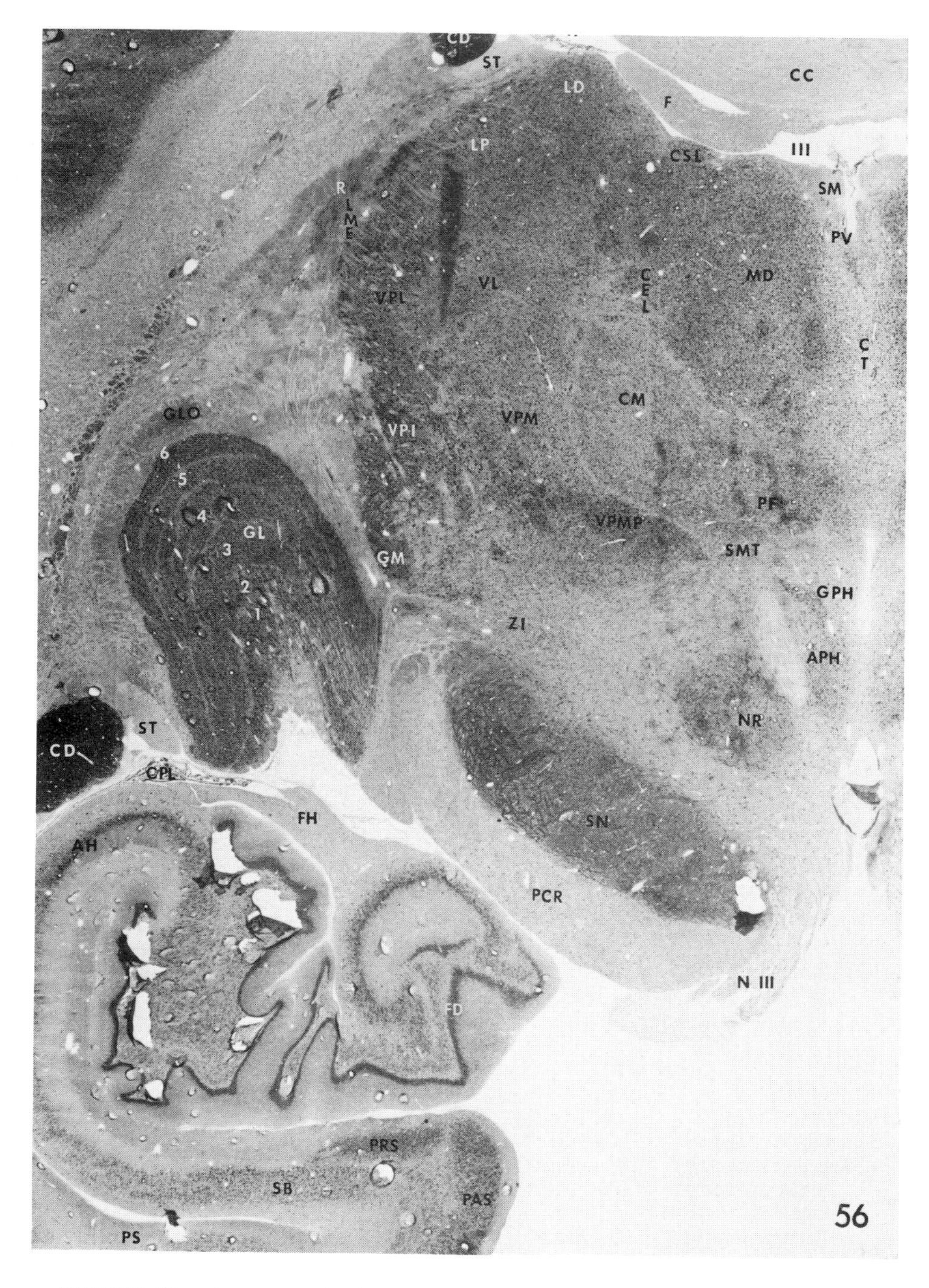

FIGURE 56. Section at the level of the corpus geniculatum laterale showing SE activity in the various nuclei of the thalamus. ×7.

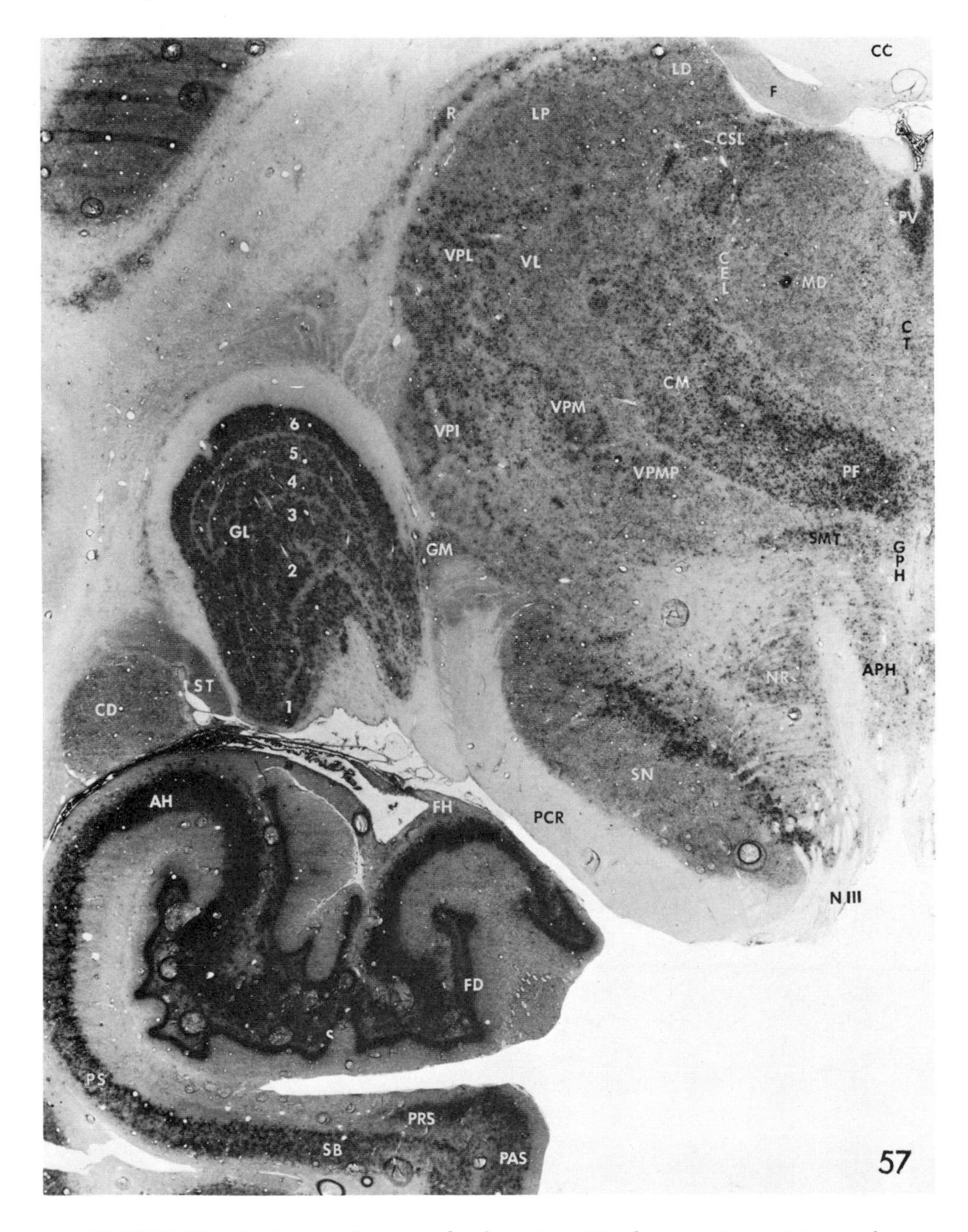

FIGURE 57. Section at the same level as Fig. 56, showing AC activity in the thalamus, corpus geniculatum laterale, and hippocampal area. Note the strong reaction in the neurons in all the areas. ×7.

The VP is situated in the ventrolateral part of the thalamus rostrally and is ventral to it posteriorly (Figs. 52–58). This nucleus is divisible into lateral (VPL), inferior (VPI), and medial (VPM) parts. The last of these shows a group of small cells which are ventromedially placed called the parvocellular part (VPMP) of the VPM. The VPL consists of small, medium, and large cells with a greater number of the last two categories. The VPM has more of the medium-sized cells with some smaller neurons. The VPMP shows small cells; the VPI also shows predominantly small cells with some medium-sized neurons. In the LDH preparations, the neurons are moderately strong to strong in the VPMP and VPI and very strong in the VPL and VPM, and the enzyme activity of the same intensity as the perikarya passes on to the cell processes for some distance. The glial cells are moderately strong, and the neuropil is moderate to moderately strong except in the VPI, where it is mild. In SDH, the neurons and neuropil are strong in the VPL and VPM, moderately strong in the VPMP, and moderate in the VPI. The SDH activity passes into the cell processes as a moderate reaction. The AC and SE activity are strong to very strong in the cells and mild to moderate in the neuropil and cell processes. The ATPase, AChE, and MAO reactions are almost uniform in the various parts. The ATPase is moderate to moderately strong in cells, moderate in neuropil and glial cells, mild to moderate in the nerve fibers passing through this area, and strong in blood vessels. The AChE is mild to moderate in the neuropil and perikarya and similar reactions in the cell processes have been observed in some neurons. The blood vessels give a moderate reaction as seen in the AK preparations and are more numerous in the VPM than in the other parts of this nucleus. The MAO shows mild positive activity throughout the ventral posterior nucleus. A number of long nerve fibers running through this area show moderate MAO activity.

Posterior nuclei

This group represents the posterior part of the thalamus and includes the nucleus pulvinaris, nucleus limitans thalami (LMT), and the regio praetectalis (APT) [Fig. 59]. The nucleus pulvinaris occupies a large area in the posterior region of the thalamus. It shows a uniform structure throughout with similar histochemistry and consists of mostly medium-sized and multipolar neurons; yet from a topographical point of view it is divided into medial (PUM), lateral (PUL), and inferior (PUI) parts (Fig. 59). In its oral extent, when the above divisions are not clearly defined, it is generally referred to as pars oralis (PUO). The PUM is the largest of all and shows evenly dispersed neurons with no intermixing of larger cells as seen in the PUL. The PUI is well

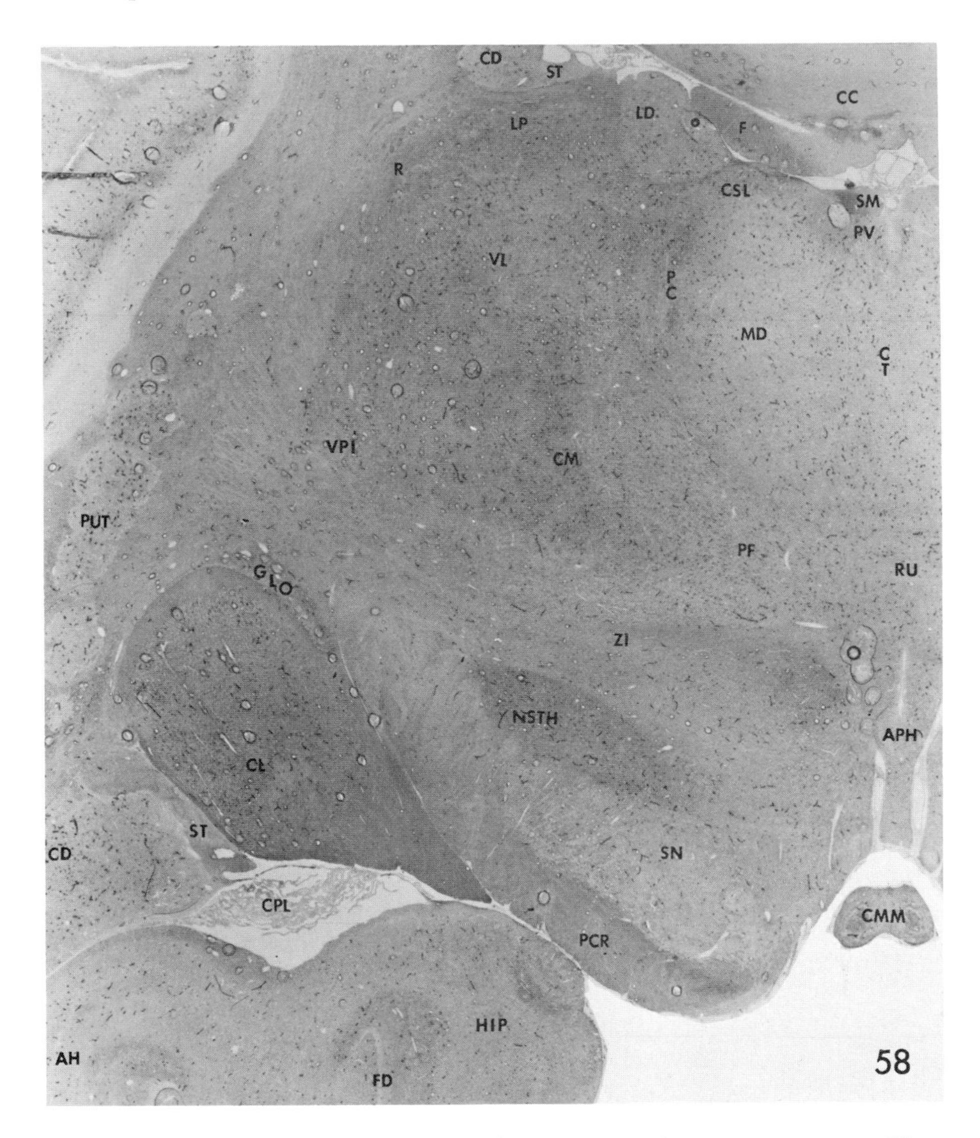

FIGURE 58. Section at the same level as Fig. 54, showing AK activity. The enzyme activity is exclusively localized in the blood vessels. ×7.

marked from the PUL and PUM by a number of fibers (brachium of colliculus superior) passing dorsally. For details see Olszewski (1952).

The neurons show strong SE and AC, moderately strong LDH, moderate SDH and ATPase, mild AChE and MAO, and negative AK reaction. The blood vessels are uniformly distributed with a moderate AK activity and also react positively with AChE and ATPase. The PUM, however, shows a larger number of blood vessels than the PUL, PUI, or PUO. The neuropil is moderately strong in LDH and ATPase; moderate in SE and SDH; and mild in AC, AChE, and MAO.

The LMT or the nucleus of the optic tract of Aronson and Papez (1934) consists of a narrow band of oval or elongate spindle-shaped cells located on the dorsolateral surface of the regio praetectalis and extending between the nucleus suprageniculatus and densocellular part of the MD. The pretectal area extends between the superior colliculus and the LMT and consists of small, spindle-shaped cells which are parallel to one another. According to Olszewski (1952), it is possible that the LMT is not a distinct functional unit and may be a part of the nucleus suprageniculatus because of its morphological resemblance to the latter. The nucleus limitans and suprageniculatus in the rhesus are identical except that in the lateral part of the latter, the neurons are somewhat smaller in size and more loosely arranged. The perikarya give a very strong SE reaction which continues into the cell processes as moderate activity. The AC, LDH, and SDH also show strong activity in the perikarya and moderate to moderately strong reaction in the cell processes. The neurons are moderate in ATPase, AChE, and MAO. In AChE some neurons also give a negligible enzyme reaction. The fibers crossing these areas are moderately strong in MAO. The neuropil of these nuclei is moderate in most preparations except in LDH and ATPase, where it is somewhat stronger.

SUBTHALAMUS

This area includes the zona incerta (ZI) [Figs. 46, 47, 52–56, 58], nucleus subthalamicus (NSTH) [Figs. 60, 61], and campus Foreli and its nucleus.

The zona incerta is situated between the thalamic and lenticular fasciculi and is continuous laterally with the nucleus reticularis thalami and medially with the nucleus of the Forel's field. The latter consists of a tegmental field of Forel (prerubral field, H), fasciculus thalamicus (H_1), and ansa and fasciculus lenticularis (H_2). There are scattered neurons found among the fibers of Forel's field H, H_1, and H_2. The neurons found in the area H are called the nucleus of the prerubral or tegmental field. The neurons in the Forel's field and the ZI are similar. They are small and medium-sized and are triangular and oval in shape. In the ZI, the cells are compactly arranged. Histochemically the ZI and the neurons in the Forel's field are similar, and since the cells are compactly arranged fewer fibers pass through the ZI and the neuropil is enzymati-

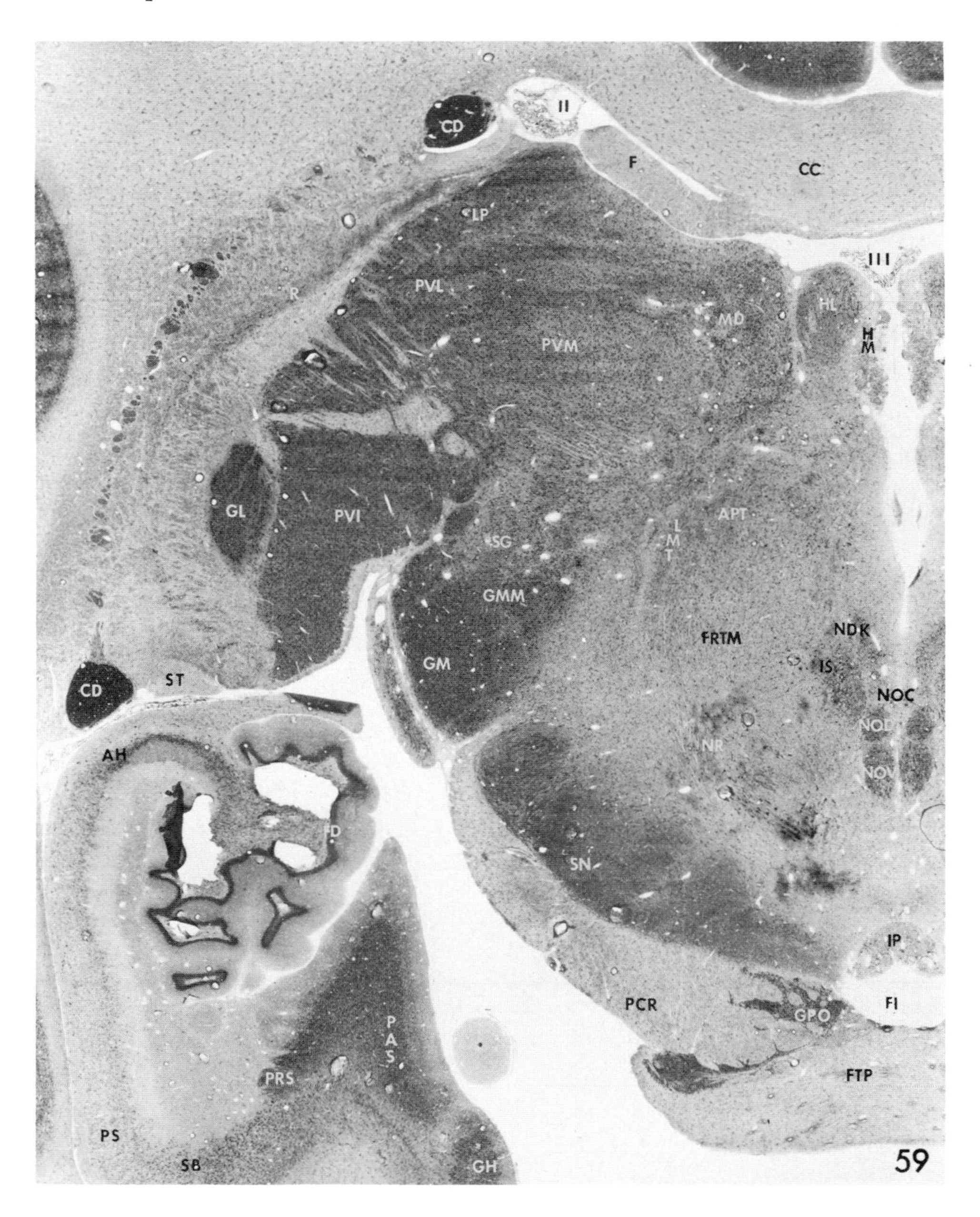

FIGURE 59. SE preparation at the level of the habenular nuclei and the nucleus n. oculomotorii. The enzyme activity in the pulvinar and the medial geniculate body is stronger than in the other nuclei. ×7.

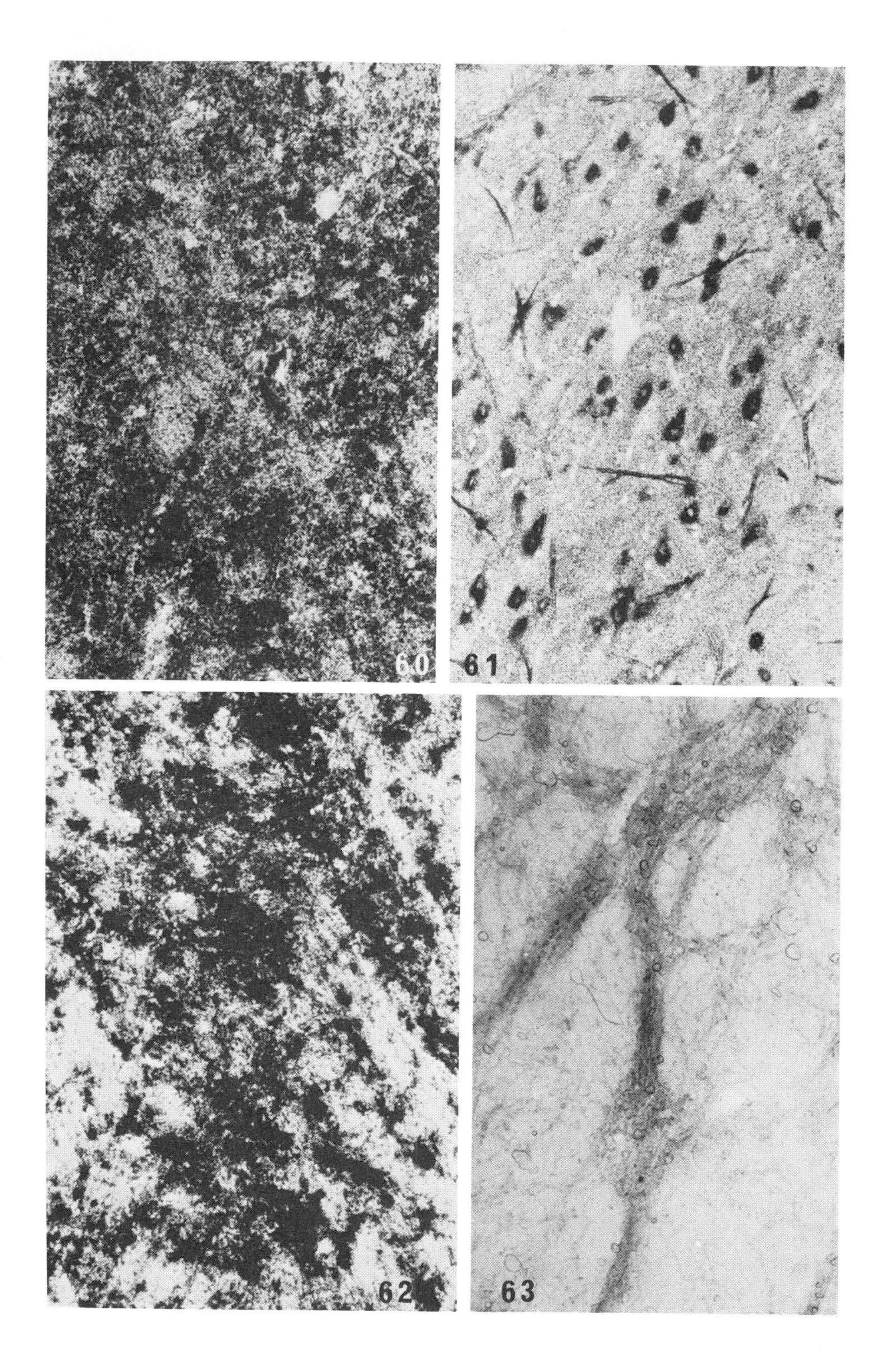
60
61
62
63

FIGURES 60, 61. SDH and SE preparations showing enzyme activity in the nucleus subthalamicus. The neurons give a similar reaction, whereas the neuropil shows a remarkable difference in the distribution of these enzymes. ×370, ×148.

FIGURES 62, 63. Nucleus reticularis thalami showing ATPase and AChE activity, respectively. The neuropil is enzymatically more active than the neurons. ×148, ×148.

cally many times more active than that of the Forel's field. The neurons exhibit strong SE, AC, and LDH; moderate SDH; mild ATPase, AChE, and MAO; and negative AK activity. The neuropil in the ZI shows moderate SE, LDH, and ATPase; mild AC, SDH, AChE, and MAO; and negative AK activity. The blood vessels show positive activity for AK, AChE, and ATPase. The ZI contains fewer blood vessels than the surrounding thalamic areas and NSTH.

The NSTH is a biconvex, lens-like mass bounded dorsomedially by the fasciculus lenticularis (H_2) and ventrolaterally by the posterior part of the capsula interna and pedunculus cerebri. The NSTH is composed mostly of medium-sized neurons, mixed with a few large and small cells which are oval, triangular, star-shaped, and fusiform. The cells are uniformly distributed throughout the nucleus and are fairly compactly arranged when compared to the ZI. Histochemically, the neurons show strong AC, SE, and LDH; moderately strong SDH; moderate ATPase; mild to moderate AChE and MAO; and negative AK activity. The neuropil shows moderately strong LDH and SDH; mild to moderate AC and MAO; moderate SE, ATPase, and AChE; and negative AK activity. The AChE activity is found in the cell processes for some distance. The NSTH, in general, stands out as a distinct nucleus in most of the histochemical preparations. The blood vessels show AK, AChE-, and ATPase-positive reaction. The glial cells in the NSTH show strong LDH; moderate SE, ATPase, and MAO; and mild AC and SDH activity.

METATHALAMUS

The metathalamus is composed of the corpus geniculatum laterale (Figs. 54–58, 64–67) and mediale (Figs. 56, 57, 59) situated on the posterolateral surface of the brain stem inferior to the pulvinar.

The lateral geniculate complex is divisible into dorsal and ventral parts (Olszewski, 1952). The dorsal part (GL) consists of six concentric layers of cells interrupted by alternating layers of fiber bundles, and these are numbered 1 to 6. Layers 1 and 2 are made up of large plump

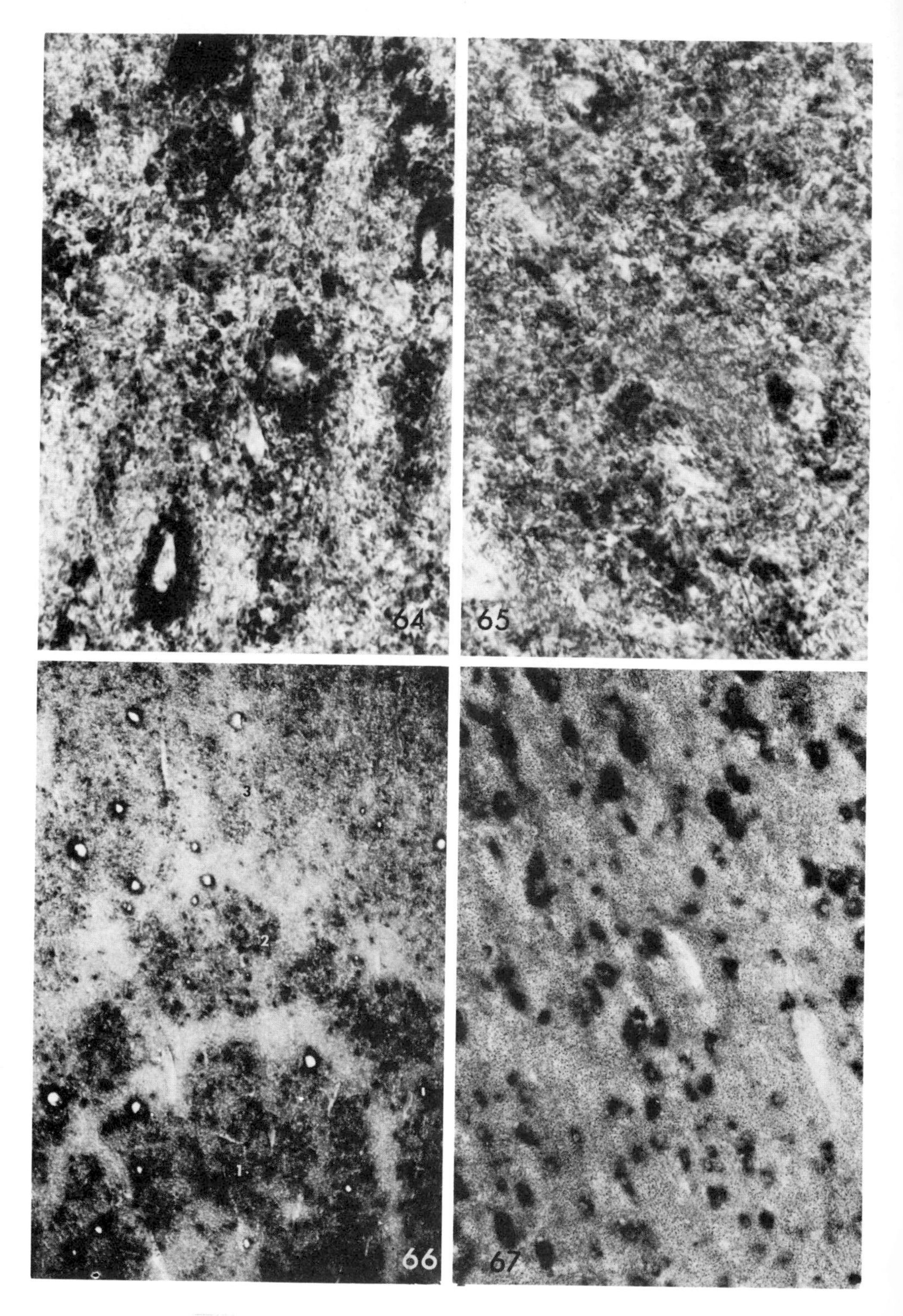
64
65
3
2
1
66
67

FIGURES 64, 65. LDH and SDH preparations of the magnocellular part of the corpus geniculatum laterale. The neurons show stronger activity in LDH compared to the SDH preparations. ×592, ×592.

FIGURES 66, 67. Magno- and parvocellular part of the corpus geniculatum laterale showing AC and SE activity, respectively. The neurons of the magnocellular part are more strongly reactive than those of the parvocellular part. ×37, ×370.

cells having ovoid and round shapes with prominent nucleoli; this area is also called the magnocellular part. Layers 3 to 6 are made up of mostly medium-sized neurons mixed with a few small cells. They are less plump, lightly stained, and oval, round, or slightly triangular with less prominent nucleoli; this region is referred to as the parvocellular part. The ventral part is further divided into oral (GLO) and caudal (GLC) parts (Olszewski, 1952). The GLO is made up of small cells, and the GLC consists of scattered large neurons. In myelin stains, the GLO stains lightly, whereas the GLC stains much darker. According to Olszewski (1952) the GLC may belong to the nucleus peripeduncularis (PD) because the former merges with the PD and both are cytoarchitecturally similar.

Histochemically, the six-layer configuration of the GL is maintained in all the enzyme preparations. The magnocellular part shows more enzyme activity than the parvocellular part and gives strong to very strong SE, LDH, and AC; strong SDH; moderate ATPase and AChE; mild MAO; and negative AK activity. The neuropil gives moderately strong LDH; moderate SE, SDH, ATPase, AChE, and MAO; mild AC; and negative AK activity. The AChE activity in the perikarya of the parvocellular part is mild compared to moderate activity in the magnocellular part. The nerve fibers in between the cell layers show moderate LDH; mild MAO, SE, ATPase, and AChE; negligible AC and SDH; and negative AK activity. The number of AK-, AChE-, and ATPase-positive blood vessels appears to be the same in both the magno- and parvocellular parts. The GL contains more blood vessels than the GLC or GLO.

The cells of the GLO are comparatively less active enzymatically than those of the GLC. On the other hand, the neuropil of the GLC is much more active enzymatically than that of the GLO. In general, the neurons give strong SE and LDH, moderately strong AC and SDH, moderate ATPase and AChE, mild MAO, and negative AK activity. The neuropil shows moderately strong LDH and ATPase; moderate SE, AChE, SDH, and MAO; mild AC; and negative AK activity. The MAO activity is greater in the neuropil than in the cells. The enzyme activity in the nucleus peripeduncularis is similar to that seen in the GLC.

The GM is located medial to the GL but, in contrast, does not show

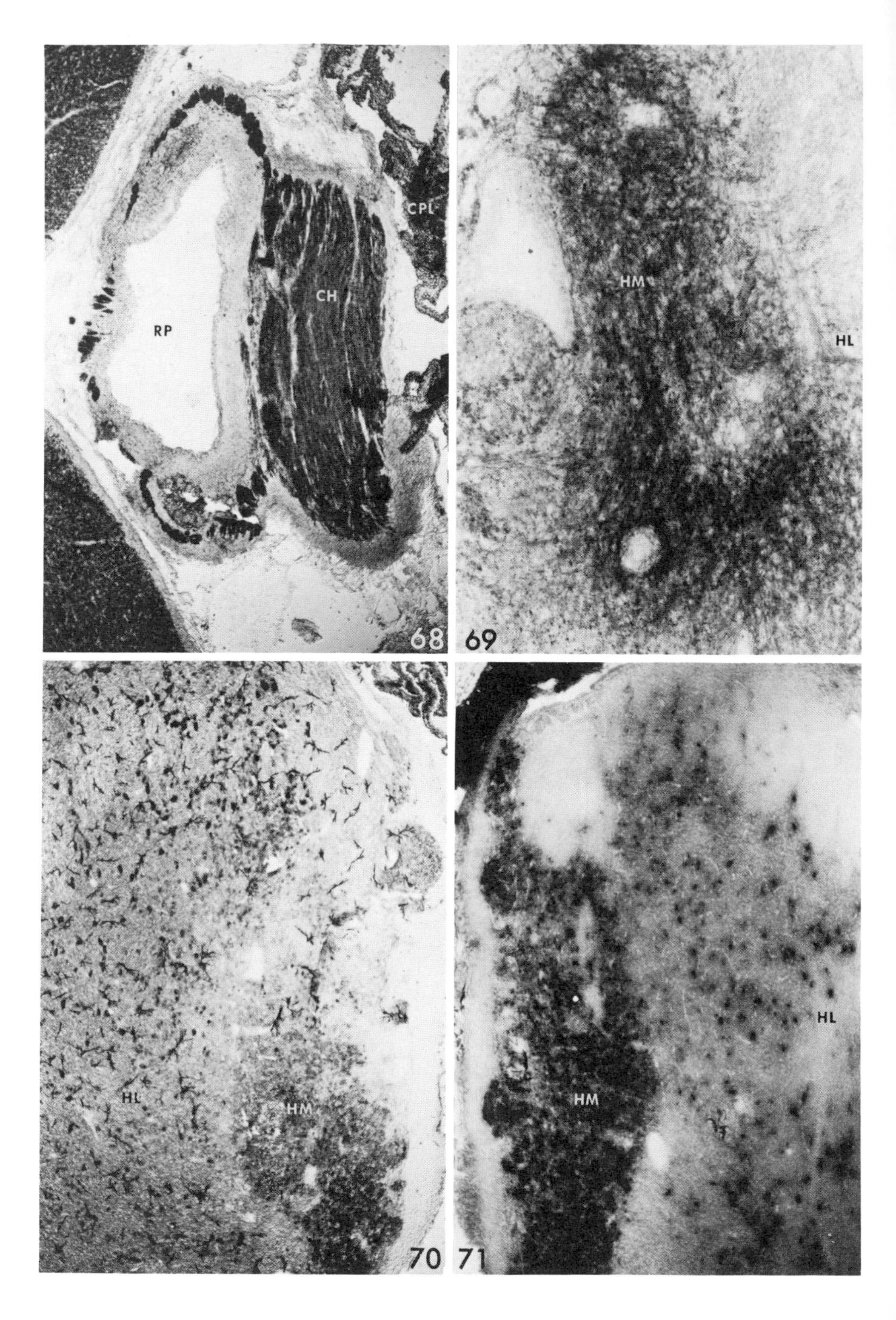
CPL
CH
RP
68
HM
HL
69
HL
HM
70
HM
HL
71

FIGURE 68. MAO preparation showing very strong reaction in the commissura habenularis. Note also the strongly positive fibers surrounding the recess pineale. ×37.

FIGURES 69–71. Lateral and medial habenular nuclei showing MAO, SE, and AC activity, respectively. Note particularly the strong activity in habenularis medialis. ×37, ×37, ×37.

any lamination. The GM is also divisible into the parvocellular and magnocellular parts; the latter is located dorsomedial to the former. The magnocellular part gives somewhat stronger enzyme reactions than the parvocellular part, but in general the enzyme activity is the same as shown by the parvocellular part of the GL.

EPITHALAMUS

Under this heading, the habenula and the epiphysis cerebri will be described.

Habenula

Situated in the dorsomedial angle of the diencephalon (Figs. 59, 69–71), the habenula is divisible into two distinct nuclei: the habenularis medialis and lateralis. The medial habenular nucleus (HM) is bounded laterally by the lateral habenular nucleus. Medially, it is separated from the third ventricle by a layer of ependymal cells and caudally, it is related to the ventral part of the commissura habenularis. The HM consists predominantly of small, closely arranged cells with relatively large nuclei and a thin cytoplasmic area. Histochemically, the neurons of the HM show moderate to moderately strong LDH, AC, and AChE; moderate SDH and SE; and strong ATPase and MAO activity. The neuropil is moderate in LDH, AC, AChE, and MAO; mild in SDH and SE; and moderately strong in the ATPase preparations.

The lateral habenular nucleus (HL) is much larger than the HM and occupies almost two-thirds of the habenular area. Based on cell size and grouping, the HL is further divisible into lateral magnocellular (HLM) and medial parvocellular (HLP) parts. The latter consists of smaller elongated cells arranged densely in the caudal part and loosely in the oral part, while the former consists of larger triangular or multipolar cells with elongated cell processes. Histochemically, the neurons of the HLM show very strong LDH and moderately strong to strong

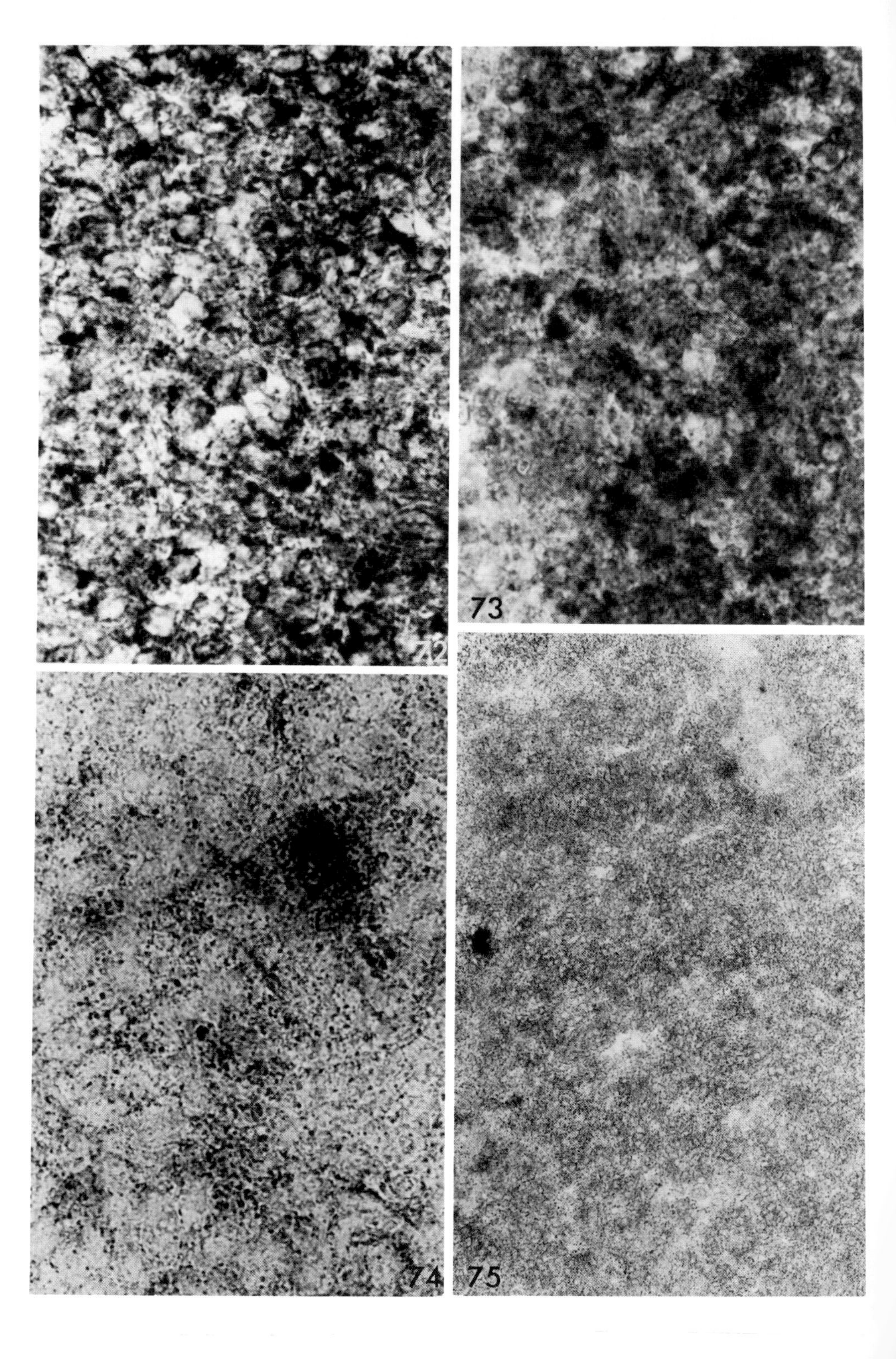

72
73
74
75

FIGURES 72–75. LDH, SDH, AK, and SE preparations, respectively, of the epiphysis cerebri. Note the variations in the enzyme localization in these figures. SDH and LDH show stronger activity than the others. ×370, ×592, ×370, ×148.

SDH activity. This compares with those of the HLP which are strong in LDH and moderate in SDH. The neuropil is also strong in LDH and moderate in SDH, whereas the glial cells are strong in LDH and negligible in SDH. A comparison of AC and ATPase reveals a moderately strong AC and strong ATPase reaction in the neurons of the HLM, and moderate AC and moderately strong ATPase activity in the HLP. The neuropil and glial cells are moderate and mild to negligible in ATPase and AC, respectively. The SE preparations show strong activity in the neurons of the HLM, a moderate reaction in those of the HLP, and a moderate reaction in the neuropil as well as glial cells. A comparison of AChE and MAO shows that the HLM exhibits moderately strong to strong MAO and moderate to moderately strong AChE in neurons, while in the HLP, the AChE and MAO activity is moderate. The neuropil of the HLM is moderate to mild in MAO, mild in AChE with moderate AChE activity in the HLP. The glial cells are negative. The AK activity is observed only in the blood vessels.

Epiphysis cerebri

Located in the dorsocaudal area of the diencephalon, the epiphysis cerebri is found above the superior colliculus (Figs. 68, 72–75). Anteriorly, the superior and inferior laminae attach the epiphysis to the habenular nuclei and posterior commissure, respectively, and between the two laminae, the cavity of the third ventricle extends to the pineal region. The parenchymal cells are strong in LDH and mild to moderate in SDH. In the former, the positive activity is found restricted to the peripheral part of the cell in the thin cytoplasm. The neurons show strong LDH and moderate to moderately strong SDH activity. In ATPase and AC, the parenchymal cells are mild to moderate and neurons moderate to moderately strong. The blood vessels give a strong reaction in ATPase as well as in AK. In AK preparations, the parenchymal cells are mild to moderate showing peripheral localization, and the neurons are moderate. The SE reaction is mild to moderate in the parenchymal cells and moderately strong in the neurons. In AChE preparations the parenchymal cells are negligible, neurons mild, and the blood vessels and sinusoids give a moderate reaction. In MAO, the parenchymal cells and the neurons are mild to moderate, and the transversely cut nerve fibers are moderately strong.

GENERAL DISCUSSION

OXIDATIVE ENZYMES AND MONOAMINE OXIDASE

A histochemical study of the hypothalamus of the rhesus monkey shows that this area contains large amounts of Embden–Meyerhof pathway enzymes and moderate amounts of TCA cycle enzymes. Small amounts of amylophosphorylase (AP) and glucose-6-phosphate dehydrogenase have also been observed. It has been suggested that anaerobic metabolism is somewhat dominant in the neurons of the magnocellular hypothalamic nuclei. The present authors have observed stronger SDH activity in the neurons of the supraoptic and paraventricular nuclei in the squirrel and rhesus monkeys (Manocha and Bourne, 1967; Iijima *et al.*, 1967) compared to that reported in rodents (Shimizu and Morikawa, 1957; Shimizu *et al.*, 1957). All the workers agree that the citric acid cycle enzymes (SDH and CYO) showed lower activity than the glycolytic pathway enzymes (Hanfeld, 1968).

The distribution of phosphorylase is mainly localized in the neuropil, and the neurons show comparatively smaller quantities. According to Shimizu and Kumamoto (1952), Shimizu (1955), and Shimizu *et al.* (1957), the adult mammalian brain exhibits deposition of glycogen in paraventricular structures such as the hypothalamus. Glycogen deposition may be correlated with a poor oxygen supply or high anaerobic metabolism. Wolff (1968) showed that the large neurons of the supraoptic and paraventricular nuclei of the hedgehog and some other hibernating mammals store a large quantity of glycogen during the hibernation period. He also showed that in the tuberal nuclei of hibernating mammals, glycogen is predominantly localized in the neuropil while the neurons are almost devoid of histochemically detectable glycogen. An increase in body temperature leads, in all parts of the hypothalamus, to a remarkable glycogen depletion. It was observed in the squirrel and rhesus monkeys that the glial cells of the magnocellular hypothalamic nuclei show strong activity for AP and LDH and moderate activity for aldolase, glucose-6-phosphate dehydrogenase, and SDH. Since the glial cells are rich in AP and glycogen, they may serve as energy donors to the neurosecretory neurons.

The results obtained from the rhesus monkey on the distribution of oxidative enzymes in the diencephalon are essentially those obtained in the squirrel monkey (Manocha and Bourne, 1967; Manocha *et al.*, 1967). A comparative assessment of various oxidative enzymes in the different nuclear groups of the thalamus has been made earlier by a number of workers (Friede, 1960; Grimmer, 1960; Friede and Fleming, 1962, 1963; Manocha and Bourne, 1967). In general, the LDH activity is stronger than that of SDH and CYO. Whereas the LDH is stronger in the cytoplasm of the neurons, passing on for some distance into the cell processes, the SDH activity is more apparent in the dendrites of the neuropil. The oligodendroglial cells in the neuropil, however, show a very strong LDH reaction compared to mild SDH activity. Moderately strong SDH and CYO are observed in the lateral and medial geniculate, habenular, pulvinar, anterior dorsalis and ventralis, lateralis dorsalis and posterior, and ventral posterior thalamic nuclei. Similar observations on the lateral geniculate body of the monkey brain have been made by Cotlier *et al.* (1965). They showed high activity for succinic, malic, isocitric, glutamic, and lactic dehydrogenase and DPN and TPN diaphorases and less activity of glucose-6-phosphate, 6 phosphogluconate and β-hydroxybutyrate dehydrogenase in the lateral geniculate body.

There are great similarities between the rhesus and human brains with regard to their oxidative enzyme content, except for minor differences in individual nuclei (such as a stronger reaction in the substantia nigra) as shown by Friede and Fleming (1962, 1963). However, as described by these authors: "Gradations among the thalamic nuclei were much less accentuated in man than in the guinea pig (Friede, 1960) where individual nuclei found quite contrasting patterns. Considerable gradations of enzymic activity were observed with certain human thalamic nuclei, such as the ventral and lateral nuclei and the pulvinar nuclei. These were characterized by bizarre patches of strong enzyme activity. These patterns were too diversified to permit one to distinguish typical findings from individual variations. Two areas showing distinct species differences, presumably phylogenetic, were the center median and the dorsomedial nucleus. The human dorsomedial nucleus had relatively strong enzyme activity in contrast to weak activity of a homologous region in the guinea pig. The extrathalamic diencephalic centers had similar enzyme patterns in both species, such as weak activity in the hypothalamus or the contrasting differentiation of the corpus subthalamicum (strong reaction) from the substantia nigra (weak reaction). These observations indicate a tendency of caudocranial differentiation of the chemoarchitecture of the brain. The human thalamus resembled only general outlines of the guinea pig pattern, while the patterns in the medulla oblongata were alike" (Friede and Fleming, 1962).

Our results are in agreement with some previous histochemical studies on the mamillary body. Nakajima and Kishi (1967) investigated the distribution of AP, AD, G6PD, SDH, CYO, MAO, and AChE in the rabbit mammillary body and described the presence of these enzymes in almost all the neurons. Presence of strong SDH activity in neurons of the MM with a compact and highly active neuropil has also been reported by Friede (1960) in the guinea pig. Although the ML shows high SDH activity in neurons, the neuropil gives very little activity. Similar reports have been made by Felgenhauer and Stammler (1962) on NAD-dependent dehydrogenases in the guinea pig and by Friede and Fleming (1962, 1963) on NAD-diaphorase of man and on LDH in the rhesus monkey. Shimizu and Abe (1966) described the lateral part of the MM in the rat as being rich in G6PD. Okinaka *et al.* (1960) reported the presence of large amounts of AChE in human mammillary bodies, but de Giacomo (1962) reported very little activity. Our studies on the rhesus monkey indicate that the neurons are only mildly to moderately active, in spite of incubating the sections for 12 to 18 hours.

The activity of monoamine oxidase (MAO) in the diencephalon is much higher in the hypothalamus than in the thalamus. The present study on the rhesus monkey indicates that, although both AChE and MAO are present in the hypothalamus in considerable quantities, their sites of activity may be quite discrete, and MAO may have a synaptic role in the adrenergic endings quite distinct from the part played by AChE in the cholinergic synapses. The involvement of MAO in the physiological inactivation or oxidative deamination of some biologically active amines is supported by the fact that quantities of adrenaline and noradrenaline increase after MAO inhibition (Spector *et al.*, 1958; Gey and Pletscher, 1961). In the brain, noradrenaline, dopamine, and 5-hydroxytryptamine are the only substrates found in abundance, and histochemical evidence suggests that MAO regulates the action of serotonin in the nerve cells. In addition, MAO may be involved in the metabolism of the visceral regions of the brain rather than exclusively participating in the function of adrenergic neurons (Shimizu *et al.*, 1959).

Carlsson *et al.* (1962) and Dahlström and Fuxe (1965), using fluorescent methods, also confirmed the presence of large quantities of monoamines in the hypothalamus. The former workers identified noradrenaline in the preoptic region, supraoptic nucleus, and paraventricular nucleus of the hypothalamus. They believe that noradrenaline is a possible synaptic transmitter in the neurons (Adams, 1965). By using the fluorescent methods, Csillik and Erulkar (1964) have described some labile stores of monoamines in the CNS. They pointed out that if the tissues were removed from animals several minutes previously, the distribution of monoamines was more widespread than earlier described (Carlsson *et al.*, 1962).

The fluorescent studies of Carlsson *et al.* (1960), Fuxe (1965), and Fuxe and Ljunggren (1965) have shown that the paraventricular nucleus exhibits many catecholamine-containing nerve terminals although the neurons themselves are negative. The supraoptic nucleus contains comparatively few fluorescent terminals, but there is no direct evidence of synaptic contact. Similarly, a dense network of catecholamine-containing fibers is observed in the preoptic recess (Sharp and Follett, 1968).

The distribution of catecholamines in the hypothalamus suggests the possibility that these compounds play some role in controlling the secretions of the pars distalis (Sharp and Follett, 1968). Donoso *et al.* (1967) also showed their role in gonadotrophin release by the measurement of catecholamines and the enzymes associated with their metabolism in the hypothalamus. Sharp and Follett suggested that there is indirect evidence that these drugs, which interfere with catecholamine metabolism, can affect peripheral endocrine organs, notably the gonads, by acting through the central nervous system. The tubero-infundibular tract in mammals contains dopamine (Fuxe and Hokfelt, 1966), and 5-hydroxytryptamine-containing nerve fibers are found in the lateral forebrain bundle.

The differences in the distribution of MAO in the various regions of the thalamus are obvious. Though the MAO reaction is much weaker in the thalamic area than in the hypothalamus, the nucleus centrum medianum thalami and epithalamus show MAO activity as strong as that in the hypothalamus. By using tritiated catecholamines and norepinephrine, Ishii and Friede (1967) demonstrated the catecholamine binding sites located diffusely in the neuropil and in pericellular aggregates on the nerve cells, thereby suggesting synaptic buttons. Lenn (1967) suggested that exogenous norepinephrine is accumulated by brain axons and nerve endings which contain endogenous norepinephrine.

There has been some controversy with regard to the type of cells in the epiphysis. Wislocki and Leduc (1952) believed that the principal cell elements are specific cells admixed with neuroglial cells, microglia, and an occasional neuron. In our histochemical preparations, the parenchymal cells are strong in LDH, moderate in SDH and MAO, and show a negligible AChE activity. The neuronal cells give a strong LDH, moderately strong SDH, moderate MAO, and mild AChE reaction. A strong reaction of LDH in the epiphysis points to the ability of this structure to carry out anaerobic glycolysis in addition to its participation in aerobic respiration. These results are interesting from the point of view of the investigations of Smith (1963), Håkanson *et al.* (1967), La Bella and Shin (1968), and Otani *et al.* (1968) on the pineal body. La Bella and Shin (1968) showed that the proportion of total glucose utilized via the hexosemonophosphate shunt in the pineal body is much greater than the brain and may be related to the secretory role of the epiphysis.

As suggested by Shimizu *et al.* (1957), the pineal body, like other paraventricular structures (area postrema, subcommissural organ, supraoptic crest), may be the site of glycogen deposition. The pineal gland has also been shown to contain postganglionic fibers of the cervical sympathetic system (Ariens–Kappers, 1960; Smith, 1963). It is also very rich in 5-hydroxytryptamine (5-HT) found in parenchymal cells and the sympathetic nerve terminals (Håkanson *et al.*, 1967). Otani *et al.* (1968) stated that the pineal gland is very rich in indoles and contains enzymes necessary for (a) tryptophane hydroxylation, (b) the decarboxylation of 5-hydroxytryptophane to form serotonin, (c) monoamine oxidation of serotonin, and enzymes that yield (d) 5-hydroxyindole acetaldehyde and 5-hydroxytryptophol and *O*-methylate 5-hydroxyindoles to form melatonin or 5-methoxytrytophol. They suggested that the pineal gland plays an active role in adrenal functions, carbohydrate metabolism, and probably some other endocrine functions as well.

The glial cells, particularly the oligodendroglial cells, as determined by their shape and branching characteristics, show much stronger LDH activity compared to mild SDH activity, and this may mean that oligodendroglia depend more on anaerobic glycolytic metabolism than on the citric acid cycle. Potanos *et al.* (1959) and Rubinstein *et al.* (1962) stated that, although the SDH activity was somewhat variable in the glial cells, oligodendrocytes and astrocytes showed similar enzyme activity.

CHOLINESTERASES

In agreement with the present observations on the rhesus monkey, Iijima *et al.* (1967) showed in the squirrel monkey that neurons of the hypothalamic magnocellular nuclei (supraoptic and paraventricular nuclei) showed moderately positive AChE activity and a negative AChE reaction in their axons. Moderate to strong AChE reactions in the neurons of these nuclei have also been reported in rodents and dogs (Abrahams *et al.*, 1957; Ishii, 1957). AChE activity has been observed not only in the neuronal cytoplasm and on the neuronal membranes, but also in high concentrations in some glia, protoplasmic astrocytes in the dorsal hypothalamus, and in the subependymal cell plate (Abrahams, 1963). Kiernan (1964) described the highest concentration of AChE in the cell bodies of supraoptic and paraventricular nuclei and other nerve cells of the hypothalamus, including the nucleus infundibularis and scattered neurons of the tuber cinereum. Bennett *et al.* (1966) discussed an interesting species difference: whereas in the dog the hypothalamus has greater activity than the thalamus, the rat shows more AChE in the thalamus than in the hypothalamus. Pickford (1947) presented direct physiological evidence that the neurons of the supraoptic nucleus are sensitive to acetylcholine and may be a typical example of noncholinergic neurons which are acetylcholine-sensitive. Okinaka *et al.* (1960) showed strong AChE activity in the ganglionic cells of the magnocellular hypothalamic nuclei, mammillo-infundibular nucleus as well as the fibers of these cells and believed that probably these ganglionic cells give rise to cholinergic axons. Abrahams *et al.* (1957) showed that since the distribution of choline acetylase and AChE in the hypothalamus is similar, it may suggest the presence in the magnocellular hypothalamic nuclei of true cholinergic neurons.

Kiernan (1964) stated that, in smaller quantities, AChE is present in the hypothalamic neuropil, the fibers of the optic chiasma, and the tract of fibers leaving the chiasma to form a plexus around the cells of the suprachiasmatic nucleus. The cells of the latter do not have any AChE activity and show only negligible activity in the neuropil. These observations are in agreement with those of de Giacomo (1962), except that he reports very little activity in the mammillary bodies in contrast to the intense activity described by Okinaka *et al.* (1960). Our observations on the rhesus monkey agree with those of Ishii and Friede (1967) that in the mammillary complex, AChE activity is somewhat higher in the dorsomedial portion of the medial mammillary nucleus.

Kivalo *et al.* (1958) showed that strong AChE, AC, and SDH activity exists in the magnocellular nuclei of the hypothalamus, where neurosecretory material is formed. In the squirrel and rhesus monkey only a few neurons showed Gomori's chrome haematoxylin-positive neurosecretory material in the cytoplasm. It is suggested that these primates may be unsuitable species for the study of neurosecretory phenomena. Nissl substance in the neurons of these hypothalamic nuclei is located in the peripheral zone of the cytoplasm, which is in agreement with previous observations (Scharrer and Scharrer, 1945).

It has been suggested that AChE plays an important role in the activation of the hypothalamo-neurohypophyseal system at the level of the supraoptic nucleus (Pickford, 1947; Duke *et al.*, 1950) and that this transmitter substance is involved in the release of hormones from the neurohypophyseal nerve terminals (Koelle, 1961, 1962). Pepler and Pearse (1957) showed that by stimulating the anterior or posterior lobe of the pituitary, an increase in the AChE content of the hypothalamic nuclei is observed. The majority of the midline nuclei and medial habenular nuclei give stronger AChE reaction than other thalamic nuclei. Next in order are the intralaminar group and anterior group. Medial, ventral, and anterior groups showed similar activity. Gerebtzoff (1959) showed that most of the thalamic nuclei have, in general, low AChE activity. The arcuate and magnocellular nuclei, reticular nuclei, and the ventral nucleus of the lateral geniculate body do show somewhat stronger AChE reaction than the other thalamic nuclei. The detailed studies of Curtis (1963, 1965), Andersen and Curtis (1964a, 1964b) and McCance *et al.* (1966) have confirmed the presence of acetylcholine-sensitive cells in the thalamus.

It is interesting to note that AChE activity in the fibers of the capsula interna is much higher than in the subcortical white matter (three times higher: Foldes *et al.*, 1962). Girgis (1967) showed that not only are the AChE-positive fibers better developed in older parts, but also they are more deeply stained in the brains of simpler animals (Shen *et al.*, 1956). Concerning the AChE reaction of other fiber systems, Ishii and Friede (1967), in a review of the AChE content of the human brain, pointed out (on the basis of the observations of Lewis and Shute, 1963) that the axons with marked AChE activity arise from neurons having AChE. The axonal AChE contributed substantially to the activity in the neuropil of the nuclei in which AChE-containing fibers terminate. This concept has been supported by transection experiments.

In a study of the developing forebrain, Krnjevic and Silver (1966) observed a very selective distribution of AChE throughout the foetal period. From this study they concluded that only certain types of fibers (e.g., corpus callosum, internal capsule, and anterior capsule) contain AChE, a property independent of the degree of myelination.

Butyrylcholinesterase (BChE) is concentrated in the blood vessels and the glial cells, particularly around the supraoptic and ventricular side of the nucleus paraventricularis. Detailed studies made by Koelle (1950, 1952, 1954) on the rat and cat have also indicated that the BChE is distributed on the walls of the blood vessels and in the cytoplasm of gliocytes, especially fibrous astrocytes of the nerve tracts. BChE is, however, mainly concentrated in the glial cells. In the diencephalon of the cat, strong BChE activity has been observed in the cytoplasm of the cells of the nucleus centrum medianum, paracentralis thalami, and subthalamicus (Abrahams, 1963). Abrahams also showed moderate enzyme activity in the fibers of the optic nerve, optic chiasma, mammillothalamic tract, descending columns of the fornix, and the cerebral peduncles. BChE is often described as a "neurohumoral scavenger" and may hydrolyze the acetylcholine which has escaped splitting by AChE, at the synapses, in the vessel walls, and the neuroglial cell membranes (Koelle, 1950; Brightman and Albers, 1959). The BChE in blood vessels may also be responsible for maintaining the blood-brain barrier (Grieg and Holland, 1949; Grieg and Mayberry, 1951). It has also been suggested that BChE in the satellite cells may be involved in some way in influencing the electrical activity of the related neuron. Earl and Thompson (1952) found that paralysis and demyelination result in a marked decrease in the activity of BChE in the brain and spinal cord, indicating its involvement in myelin metabolism. Later reports, however, do not substantiate the above finding. Myers and Mendel (1953) in the rat and Barnes and Denz (1953) in the rabbit inhibited BChE by tri-ortho-Cresyl phosphate, but could not confirm that prolonged inhibition of BChE is accompanied by paralysis.

Feldberg and Sherwood (1954) showed that small amounts of acetylcholine (ACh) in the ventricles of the cat would cause prolonged stupor. Abrahams *et al.* (1957) pointed out the BChE in the subependymal tissue adjacent to the nucleus paraventricularis may be involved in preventing the diffusion of acetylcholine from the hypothalamus to the cerebrospinal fluid in the ventricle, and to destroy the excess amounts of ACh even at the site of its release during neuronal activity.

PHOSPHATASES

The activity of AK and AC in the hypothalamus has been reviewed by Samorajski and Fitz (1960). The AK is particularly strong in the capillary network of the hypothalamus, especially the supraoptic and paraventricular nuclei. These nuclei are more vascular than other areas of the brain, with the exception of the nucleus colliculus inferior. The nucleus supraopticus is more richly supplied with blood vessels than the nucleus paraventricularis. The neurons do not show any AK activity when Burstone's Naphthol AS Phosphate method is used. This is in contrast to the positive reaction observed by Shimizu (1950) in the hypothalamic perikarya and nucleoli of these neurons in rodent brain. Iijima *et al.* (1968) observed mild AK activity on the peripheral part of the cells in the squirrel monkey, and Samorajski and Fitz (1960) observed negative reaction in these neurons.

Shimizu (1950) suggested that the rich AK activity of the blood vessels might control the passage of various substances between the blood and the brain tissue. On the basis of the neurosecretory function of these neurons, Iijima *et al.* (1968) suggested that these neurons might be physiochemical receptors appropriating enough energy and precursor material to synthesize protein from the rich blood supply. According to them, ATPase and AMPase (present in these neurons, glial cells, and blood vessels) also help this process.

In a histochemical study of the developing hypothalamus, Rogers *et al.* (1960) observed that the highest level of AK activity occurred before birth, and maximum AK activity accompanies differentiation of the hypothalamus. Cohn and Richter (1956), Borghese (1957), and Duckett and Pearse (1967) observed that AK activity rose from its level at birth to a peak value between the tenth and fifteenth days and subsequently fell again. These studies seem to agree with the belief of Smiechowska (1964) that there is a distinct correlation between AK activity and an increase in RNA on the one hand and an increased differentiation of the hypothalamic neuroblasts on the other.

The activity of AC is strong in the neurons of the hypothalamic nuclei. The large neurons of the supraoptic and paraventricular nuclei give a particularly strong AC reaction. Occasionally, some neurons show perinuclear aggregation of dense AC-reactive material, but in most of them the reaction product is uniformly distributed. The pericapillary glial cells in this region show moderate AC activity as also observed by Osinchak (1964). Iijima *et al.* (1968) made similar observations in the squirrel monkey and are in agreement with earlier studies on a number of animals (Eranko, 1951; Scharrer and Scharrer, 1954; Cohn and Richter, 1956; Shimizu *et al.*, 1957, 1959; Talanti *et al.*, 1958; Kivalo *et al.*, 1958; Osinchak, 1964; Pilgrim, 1967; Jongkind and Swaab, 1967). These studies support the fact that very active metabolic processes take place in the magnocellular hypothalamic nuclei. The reaction is particularly strong in the larger cells which are most probably neurosecretory in nature. Osinchak (1964) also observed a sparse reaction product in the more typical neurons surrounding hypothalamic areas. The close relationship of acid phosphatase to the cytotic mechanism, because of its association with lysosomes (Novikoff, 1961; Koenig *et al.*, 1964), facilitates the morphological investigation of important intracellular processes because it can be easily detected by histochemical methods. (Schiffer *et al.*, 1967). Lipofuscins are probably produced by the transformation of the lysosomal bodies, as is evident from the studies of a number of workers who have shown transitional forms having the histochemical properties intermediate between lysosomes and lipofuscin granules (Essner and Novikoff, 1960; Koenig and Barron, 1962). Brun and Brunk (1967) investigated histochemically the effects of experimental lead poisoning in the brain and found an apparent increase in the AC activity which may be due either to an effort to cope with the abnormal situation or as a consequence of liberation of the enzyme in the cytoplasm for the ruptured lysosomes.

In the developing rat, the AC and SE levels increase sharply after birth so that in the neurosecretory cells of a three-month-old rat hypothalamus, the reactions for AC and SE were very strong compared to a mild reaction for AK (Smiechowska, 1964). Talanti *et al.* (1958) observed in the third intrauterine month of the cow that there is a correlation between the AC activity and the aldehyde fuchsin-positive material in the hypothalamic magnocellular nuclei.

Pilgrim (1967) and Jongkind and Swaab (1967) studied the magnocellular neurosecretory nuclei in rats subjected to varying periods of water withdrawal. Pilgrim observed a marked increase in the activity of AC, thiamine pyrophosphatase (TPPase), and glucose-6-phosphate dehydrogenase in the thirsting animals. Jongkind and Swaab (1967) noted a 40% increase in TPPase in the supraoptic nucleus after two days of water deprivation, with a progressive increase in TPPase-positive material following longer dehydration periods, indicating a direct relationship between the amount of Golgi apparatus and the phenomenon of neurosecretion.

In the normal animals, deeply stained vesicular enlargements of the TPPase-positive network interconnected by lightly stained thin strands are observed in most neurons of the hypothalamus and thalamus (Shantha, unpublished data). Some groups of cells, however, show highly irregular TPPase-positive fenestrated plates forming a Golgi network. Some cells show simple TPPase-positive Golgi networks, and others (especially very small neurons) have separate TPPase-positive vesicles and granules as their Golgi apparatus.

REFERENCES

Abrahams, V. C. Histochemical localization of cholinesterases in some brain stem regions of the rat. *J. Physiol.* (*London*) **165**:55P (1963).

Abrahams, V. C., Koelle, G. B., and Smart, P. Histochemical demonstration of cholinesterases in the hypothalamus of the dog. *J. Physiol.* (*London*) **139**:137–144 (1957).

Adams, C. W. M. (ed.) Histochemistry of the cells in the nervous system. In: *Neurohistochemistry*, pp. 253–331. Elsevier, Amsterdam, 1965.

Anderson, P., and Curtis, D. R. The pharmacology of the synaptic and acetylcholineline. *Acta Physiol. Scand.* **61**:85–99 (1964a).

Andersen, P., and Curtis, D. R. The pharmacology of the synaptic and acetylcholine-induced excitation of ventrobasal thalamic neurones. *Acta Physiol. Scand.* **61**: 100–120 (1964b).

Ariens-Kappers, J. The development, topographical relations and innervation of the epiphysis cerebri in the albino rat. *Z. Zellforsch.* **52**:163–215 (1960).

Aronson, L. R., and Papez, J. W. Thalamic nuclei of *Pithecus* (*Macacus*) *rhesus*. II. Dorsal thalamus. *Arch. Neurol. Psychiat.* **32**:27–44 (1934).

Barnes, J. M., and Denz, F. A. Experimental demyelination with organophosphorus compounds. *J. Pathol. Bacteriol.* **65**:597–605 (1953).

Bennett, E. L., Diamond, M. C., Morimoto, H., and Herbert, M. Acetylcholinesterase activity and weight measures in fifteen brain areas from six lines of rats. *J. Neurochem.* **13**:563–572 (1966).

Borghese, E. Recent histochemical results of studies on embryos of some birds and mammals. *Intern. Rev. Cytol.* **6**:289–341 (1957).

Brightman, N. W., and Albers, R. W. Species differences in the distribution of extraneuronal cholinesterases within the vertebrate central nervous system. *J. Neurochem.* **4**:244–250 (1959).

Brun, A., and Brunk, U. Histochemical studies on brain phosphatases in experimental lead poisoning. *Acta Pathol. Microbiol. Scand.* **70**:531–536 (1967).

Carlsson, A., Lindqvist, M., and Magnusson, T. On the biochemistry and possible functions of dopamine and noradrenalin in brain. *Ciba Found. Symp., Adrenergic Mechanisms,* pp. 432–439 (1960).

Carlsson, A., Falck, B., Hillarp, N., and Torp, A. Histochemical localization at the cellular level of hypothalamic noradrenalin. *Acta Physiol. Scand.* **54**:385–386 (1962).

Cohn, P., and Richter, D. Enzymic development and maturation of the hypothalamus. *J. Neurochem.* **1**:166–172 (1956).

Cotlier, E., Lieberman, T. W., and Gay, A. J. Dehydrogenases and diaphorases in monkey lateral geniculate body. *Arch. Neurol.* **12**:294–299 (1965).

Crosby, E. C., Humphrey, T., and Lauer, E. W. *Correlative Anatomy of the Nervous System.* Macmillan, New York, 1962.

Crouch, R. L. Nuclear configuration of the hypothalamus and subthalamus of *Macacus rhesus. J. Comp. Neurol.* **59**:431–450 (1934a).

Crouch, R. L. The nuclear configuration of the thalamus of *Macacus rhesus. J. Comp. Neurol.* **59**:451–484 (1934b).

Csillik, B., and Erulkar, S. D. Labile stores of monoamines in the central nervous system. A histochemical study. *J. Pharmacol. Exptl. Therap.* **146**:186–193 (1964).

Curtis, D. R. Acetylcholine as a central transmitter. *Can. J. Biochem. Physiol.* **41**: 2611–2618 (1963).

Curtis, D. R. Actions of drugs on single neurones in the spinal cord and thalamus. *Brit. Med. Bull.* **21**:5–9 (1965).

Dahlström, A., and Fuxe, K. Evidence for the existence of monoamine-containing neurons in the central nervous system. *Acta Physiol. Scand.* **62**: Suppl. 232 (1965).

de Giacomo, P. Distribution of cholinesterase activity in the human central nervous system. *Proc. Fourth Intern. Congr. Neuropath., Munich, 1961.* Vol. 1, pp. 198–205. Thieme, Stuttgart, 1962.

Donoso, A. O., Stefano, F. J., and Biscardi, A. M. Effects of castration on hypothalamic catecholamines. *Am. J. Physiol.* **212**:737–739 (1967).

Duckett, S., and Pearse, A. G. E. Histoenzymology of the developing human basal ganglia. *Histochemie* **8**:334–341 (1967).

Duke, H. N., Pickford, M., and Watt, J. A. The immediate and delayed effects of diisopropyl fluorophosphate injected into the supraoptic nuclei of dogs. *J. Physiol. (London)* **111**:81–88 (1950).

Earl, C. J., and Thompson, R. H. S. Cholinesterase levels in the nervous system in tri-ortho-cresyl phosphate poisoning. *Brit. J. Pharmacol.* **7**:685–694 (1952).

Eränkö, O. Histochemical evidence of intense phosphatase activity in the hypothalamic magnocellular nuclei of the rat. *Acta Physiol. Scand.* **24**:1–6 (1951).

Essner, E., and Novikoff, A. B. Human hepatocellular pigments and lysosomes. *J. Ultrastruct. Res.* **3**:374–391 (1960).

Feldberg, W., and Sherwood, S. L. Injections of drugs into lateral ventricle of cat. *J. Physiol.* (*London*) **123**:148–167 (1954).

Felgenhauer, K., and Stammler, A. Das Verteilungmuster der Dehydrogenasen und Diaphorasen im Zentralnervensystem des Meerschweinchens. *Z. Zellforsch.* **58**: 219–233 (1962).

Foldes, F. F., Zzigmond, E. K., Foldes, V. M., and Erdos, E. G. The distribution of acetylcholinesterase and butyrylcholinesterase in the human brain. *J. Neurochem.* **9**:559–572 (1962).

Friede, R. L. Histochemical investigations of succinic dehydrogenase in the central nervous system. V. The diencephalon and basal telencephalic centers of the guinea pig. *J. Neurochem.* **6**:190–199 (1960).

Friede, R. L., and Fleming, L. M. A mapping of oxidative enzymes in the human brain. *J. Neurochem.* **9**:179–198 (1962).

Friede, R. L., and Fleming, L. M. A mapping of the distribution of lactic dehydrogenase in the brain of the rhesus monkey. *Am. J. Anat.* **113**:215–234 (1963).

Fuxe, K. Evidence for the existence of monoamine neurons in the central nervous system. IV. Distribution of monoamine nerve terminals in the central nervous system. *Acta Physiol. Scand.* **64**: Suppl. 247 (1965).

Fuxe, K., and Hökfelt, T. Further evidence for the existence of tuberoinfundibular dopamine neurons. *Acta Physiol. Scand.* **66**:245–246 (1966).

Fuxe, K., and Ljunggren, L. Cellular localization of monoamines in the upper brain stem of the pigeon. *J. Comp. Neurol.* **125**:355–388 (1965).

Gerebtzoff, M. A. *Cholinesterases.* International Series of Monographs on Pure and Applied Biology (Division: Modern Trends in Physiological Sciences), (P. Alexander and Z. M. Bacq, eds.). Pergamon Press, New York, 1959.

Gey, K. H., and Pletscher, A. Activity of monoamine oxidase in relation to the 5-hydroxytryptamine and norepinephrine content of the rat brain. *J. Neurochem.* **6**:239–243 (1961).

Girgis, M. Distribution of cholinesterase in the basal rhinencephalic structures of the coypu (*Myocastor coypus*). *J. Comp. Neurol.* **129**:85–95 (1967).

Grieg, M. E., and Holland, W. C. Increased permeability of the mematoencephalic barrier produced by physostigmine and acetylcholine. *Science* **110**:237 (1949).

Grieg, M. E., and Mayberry, T. C. The relationship between cholinesterase activity and brain permeability. *J. Pharmacol. Exptl. Therap.* **102**:1–4 (1951).

Grimmer, W. Die Verteilung der Succinodehydrogenase-Aktivität in Gehirn von *Macaca mulatta. Z. Anat. Entwicklungsgeschichte* **122**:414–440 (1960).

Grünthal, E. Der Zellbau im thalamus der Säuger und des Menschen. *J. F. Psychol. u. Neur.* **46**:41–112 (1934).

Håkanson, R., Lombard des Gouttes, M. N., and Owman, C. Activities of tryptophan hydroxylase, dopa decarboxylase and monoamine oxidase as correlated with the appearance of monoamines in developing rat pineal gland. *Life Sci.* **6**:2577–2585 (1967).

Hanefeld, F. Histochemical enzyme demonstration in the neurosecretory nuclei of rat (in German). *Acta Neuropathol.* **10**:91–94 (1968).

Heiner, J. R. A reconstruction of the diencephalic nuclei of the chimpanzee. *J. Comp. Neurol.* **114**:217–238 (1960).

Iijima, K., Shantha, T. R., and Bourne, G. H. Enzyme-Histochemical studies of the hypothalamus with special reference to the supraoptic and paraventricular nuclei of squirrel monkey (*Saimiri sciureus*). *Z. Zellforsch.* **79**:76–91 (1967).

Iijima, K., Shantha, T. R., and Bourne, G. H. Histochemical studies on the nucleus basalis Meynert of the squirrel monkey (*Saimiri sciureus*). *Acta Histochem.* **30:** 46–108 (1968).

Ishii, T., and Friede, R. L. A comparative histochemical mapping of the distribution of acetylcholinesterase and nicotinamide adenine-dinucleotide-diaphorase activities in the human brain. *Intern. Rev. Neurobiol.* **10**:231–275 (1967).

Ishii, Y. The histochemical studies of cholinesterase in the central nervous system. I. Normal distribution in rodents (in Japanese). *Arch. Histol. Jap.* **12**:587–611 (1957).

Jongkind, J. F., and Swaab, D. F. The distribution of thiamine diphosphate-phosphohydrolase in the neurosecretory nuclei of the rat following osmotic stress. *Histochemie* **11**:319–324 (1967).

Kesarev, V. S. Structural features of hypothalamus in man and other primates (chimpanzee, macaque). *Zh. Nevropathol. Psikhiatr. Korsakov.* **65**:696–702 (1965).

Kiernan, J. A. Carboxylic esterases of the hypothalamus and neurohypophysis of the hedgehog. *J. Roy. Microscop. Soc.* **83**:297–306 (1964).

Kivalo, E., Rinne, U. K., and Mäkelä, S. Acetylcholinesterase, acid phosphatase, and succinic dehydrogenase in the hypothalamic magnocellular nuclei after chlorpromazine administration: histochemical studies. *Experientia* **14**:293–294 (1958).

Koelle, G. B. The histochemical differentiation of types of cholinesterases and their localization in tissues of the cat. *J. Pharmacol. Exptl. Therap.* **100**:158–179 (1950).

Koelle, G. B. Histochemical localization of cholinesterases in the central nervous system of the rat. *J. Pharmacol. Exptl. Therap.* **106**:401 (1952).

Koelle, G. B. The histochemical localization of cholinesterases in the central nervous system of the rat. *J. Comp. Neurol.* **100**:211–228 (1954).

Koelle, G. B. A proposed dual neurohumoral role of acetylcholine: its functions at the pre- and post-synaptic sites. *Nature* **190**:208–211 (1961).

Koelle, G. B. A new general concept of the neurohumoral functions of acetylcholine and acetylcholinesterase. *J. Pharmacol.* (*London*) **14**:65–90 (1962).

Koenig, H., and Barron, K. D. Morphological and enzymic alterations in reacting glia. *Acta Neurol. Scand.* **38**: Suppl. 1, 72–73 (1962).

Koenig, H., Gaines, D., McDonald, T., Gray, R., and Scott, J. Studies of brain lysosomes. I. Subcellular distribution of fine acid hydrolases, succinate dehydrogenase and gangliosides in rat brain. *J. Neurochem.* **11**:729–743 (1964).

Krieg, W. J. S. A reconstruction of the diencephalic nuclei of *Macacus rhesus*. *J. Comp. Neurol.* **88**:1–51 (1948).

Krieg, W. J. S. *Functional Neuroanatomy*, pp. 506–514. Blakiston's Son, Philadelphia, 1953.

Krnjevic, K., and Silver, A. Acetylcholinesterase in the developing forebrain. *J. Anat.* (*London*) **100**:63–89 (1966).

La Bella, F. S., and Shin, S. Estimation of cholinesterase and choline acetyltransferase in bovine anterior pituitary, posterior pituitary, and pineal body. *J. Neurochem.* **15**:335–343 (1968).

Lenn, N. J. Localization of uptake of tritiated norepinephrine by rat brain *in vivo* and *in vitro* using electron microscopic autoradiography. *Am. J. Anat.* **120**:377–390 (1967).

Lewis, P. R., and Shute, C. C. D. Tracing presumed cholinergic fiber in rat forebrain. *J. Physiol.* (*London*) **168**:33P–35P (1963).

Manocha, S. L., and Bourne, G. H. Histochemical mapping of succinic dehydrogenase and cytochrome oxidase in the diencephalon and basal telencephalic centers of the brain of squirrel monkey (*Saimiri sciureus*). *Histochemie* **9**:300–319 (1967).

Manocha, S. L., Shantha, T. R., and Bourne, G. H. Histochemical mapping of the distribution of monoamine oxidase in the diencephalon and basal telencephalic centers of the brain of squirrel monkey (*Saimiri sciureus*). *Brain Res.* **6**:570–586 (1967).

Manocha, S. L., Shantha, T. R., and Bourne, G. H. *A Stereotaxic Atlas of the Brain of the Cebus Monkey* (*Cebus apella*). Oxford University Press, London, 1968.

McCance, I., Phillis, J. W., and Westerman, R. A. Responses of thalamic neurones to iontophoretically applied drugs. *Nature* **209**:715–716 (1966).

Myers, D. K., and Mendel, B. Studies on di-esterases and other lipid hydrolyzing enzymes; inhibition of esterases and acetoacetate production of liver. *Biochem. J.* **53**:16–25 (1953).

Nakajima, Y., and Kishi, K. Histochemical studies on the mammillary body of the rabbit. *Bull. Tokyo Med. Dent. Univ.* **14**:279–292 (1967).

Novikoff, A. Lysosomes and related particles. In: *The Cell* (J. Brachet and A. Mirsky, eds.), Vol. II, pp. 423–488. Academic Press, New York and London, 1961.

Okinaka, S., Yoshikawa, M., Uono, M., Mozai, T., Toyota, M., Muro, T., Igata, T., Tanabe, H., and Ueda, T. Histochemical study on cholinesterase of the human hypothalamus. *Acta Neuroveget.* (*Vienna*) **22**:53–62 (1960).

Olszewski, J. *The Thalamus of the Macaca mulatta. An Atlas for Use with the Sterotaxic Instrument.* S. Karger, Basel and New York, 1952.

Osinchak, J. Electron microscopic localization of acid phosphatase and thiamine pyrophosphatase activity in hypothalamic neurosecretory cells of the rat. *J. Cell Biol.* **21**:35–49 (1964).

Otani, T., Györkey, F., and Farrell, G. Enzymes of the human pineal body. *J. Clin. Endocrinol. Metabol.* **28**:349–355 (1968).

Papez, J. W., and Aronson, L. R. Thalamic nuclei of Pithecus (*Macacus*) rhesus. I. Ventral thalamus. *Arch. Neur. Psych.* **32**:1–26 (1934).

Peele, T. L. *The Neuroanatomic Basis for Clinical Neurology.* McGraw-Hill, New York, 1961.

Pepler, W. J., and Pearse, A. G. E. The histochemistry of the esterases of rat brain, with special reference to those of the hypothalamic nuclei. *J. Neurochem.* **1**:193–202 (1957).

Pickford, M. The action of acetylcholine in the supraoptic nucleus of the chloralosed dog. *J. Physiol.* (*London*) **106**:264–270 (1947).

Pilgrim, C. Enzyme histochemistry studies on thirst-activated neurosecretory cells in rats. *Experientia* **23**:943 (1967).

Potanos, J. N., Wolf, A., and Cowen, D. Cytochemical localization of oxidative enzymes in human nerve cells and neuroglia. *J. Neuropathol. Exptl. Neurol.* **18**:627–635 (1959).

Riley, H. L. *An Atlas of the Basal Ganglia, Brain Stem, and Spinal Cord.* Hafner, New York, 1960.

Rogers, K. T., DeVries, L., Kepler, J. A., Kepler, C. R., and Speidel, E. R. Studies on chick brain of biochemical differentiation related to morphological differentiation and onset of function. I. Morpohological development. II. Alkaline phosphatase and cholinesterase levels, and onset of function. *J. Exptl. Zool.* **144**:77–103 (1960).

Rose, J. E. The cortical connections of the reticular complex of the thalamus. *Res. Publ. Ass. Res. Nerv. Ment. Dis.* **30**:454–479 (1952).
Rose, J. E., and Woolsey, C. N. A study of thalamocortical relations in the rabbit. *Bull. Johns Hopkins Hosp.* **72**:65–128 (1943).
Rubinstein, L. J., Klatzo, I., and Miguel, J. Histochemical observations on oxidative enzyme activity of glial cells in a local brain injury. *J. Neuropathol. Exptl. Neurol.* **21**:116–136 (1962).
Samorajski, T., and Fitz, G. R. Histochemical analysis of phosphomonoesterase in the hypothalamus and pituitary gland of the rat. *Lab. Invest.* **9**:517–534 (1960).
Scharrer, E., and Scharrer, B. Neurosecretion. *Physiol. Rev.* **25**:171–181 (1945).
Scharrer, E., and Scharrer, B. Hormones produced by neurosecretory cells. *Recent Progr. Hormone Res.* **10**:183–232 (1954).
Scheibel, M. E., and Scheibel, A. B. The organisation of the nucelus reticularis thalami: a Golgi study. *Brain Res. Netherl.* **1**:43–62 (1966).
Schiffer, D., Fabiani, A., and Monticone, G. F. Acid phosphatase and nonspecific esterase in normal and reactive glia of human nervous tissue. A histochemical study. *Acta Neuropath.* **9**:316–327 (1967).
Shantha, T. R., and Manocha, S. L. Basal ganglia, septal area, epithalamus, dorsal thalamus, metathalamus, subthalamus, and hypothalamus. In: *Handbook of Chimpanzee* (G. H. Bourne, ed.), Vol. I, pp. 250–317. S. Karger, Basel and New York, 1969.
Shantha, T. R., Manocha, S. L., and Bourne, G. H. *A Stereotaxic Atlas of the Java Monkey Brain* (*Macaca irus*). S. Karger, Basel and New York, 1968.
Sharp, P. J., and Follett, B. K. The distribution of monoamines in the hypothalamus of the Japanese quail, *Corturnix coturnix japonica. Z. Zellforsch.* **90**:245–263 (1968).
Shen, S. C., Greenfield, P., and Boell, E. J. Localization of acetylcholinesterase in chick retina during histogenesis. *J. Comp. Neurol.* **106**:433–461 (1956).
Shimizu, N. Histochemical studies on the phosphatase of the nervous system. *J. Comp. Neurol.* **93**:201–218 (1950).
Shimizu, N. Histochemical studies of glycogen of the area postrema and the allied structures of the mammalian brain. *J. Comp. Neurol.* **102**:323–339 (1955).
Shimizu, N., and Abe, T. Histochemical studies of the brain with reference to glucose metabolism. *Progr. Brain Res.* **21A**:197–216 (1966).
Shimizu, N., and Kumamoto, T. Histochemical studies on the glycogen of the mammalian brain. *Anat. Rec.* **114**:479–498 (1952).
Shimizu, N., and Morikawa, N. Histochemical studies of succinic dehydrogenase of the brain of mice, rats, guinea pigs, and rabbits. *J. Histochem. Cytochem.* **5**:334–345 (1957).
Shimizu, N., Morikawa, N., and Ishii, Y. Histochemical studies of succinic dehydrogenase and cytochrome oxidase of the rabbit brain, with special reference to the results in the paraventricular structures. *J. Comp. Neurol.* **108**:1–21 (1957).
Shimizu, N., Morikawa, N., and Okada, M. Histochemical studies of monoamine oxidase in the brain of rodents. *Z. Zellforsch.* **49**:389–400 (1959).
Smiechowska, B. Badania histochemiczne podwzgorza szczura bialego podczas ontogenzy. *Folia Morphol.* **23**:227–256 (1964).
Smith, B. Monoamine oxidase in the pineal gland, neurohypophysis, and brain of the albino rat. *J. Anat.* (*London*) **97**:81–86 (1963).
Spector, S., Prockop, D., Jr., Shore, P. A., and Brodie, B. B. Effect of Iproniazid on brain levels of norepinephrine and serotonin. *Science* **127**:704 (1958).

Talanti, S., Kivalo, E., and Kivalo, A. The acid phosphatase activity in the hypothalamic magnocellular nuclei of the cow embryo. *Acta Endocrinol.* **29**:302–306 (1958).

Walker, E. A. A note on the thalamic nuclei of *Macaca mulatta*. *J. Comp. Neurol.* **66**:145–155 (1937).

Walker, E. A. *The Primate Thalamus* (F. C. McLean, A. J. Carlson, H. G. Wells, eds.). The Univ. of Chicago Press, Chicago, 1938.

Wislocki, G. B., and Leduc, E. H. Vital staining of the hematoencephalic barrier by silver nitrate and trypan blue, and cytological comparison of the neurohypophysis, pineal body, area postrema, intercolumnar tubercle and supraoptic crest. *J. Comp. Neurol.* **96**:371–413 (1952).

Wolff, H. Histochemical and electronmicroscopic observations on the distribution of glycogen in the hypothalamus of some hibernating mammals. (With quantitative comments). *Z. Zellforsch.* **88**:228–262 (1968).

VI

BRAIN STEM

CRANIAL NERVE NUCLEI

Nucleus n. oculomotorii

The oculomotor nucleus is located ventral to the central gray mass of the midbrain below the aqueductus Sylvii and is divisible anatomically as well as histochemically into (a) lateral part, (b) central part, and (c) nucleus Edinger–Westphal (Figs. 59, 76, 77). The first is the main part of the oculomotor complex and is further divided into the dorsal (NOD) and ventral (NOV) areas by a small space containing some nerve fibers with only a few scattered neurons. Most of the cells of the lateral part are medium-sized and large and multipolar with elongate processes. The neurons of the NOV are, however, more compactly arranged than those of the NOD. The neuropil of the lateral part of the oculomotor complex gives moderate SE, AC, AChE, and MAO; moderately strong ATPase; and a strong LDH reaction. The SDH reaction, although moderate in general, is a bit stronger in the NOD than in the NOV. The neurons are usually stronger than the neuropil, with some exceptions such as in ATPase, AChE, and MAO preparations. Most neurons show strong to very strong SE and LDH; moderately strong and strong AC, SDH, and ATPase; and mild to moderate AChE and MAO. A slightly weaker enzyme reaction than that of the perikarya has been observed in the cell processes in SE, AC, LDH, SDH, and ATPase preparations. The blood vessels are positive in AK, ATPase, and AChE.

The central part consists of a central nucleus (NOC) made up of small, compactly arranged neurons located dorsomedial to the lateral part and an elongate group of small to medium-sized, oval and spindle-shaped neurons called the nucleus of Perlia (NOP).

The NOP is much less cellular than the other components of the oculomotor complex. The NOC gives a moderate to moderately strong SE reaction in the neurons and shows mild activity in the neuropil. The strength of this SE reaction is in contrast to the generally mild one in the neurons and a negligible reaction in the neuropil of the NOP. The AC, LDH, SDH, and ATPase reactions are stronger. The neurons of the NOC are strong in AC, LDH, and SDH, and the cells of the NOP give a moderate AC, SDH, and ATPase and moderately strong to strong LDH activity. The neuronal activity in the AChE and MAO reactions is mostly mild and moderate in a few cells. The enzyme activity in the cell processes has been observed mostly in the SE, LDH, and SDH preparations. The neuropil in the NOC gives mild SE and AChE; moderate AC, ATPase, and SDH; and moderately strong LDH reactions; in the NOP, the neuropil is somewhat weak compared to that of the NOC. The cut nerve fibers give a mild to moderate AChE and moderate MAO reaction. The AK preparations are negative and show only the blood vessels. The NOP shows a lesser number of blood vessels than the other parts of the third nerve nucleus.

The nucleus Edinger–Westphal (NOW) consists of closely arranged small and medium-sized neurons of oval, fusiform, and triangular outlines. The neurons give mild AChE and MAO; moderate SE, AC, and ATPase; moderately strong SDH; and strong LDH reactions, whereas the neuropil is mild in AC and MAO and moderate in SE, SDH, LDH, ATPase, and AChE preparations.

Nucleus n. trochlearis (NT) and nucleus supratrochlearis (NSTT)

The trochlear nucleus is represented by an oval mass of medium-sized and large neurons located in the midbrain at the level of the colliculus inferior (Figs. 78, 79, 83, 108–109, 114). A few cells scattered in the vicinity of the main mass or partially embedded in the fasciculus longitudinalis medialis also belong to the NT, based on their histochemical reactions. A group of closely arranged neurons dorsal and dorsomedial to the NT is generally called the nucleus supratrochlearis and is histochemically identical to the NT. The motor type neurons of the NT and NSTT give a strong reaction in SE, AC, LDH, and SDH; mild to moderate in ATPase and AChE; mild in MAO; and negative in AK. A weaker reaction than the perikarya is seen in the cell processes in the SE, LDH, and SDH. The neuropil shows strong LDH; moderately

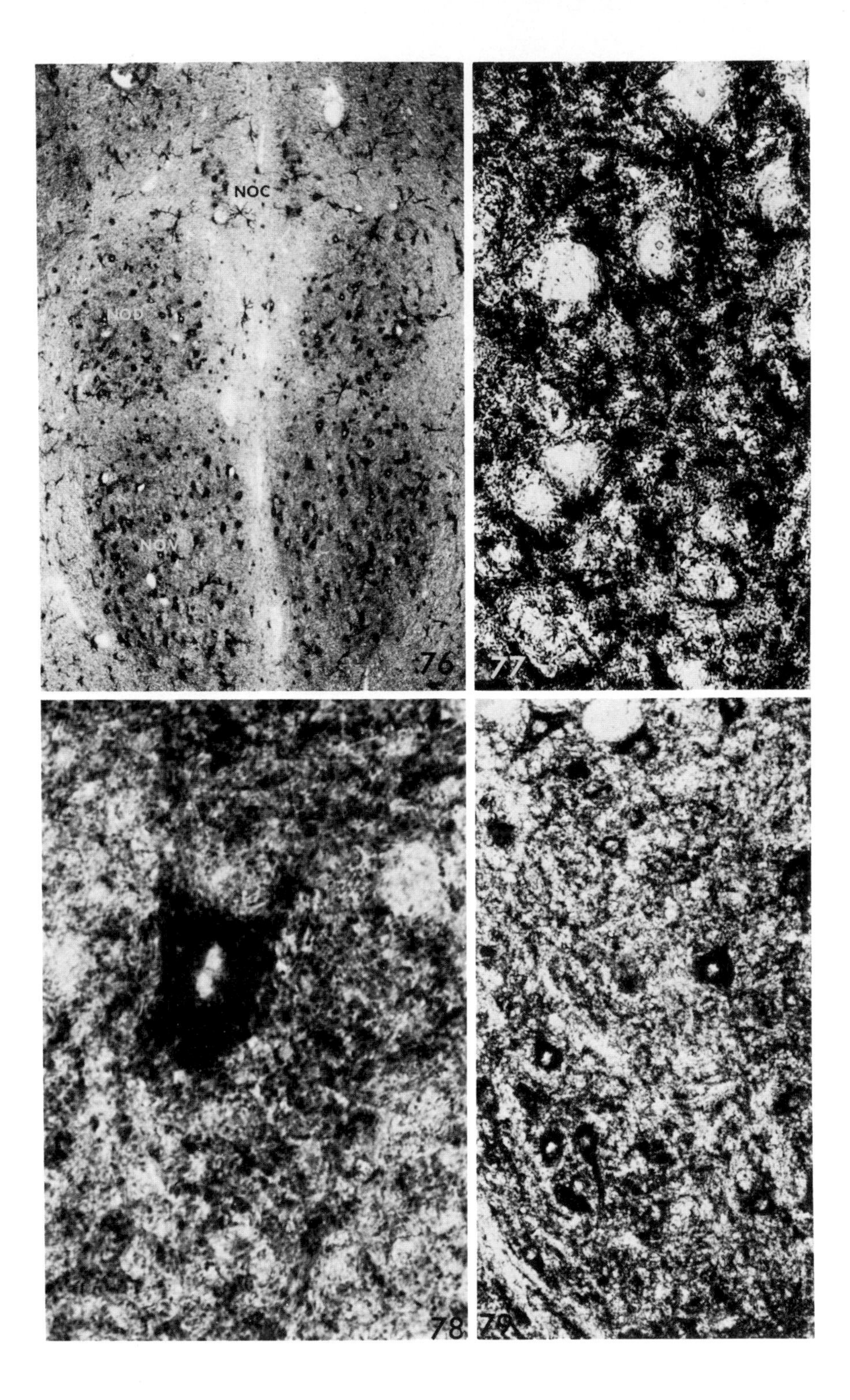
NOC
NOD
NOV
76
77
78
79

FIGURES 76, 77. SE and ATPase preparations of the oculomotor complex. ×37, ×370.

FIGURES 78, 79. SDH and LDH preparations of the nucleus n. trochlearis. ×592, ×148.

strong SDH and ATPase; moderate AC, SE, and MAO; and mild to moderate AChE activity. The glial cells are SE-, LDH-, and SDH-positive, and the blood vessels are positive in ATPase, AChE, and AK. As observed in the latter preparations, there are more blood vessels in the NT and NSTT than in the adjoining gray matter.

Nucleus n. trigemini

The trigeminal nerve nuclear complex comprises the following four separate sensory and motor nuclei located in the different parts of the brain stem.

Nucleus tractus mesencephali n. trigemini (MV)

The MV is located in the lateral part of the griseum centrale of the midbrain and pons and is represented by groups of scattered neurons, most of which are large and oval or round (Figs. 80–82, 84, 85, 108, 109, 119, 120, 125, 126, 131). The number of cells in the sections at various levels in the midbrain and pons differs greatly from a few to many. The MV is more conspicuous at the level of the nucleus locus coeruleus (LCR), with which the MV cells sometimes intermingle. The MV cells can, however, be distinguished from those of the LCR by their ganglion-type characteristics and lack of melanin pigment which is often seen in the cells of the LCR. The MV cells show strong to very strong SE, AC, LDH, and SDH; moderate ATPase; mild to moderate AChE and MAO; and negative AK activity. A somewhat weaker enzyme reaction in the cell processes has also been observed for some distances in the LDH and SDH preparations.

Nucleus motorius n. trigemini (NMV)

The NMV is situated in the lateral part of the midpontine tegmentum at about the same level as the olivaris superior rostrally and nucleus n. facialis caudally (Figs. 86, 87, 125, 131, 133–136). Most neurons are medium-sized and large, with a few small cells, and are not compactly arranged. They are round, oval, triangular, and elongate, and some of them have large cell processes. Histochemically, the smaller cells show weaker enzyme activity than the larger ones; e.g., small cells show moderate SE, AC, and SDH, whereas the large ones show strong

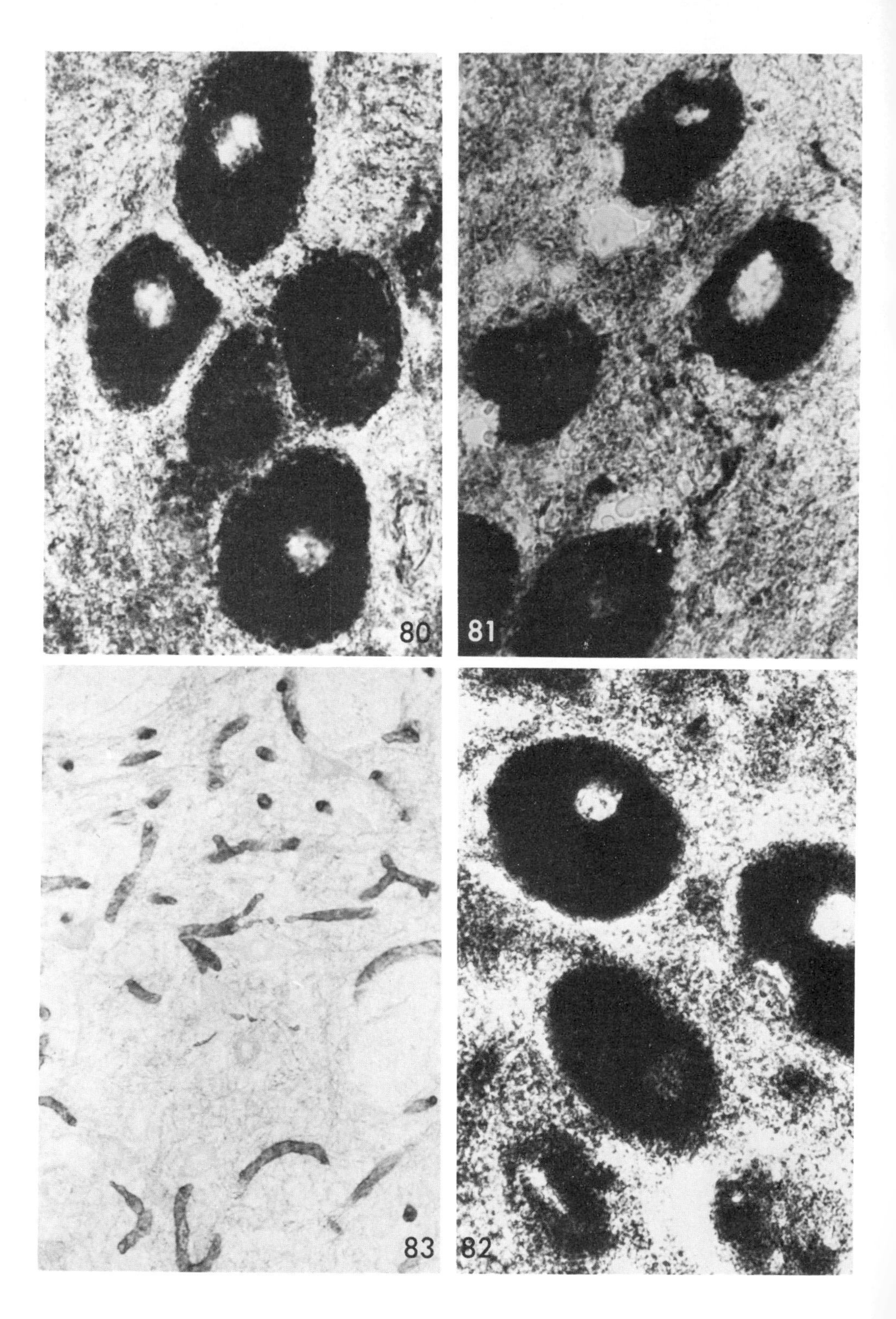
80
81
83
82

FIGURES 80–82. Neurons of the nucleus mesencephalicus n. trigemini showing LDH, SDH, and SE activity. Note the strong reaction shown by these cells for all these enzymes, the activity being uniformly distributed throughout the cytoplasm. ×592, ×592, ×592.

FIGURE 83. AChE activity in the nucleus n. trochlearis showing negligible to mildly positive neurons and mildly positive neuropil. Also note the positive reaction in the blood vessels. ×148.

enzyme activity. In general, the neurons tend to be strongly reactive in SE, AC, LDH, and SDH, whereas they are moderate in ATPase, AChE, and MAO preparations. The cell processes and the neuropil are weaker than the perikarya in the first four preparations (SE, AC, LDH, and SDH). In the ATPase, AChE, and MAO preparations the reverse is true, and the neuropil shows moderately strong enzyme activity. The blood vessels are positive in AChE, ATPase, and AK. A numerical count in AK shows that the NMV has a larger number of blood vessels than the adjoining principal nucleus described below.

Nucleus principalis n. trigemini (PVA)

The PVA is made up of compactly arranged medium-sized and small neurons; the former are mostly round and oval, whereas the smaller cells are oval, fusiform, and elongate (Figs. 131, 133, 135). The PVA is located ventral and lateral to the NMV and can be easily differentiated from the latter by the smaller cells. Histochemically, the small cells of the PVA show less enzyme activity than the medium-sized ones. The latter give moderate to moderately strong reactions in SE, AC, LDH, and SDH; moderate in ATPase; and mild in AChE and MAO preparations. The neuropil shows slightly weaker activity except in the ATPase and AChE preparations. In the former case the neurons are moderate and the neuropil moderately strong, whereas in AChE the neurons are mild and the neuropil reacts moderately.

Nucleus tractus spinalis n. trigemini (NSV)

The NSV is a very long nucleus and covers the entire length of the medulla oblongata, going into the pons to approximately the caudal level of the PVA (Figs. 132, 134, 136–143, 155–160, 171, 172). As in the chimpanzee brain (Shantha and Manocha, 1969), the NSV in the rhesus monkey can be subdivided into (a) caudal part, extending from the cervical part of the spinal cord to the level of the cuneate nucleus, (b) intermediate part, extending to the middle of the nucleus olivaris inferior, and (c) the oral part. Histochemically, however, it is only the caudal part which shows variations and can be divided into three distinct areas:

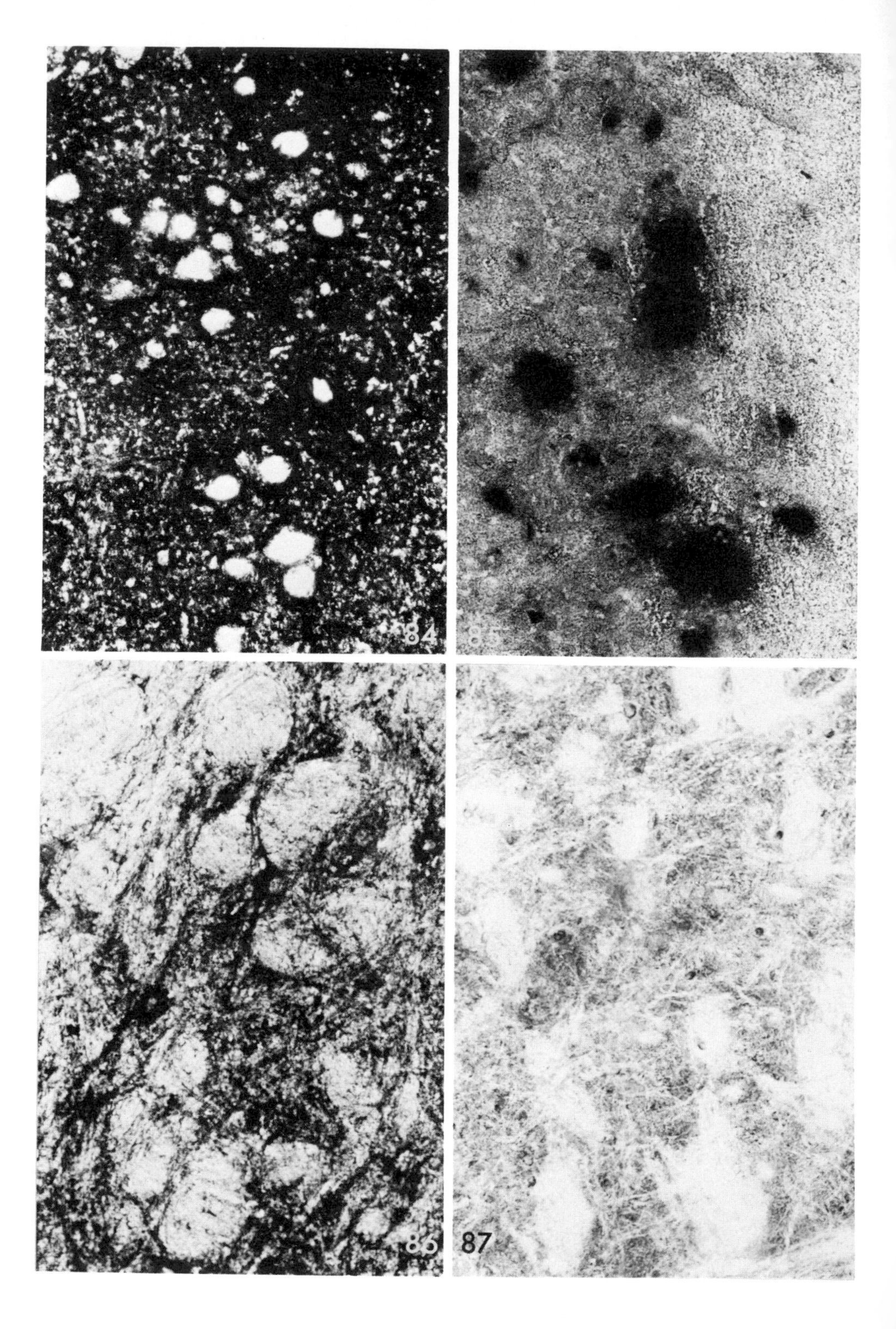
84
87

FIGURES 84, 85. The neurons of the nucleus mesencephalicus n. trigemini showing ATPase and AC activity. Note the stronger reaction in the neuropil for ATPase compared to the AC reaction. ×148, ×148.

FIGURES 86, 87. LDH and AChE preparations of the nucleus motorius n. trigemini showing strong LDH and negligible AChE activity in the neurons. ×148, ×148.

medial, intermediate, and outermost. These areas have been referred to as pars magnocellularis, pars gelatinosus, and pars zonalis, respectively. The medial part consists of mostly medium-sized with some large cells, whereas the intermediate part, the largest of the three areas, consists of small and medium-sized, rather closely packed oval, fusiform, and triangular cells. The outermost part can be recognized by the arrangement of medium-sized cells, whose long axes are parallel to the outer border of the NSV. The magnocellular part shows strong to very strong LDH and AC, moderately strong SDH and SE, moderate ATPase and MAO, and negligible to mild AChE activity. The neuropil gives a moderately strong reaction in LDH and ATPase; moderate in MAO; and mild in SE, AC, SDH, and AChE preparations. Histochemically, the pars gelatinosus and zonalis are similar in the AC, LDH, ATPase, AChE, and MAO preparations. The neurons are strong to very strong in LDH, moderate to moderately strong in AC, moderate in ATPase, mild in MAO, and negligible in the AChE preparations. The neuropil is moderately strong in LDH and ATPase, moderate in MAO, and mild in AC and AChE. In the MAO, the cut nerve fibers passing through this nucleus also show moderate enzyme activity. The pars zonalis gives mild to moderate SE and SDH activity in the neurons and mild activity in the neuropil; this is in contrast to moderately strong to strong neurons and moderate neuropil in the pars gelatinosus. In SDH the neurons as well as the neuropil show moderately strong activity.

Nucleus n. abducentis (NAB)

The NAB is an ovoid, well demarcated mass of large and some medium-sized neurons located beneath the facial colliculus, lateral to the fasciculus longitudinalis medialis (Figs. 88, 89, 131–138, 140). In some preparations the facial nerve fibers course either through the nucleus or somewhat behind it from the ventromedial side. Histochemically, the reactions are almost similar to those of the nucleus trochlearis. The neurons are strong in SE, AC, LDH, and SDH; mild to moderate in ATPase; and mild in AChE and MAO, whereas the neuropil is moderately strong in LDH and SDH, moderate in SE and ATPase, mild

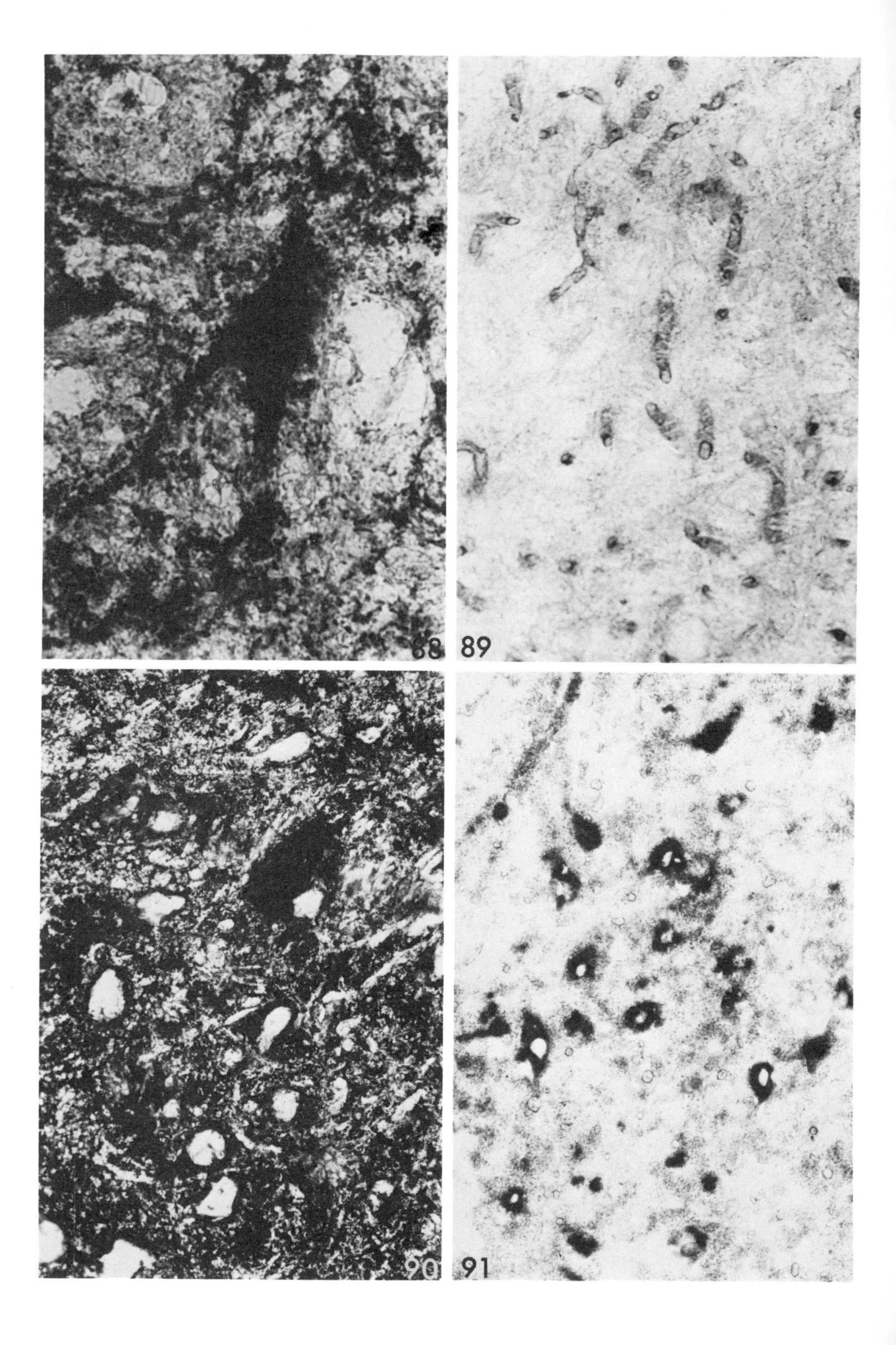
88
89
90
91

FIGURES 88, 89. A comparison of SDH and AChE activity in the nucleus abducentis. Note the strong SDH and weak AChE reactions in the neurons. ×592, ×148.

FIGURES 90, 91. ATPase and SE preparations of the nucleus n. facialis. The neuropil is stronger in the ATPase reaction and the neurons are stronger in the SE preparation. ×148, ×148.

to moderate in AC and AChE, and mild in MAO preparations. The blood vessels show moderately strong AChE and AK and strong ATPase activity.

Nucleus n. facialis (NF)

The facial nucleus appears in serial sections, close to the lower end of the nucleus abducentis. It appears as groups of large and medium-sized triangular, oval, and fusiform neurons with elongate cell processes in a position dorsolateral to the nucleus olivaris inferior (Figs. 90, 91, 139, 142, 143). Unlike the abducentis, the NF is not compact because of a number of longitudinal fibers running through the main body which divide the nucleus into certain subgroups having a varying number of cells in each. Pearson (1946) and Olszewski and Baxter (1959) in the human and Shantha and Manocha (1969) and Manocha (1969) in the chimpanzee observed six subdivisions of the NF, designated medial or ventral according to their position. Histochemically, most neurons give identical reactions. They show strong AC, SE, and LDH; moderately strong SDH; mild to moderate ATPase; and mild AChE and MAO activity. The enzyme activity in somewhat weaker form passes into the cell processes for some distance. The neuropil is moderately strong in LDH and ATPase; moderate in SDH, SE, AChE, and MAO; and mild in AC. The blood vessels are strong in ATPase and moderately strong in AChE and AK.

Nuclei cochlearis

Located at about the level of the junction of the medulla oblongata and pons, dorsolateral to the corpus restiforme, the cochlear area consists of a dorsal (CDS) and a ventral (CVS) cochlear nucleus (Figs. 94, 95, 99, 139, 142). The CDS shows smaller neurons, whereas the CVS consists of mostly medium-sized cells with a few smaller neurons. They are all oval, fusiform, or round. The histochemical reactions in the CDS and CVS are essentially identical, except for somewhat weaker enzyme reactions in the neurons of the CDS. This is because the CVS has larger

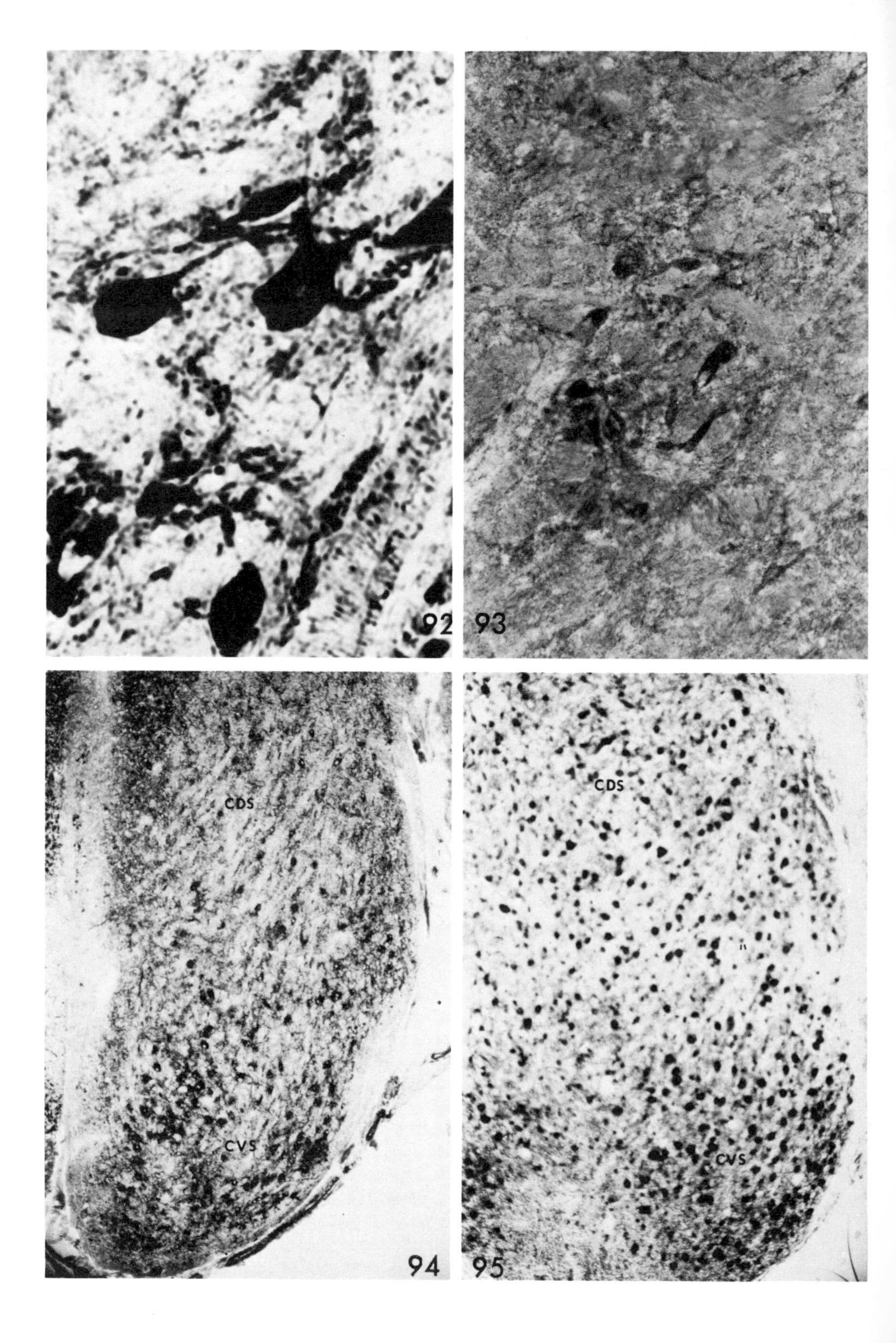
92
93
CDS
CVS
94
CDS
CVS
95

FIGURES 92, 93. LDH and SDH activity, respectively, in the neurons of the nucleus ambiguus. ×370, ×148.

FIGURES 94, 95. Low power photomicrographs of the dorsal and ventral cochlear nuclei. Note the differences in the distribution of LDH and SE activity. ×37, ×37.

cells, which show stronger SE, AC, LDH, and SDH reactions. Most of the neurons are strong in LDH, SDH, AC, and SE; moderate in ATPase; mild to moderate in MAO; and mild in AChE. The cell processes in a large number of neurons react moderately in the SE, LDH, and SDH preparations. The glial cells are SE-, LDH-, and ATPase-positive. The neuropil shows moderate LDH, SDH, MAO, and ATPase and mild AC and SE activity. In AChE the neuropil is moderate in the dorsolateral part, and mild in the medial part of the nucleus.

Nuclei vestibularis

The vestibular complex occupies a large area in the dorsolateral aspect of the upper part of the medulla oblongata and consists of four distinct nuclei, named according to their anatomical position: medial (VM), lateral (VL), inferior (VI), and superior (VS) [Figs. 96–98, 137–143, 148, 149]. The medial vestibular nucleus is located beneath the floor of the fourth ventricle in a slightly lateral position and consists of mostly medium-sized and small neurons which are more compactly arranged than the other vestibular nuclei. The neurons are round, oval, fusiform, and triangular and give a moderately strong to strong reaction in LDH, AC, SE, and SDH, with mild AChE and mild to moderate ATPase and MAO activity. The glial cells, which are fairly abundant in the VM, give strong LDH and mild to moderate ATPase, MAO, SE, AC, and SDH activity. The neuropil gives mild AChE; moderate MAO, ATPase, SDH, AC, and SE; and moderately strong LDH activity. The blood vessels show positive AChE, ATPase, and AK activity. The VM is more vascular than the other three components of the vestibular nuclei.

The lateral vestibular nucleus (VL) is lateral and dorsolateral to the VM and is related dorsally to the lateral part of the floor of the fourth ventricle, from which it is separated by a subependymal layer. Laterally the VL is related to the corpus restiforme, and ventrally to the VI. The cell population consists of mostly medium-sized and large neurons with more of the latter, and shows cell outlines varying from star-shaped to oval and triangular with elongate cell processes. The VL is not as compact as the VM, which is due partly to the passage of the fibers of the vestibular nerve. Histochemically, the neurons show strong LDH, SDH, AC, and SE; moderate ATPase; and mild MAO and AChE activity.

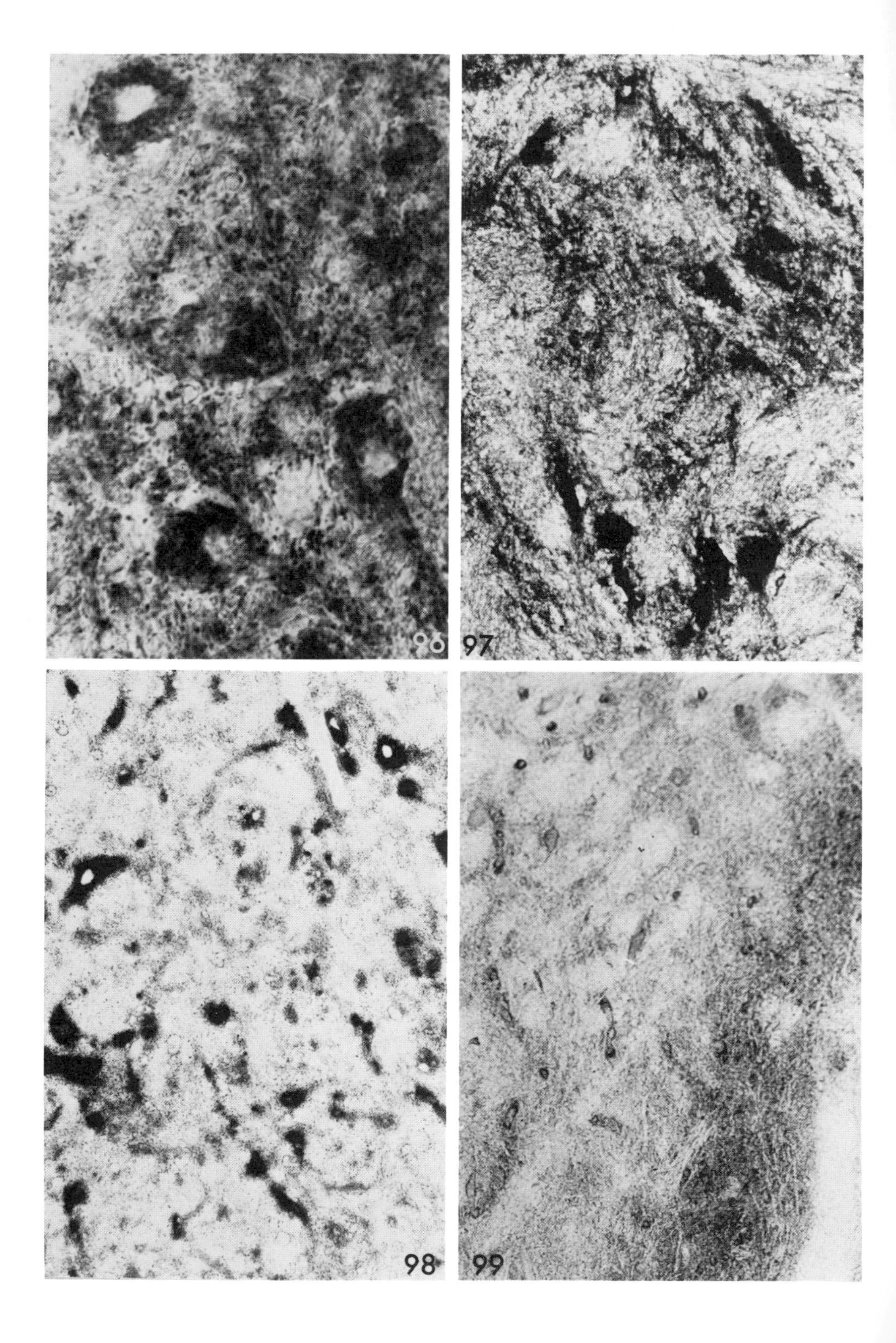
96
97
98
99

FIGURES 96–98. Photomicrographs of the nucleus vestibularis lateralis showing a comparison of SDH, LDH, and SE activity in its giant cells. LDH and SE activity is stronger than that of SDH. ×592, ×148, ×148.

FIGURE 99. Nucleus cochlearis showing AChE activity. Note the concentrations of the enzyme at the peripheral part of the nucleus. ×148.

The enzyme activity is also observed in somewhat weaker form in the cell processes in most of the larger neurons. The neuropil is strong in LDH; moderate in SDH and ATPase; and mild in AC, SE, MAO, and AChE preparations. The VI is histochemically identical to the VL, except that the large cells of the latter are enzymatically more active.

The superior vestibular nucleus is the oral part of the vestibular complex. Its neurons are more compactly arranged than those of the VL and VI but not to the extent of those of the VM. Similarly, the cell size in general is smaller than in the VL, but larger than in the VM. The histochemical reactions in the cells vary from mild AChE, ATPase, and MAO to moderately strong to strong SDH, AC, and SE and very strong LDH. The neuropil is mild in SE, AC, MAO, and AChE; moderate in SDH; and moderately strong in ATPase and LDH. The glial cells show stronger activity in LDH and ATPase than do other enzymes investigated in the present study.

Nucleus ambiguus (AB)

The AB extends through the entire length of the nucleus olivaris inferior and is located ventromedial to the nucleus tractus spinalis n. trigemini (Figs. 92, 93, 150, 155). It consists of loosely arranged, mostly medium-sized and some large cells which are flask-shaped, fusiform, elongate, and oval and have many cell processes. The neurons show strong to very strong LDH, AC, and SE; moderately strong SDH; mild to moderate ATPase; and mild AChE and MAO activity. Somewhat weaker enzyme activity is also seen in the cell processes in most of the preparations. The neuropil is moderately strong in LDH and ATPase; moderate in SE, AC, SDH, and MAO; and mild in AChE. The cut nerve fibers give a moderate MAO reaction.

Nucleus dorsalis vagi (DV)

The DV is located in the medulla oblongata between the nucleus tractus solitarius, situated laterally, and the nucleus intercalatus, situated medially and ventral to the floor of the fourth ventricle. It is separated from the latter by a layer of subependymal tissue (Figs. 102, 103, 149,

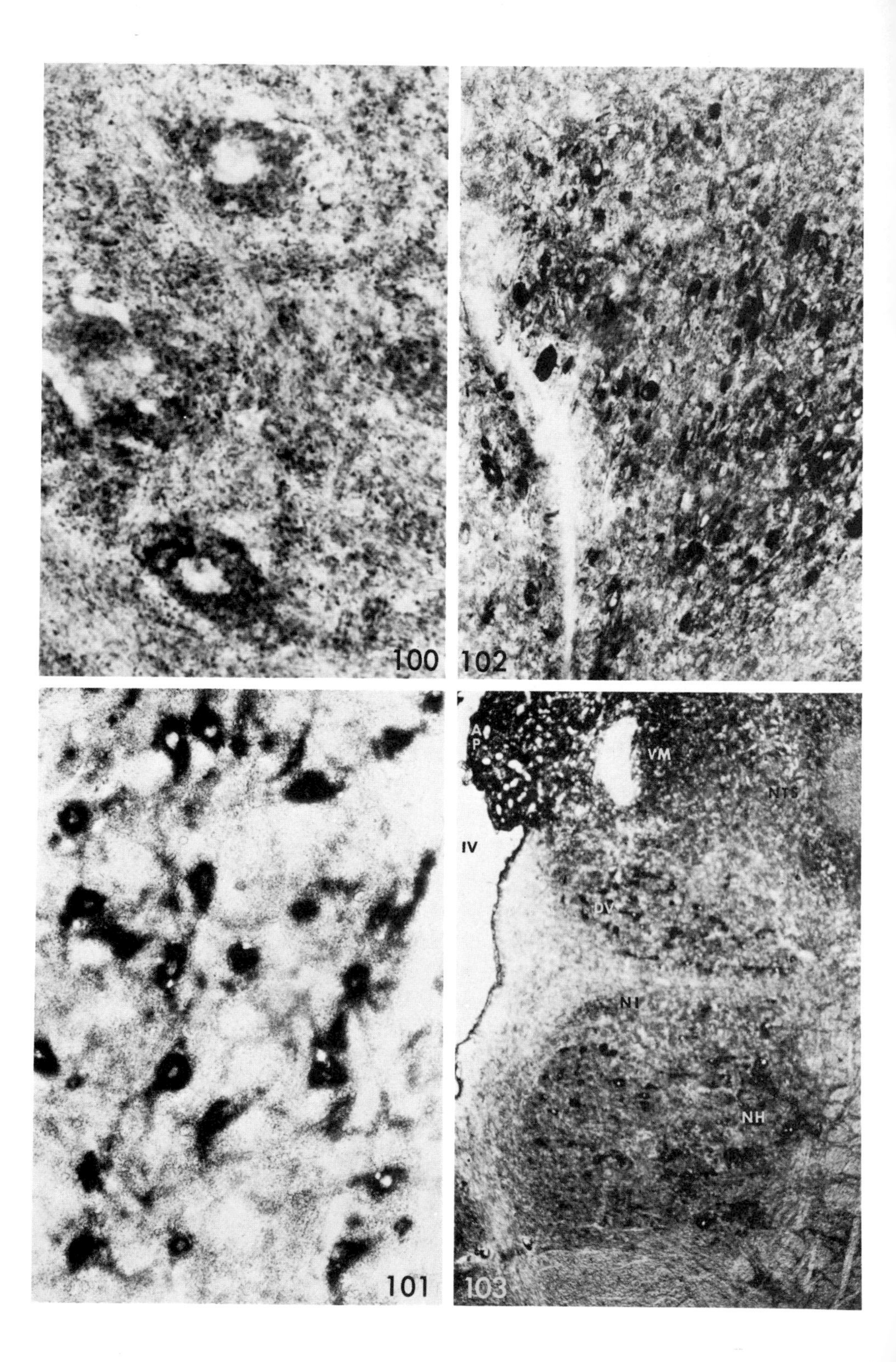
100
102
101
103
AP
VM
NTS
IV
DV
NI
NH

FIGURES 100, 101. SDH and SE activity in the neurons of the nucleus n. hypoglossi. ×592, ×148.

FIGURE 102. LDH preparation of the dorsal vagus nucleus showing very strong activity in the neurons and moderately strong activity in the neuropil. ×148.

FIGURE 103. Low power photomicrograph showing the difference in LDH localization in the nuclei n. hypoglossi, intercalatus, dorsalis vagi, tractus solitarius, vestibularis medialis, and area postrema. ×37.

150, 155, 156). It is composed of mostly medium-sized and some small and large neurons, having fusiform, round, or oval shapes. Some of the neurons show melanin pigment granules in histological preparations, but the occurrence of melanin appears to be lower in the rhesus than in the squirrel monkey, as observed by Iijima *et al.* (1968). Most of the nerve cells are strong to very strong in LDH and AC, moderate to moderately strong in SE, moderate in SDH, mild to moderate in AChE, and mild in MAO. In the ATPase some neurons are mild in the inner part of the cytoplasm and moderate at the periphery. The cell processes in most of the enzyme preparations show some activity for some distance. The neuropil of the DV is moderate in SE, AC, SDH, AChE, and MAO and moderately strong in LDH and ATPase. The blood vessels react positively in ATPase, AChE, and AK preparations. An estimate of the last preparation reveals that the DV is more vascular than the adjoining nuclei, such as nucleus tractus solitarius and nucleus n. hypoglossus.

Nucleus tractus solitarius (NTS)

The NTS is located in the medulla oblongata at the level of the nucleus dorsalis vagi (DV), although it extends orally farther than the DV to about the caudal level of the nucleus n. facialis (Figs. 103, 143, 148–150, 155). The fibers of the tractus solitarius pass through the middle of this nucleus, thus dividing the main body of the NTS into histochemically similar dorsal and ventral parts. Most neurons are medium-sized, mixed with a number of small cells, and are rather compactly arranged. They are fusiform, oval, and triangular; a number of them are multipolar and show enzyme activity in cell processes for some distance. In general, smaller neurons show a one step weaker enzyme reaction than the medium-sized ones; for example, small cells give moderate AC and SE activity compared to a moderately strong reaction in the medium-sized cells. The other enzyme preparations show moderately strong LDH, moderate SDH and ATPase, mild to moderate MAO, and negligible to mild AChE activity in the neurons. The neuropil is moderately strong in LDH and ATPase; moderate in SE, SDH, and MAO; and mild in AC and AChE activity. The fibers of the tractus solitarius, as well as the transversely cut fibers, give moderate MAO, mild to moderate ATPase, mild LDH, and negligible to negative reactions in SE, AC, SDH, and AChE.

Nucleus n. hypoglossi (NH)

The hypoglossal nucleus is a long nucleus, extending from the upper part of the pyramidal decussation to the caudal end of the nucleus praepositus (Figs. 100, 101, 103, 149, 150, 155–160). It consists of compactly arranged large and medium-sized multipolar neurons which are round, oval, or fusiform. The position of the NH in the medulla oblongata is well marked because of the distinctness of its neurons in the various histochemical preparations. Dorsally, it is related to the subependymal layer of the fourth ventricle, dorsolaterally to the nuclei intercalatus and dorsal vagus, ventrolaterally to the formatio reticularis myelencephali, and ventrally to the fasciculus longitudinalis medialis. The neurons show strong SE, AC, and LDH; moderately strong SDH; and mild to moderate AChE and ATPase activity. The neuropil is weaker than the perikarya and gives moderately strong activity in LDH, and moderate in SDH, AC, SE, MAO, AChE, and ATPase. The blood vessels are positive in ATPase and AK.

MIDBRAIN

Nucleus Ruber

This nucleus is located in the tegmental area of the midbrain, extending rostrally into the subthalamic area and caudally to the lower end of the oculomotor nerve (Figs. 56, 57, 59). The caudal part of this nucleus consists mainly of medium-sized and large cells, which are multipolar and triangular in shape. In the rostral part, the number of large cells decreases, and in this area the dorsolateral part of this nucleus contains mostly small and some medium-sized fusiform or triangular neurons. In the chimpanzee and human this area is called the parvocellular part. Histochemically, the large and medium-sized neurons show stronger enzyme activity than the small cells, but in general they show strong AC, LDH, and SDH; moderately strong SE; moderate ATPase; mild MAO; negligible AChE; and negative AK activity. The neuropil shows moderately strong LDH and ATPase; moderate SDH and SE; mild AC, AChE, and MAO; and negative AK activity. Blood vessels show AK,

AChE, and ATPase activity. The nucleus ruber has almost the same number of blood vessels as the substantia nigra.

Substantia nigra (SN)

The SN is formed by a broad band of cell masses found almost throughout the midbrain (Figs. 53–59, 104–107). Like the locus coeruleus, it is characterized by the presence of melanin pigment in the cytoplasm. The SN is situated between the pedunculus cerebri and the nucleus subthalamicus and the red nucleus; as observed in other primates, it is divisible into the pars diffusa situated ventrally and the pars compacta situated laterally. The latter occupies the dorsal and medial part of the nucleus and is composed of compactly arranged medium-sized cells which are oval or triangular, with coarse, darkly stained Nissl substance and prominent nucleoli. They give strong to very strong LDH and AC, moderately strong SE, moderate SDH and ATPase, mild to moderate AChE and MAO, and negative AK activity. The neuropil of pars compacta shows stronger enzyme activity than that of pars diffusa or the red nucleus, and shows strong LDH; moderately strong SE, SDH, and ATPase; moderate AChE and MAO; mild AS; and negative AK activity. The blood vessels show ATPase, AK, and AChE activity.

Nucleus interstitialis (Cajal) [IS]

The IS is a well delineated mass in the most oral part of the mesencephalic tegmentum at the level of the nucleus Darkschewitsch (Fig. 59). It is dorsolateral to the oral extremity of the nucleus oculomotorius, and is made up of medium-sized cells intermixed with a few large and small neurons. The cells show strong LDH, moderately strong SDH, moderate SE and AC, mild to moderate AChE and ATPase, mild MAO, and negative AK activity. The neuropil is moderately strong in LDH and SDH; moderate in ATPase, SE, and MAO; mild in AC and AChE; and negative in AK activity. The number of AK-positive vessels found in this nucleus is less than in the oculomotor complex.

Nucleus Darkschewitsch (NDK)

This fairly well-delineated nucleus is located at the same level as the nucleus interstitialis of Cajal, on the ventrolateral border of the central gray (Fig. 59). It is composed of both small and medium-sized elongate, fusiform, and triangular neurons, which are oriented in a ventromedial direction. They show strong LDH; moderately strong SE,

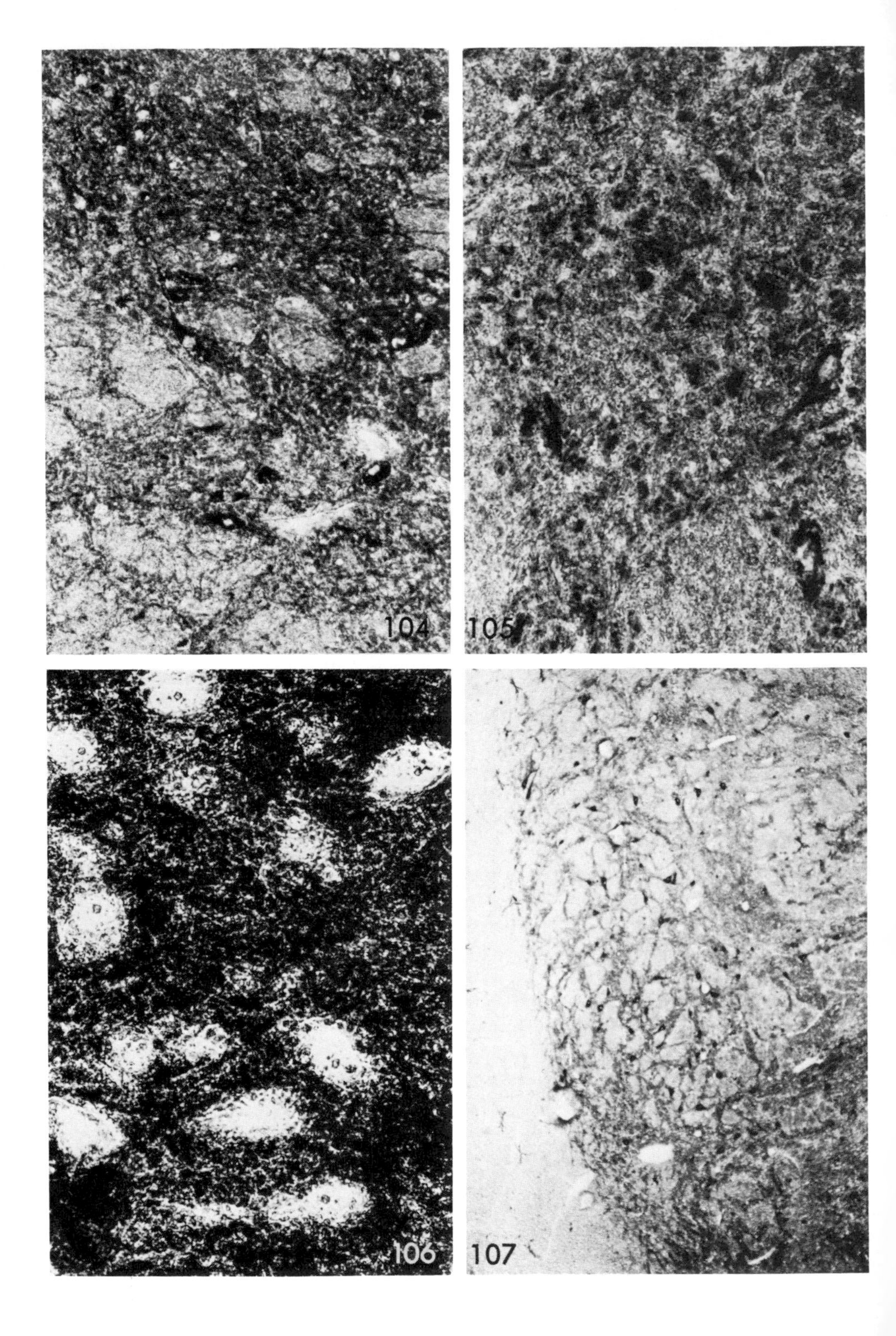

FIGURES 104–107. The photomicrographs of subsantia nigra showing LDH, SDH, ATPase, and SE activity, respectively. The neurons are strongly reactive in LDH, SDH, and SE preparations, whereas they are mild to moderate in the ATPase. The neuropil, on the other hand, is stronger in the ATPase preparation. ×148, ×370, ×370, ×37.

SDH, and AC; mild to moderate AChE and ATPase; mild MAO; and negative AK activity. The smaller neurons show less activity than the medium-sized ones. The neuropil is histochemically similar to that of the IS.

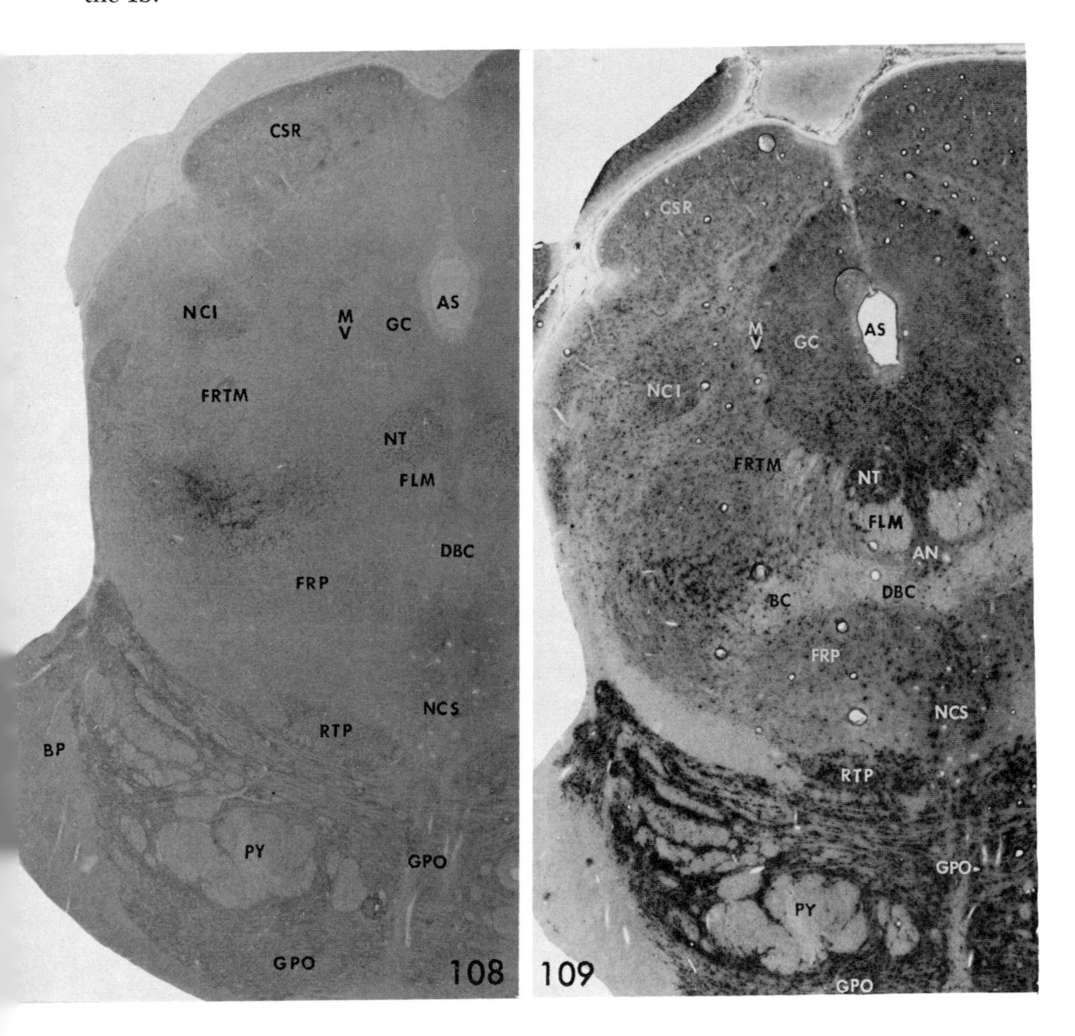

FIGURES 108, 109. AChE and AC preparations of the coronal sections at the level of the nucleus n. trochlearis showing a comparative view of the distribution of these enzymes at this level. ×7, ×7.

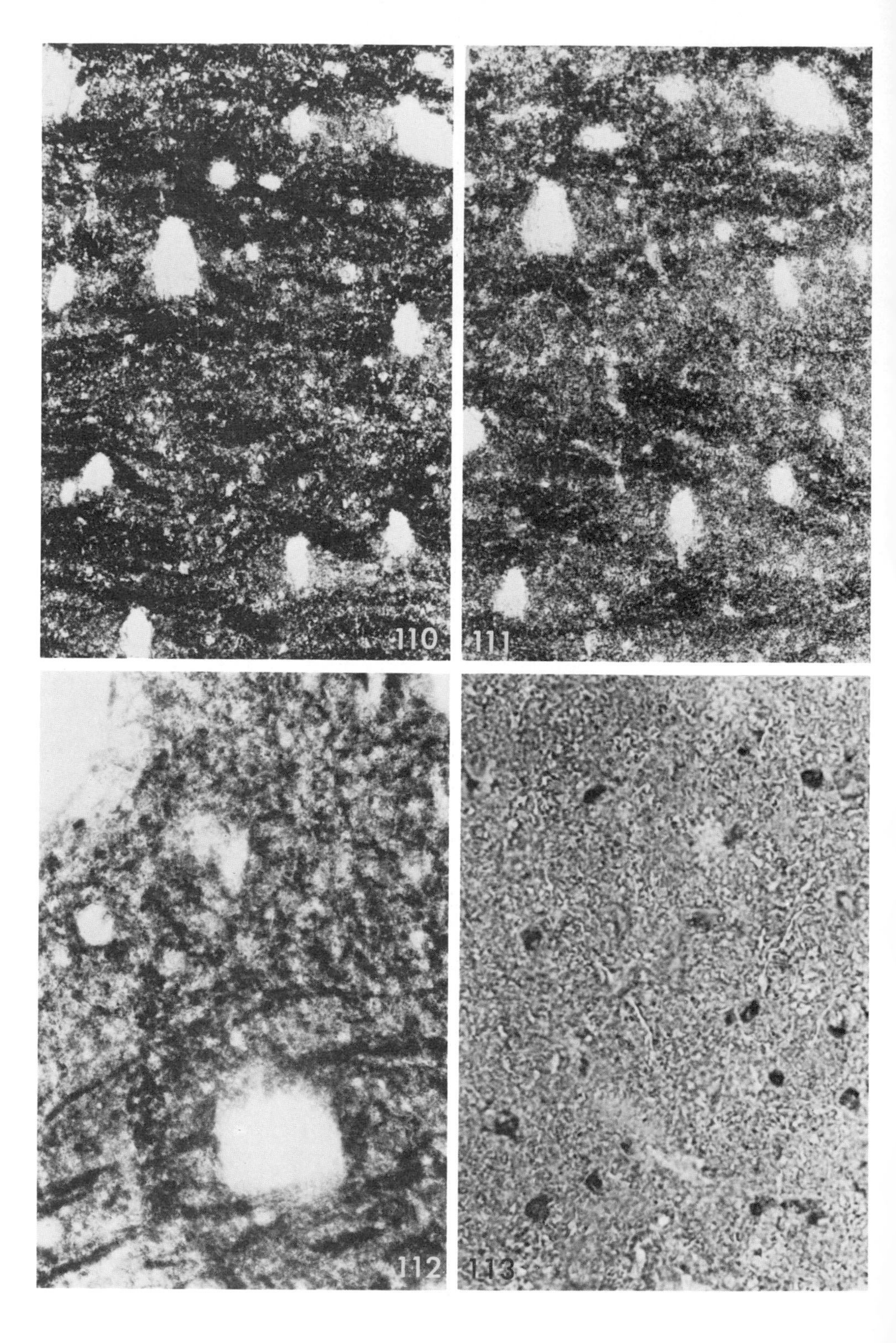
110
111
112
113

FIGURES 110–113. Photomicrographs of the nucleus interpeduncularis showing the localization of LDH, SDH, MAO, and AChE activity, respectively. The neuropil in these preparations shows stronger activity compared to the neurons, although to a varying degree. In addition, the blood vessels are strongly positive in the AChE preparations. ×148, ×148, ×148, ×592.

Nucleus interpeduncularis (IP)

The IP is an unpaired, well-circumscribed nucleus situated in the midline of the dorsal part of the interpeduncular fossa, in the caudal part of the midbrain (Figs. 59, 110–113). Caudally, the IP is located between the superior cerebellar decussation and central mass of the griseum pontis. It contains mostly small and some medium-sized elongate, fusiform, or triangular neurons, which are not very compactly arranged. The neurons are moderately strong to strong in SE, AC, LDH, and MAO; moderate in ATPase and SDH; mild in AChE; and negative in AK. The neuropil shows moderately strong LDH, MAO, and ATPase; moderate AChE and SDH; and mild to moderate AC and SE activity. Most of the small capillaries as well as blood vessels show mild to moderate AChE, AK, and ATPase activity. It is interesting that the MAO reaction in this nucleus is greater than that of any other nucleus in the brain stem; nerve fibers, crossing through the IP, also give a strong MAO reaction.

Colliculus superior (CSR)

The rostral half of the tectum is composed of the colliculus superior (CSR), which is made up of several alternating gray and white layers (Figs. 108, 109, 114). These layers do not have clear-cut boundaries in the histochemical preparations, as are seen in the Nissl and Weil preparations. The following is a description of the seven layers and their histochemical nature.

The *stratum zonale* consists of nerve fibers and a few glial cells and fusiform nerve cells. The *stratum cinereum* is composed of mostly small and medium-sized cells, whereas the *stratum opticum* is made up chiefly of nerve fibers with only a few medium-sized cells placed between the fibers. The *stratum griseum medium* consists of small and medium-sized cells with an occasional large one, while the *stratum album medium* contains longitudinal and transverse fibers with a few medium-sized cells lying in between. The *stratum griseum profundum* is the deepest gray layer, made up of medium-sized and large neurons with occasional small cells. The *stratum album profundum* is made up mostly of thick nerve fibers, between which lie a few small and medium-sized neurons. The

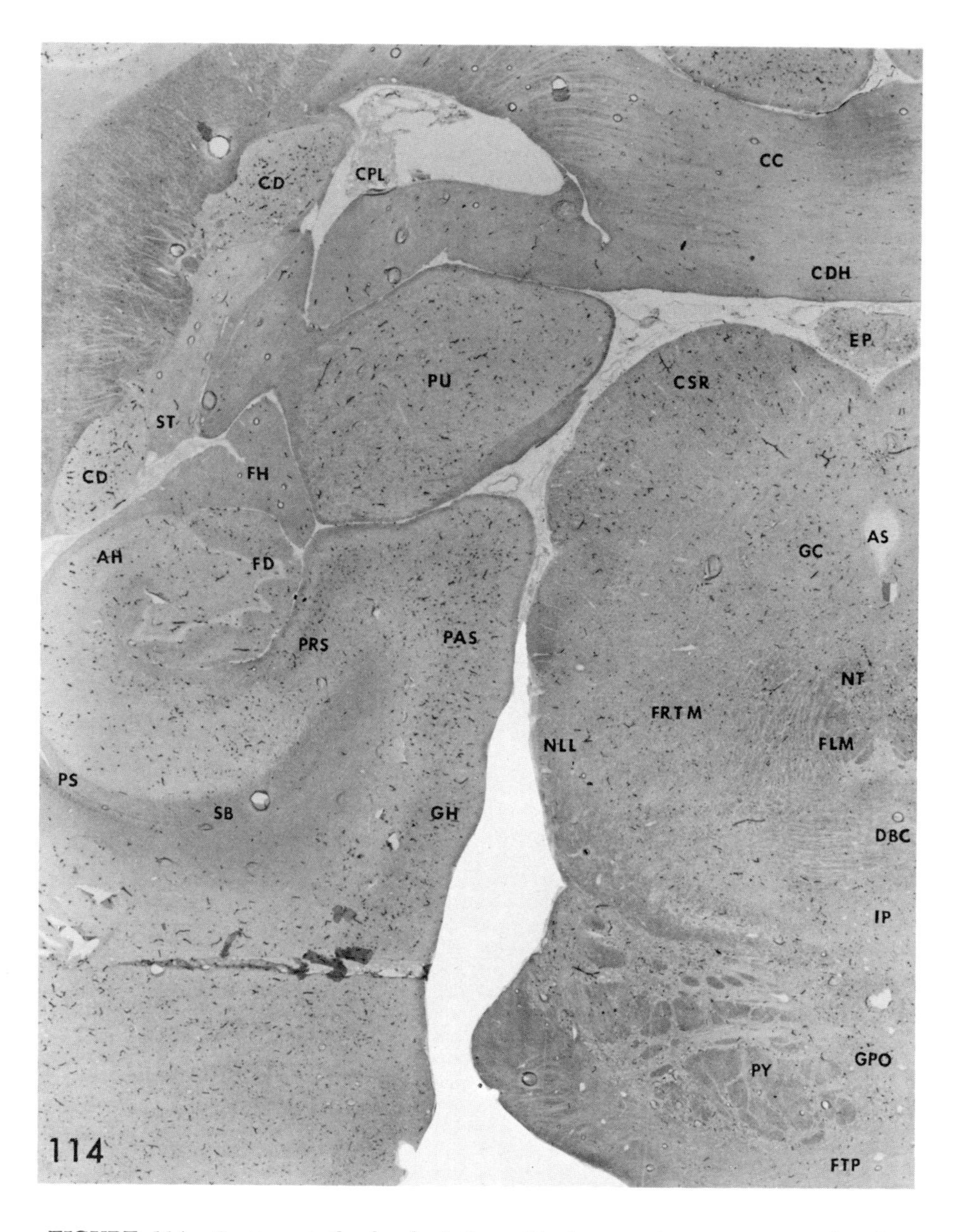

FIGURE 114. Section at the level of the colliculus superior, commissura dorsalis hippocampi, and epiphysis showing AK activity. Note the varying distribution of the blood vessels in different areas. ×7.

histochemical reactions in the cells vary with the cell size and the number of cells in a particular layer; the larger ones give stronger reactions in the various enzyme preparations. The medium-sized cells are moderately strong in LDH; moderate in SDH, SE, AC, and ATPase; and mild to moderate in AChE and MAO preparations. The enzyme activity for the large and small cells may be derived from these gradations. The neuropil shows moderately strong LDH, SDH, and ATPase; moderate MAO and AChE; and mild SE and AC activity. The glial cells give a moderately strong LDH; moderate ATPase, SE, and SDH; mild AC and MAO; and negligible AChE and AK activity. The gray layers of the CSR are more vascular than the white layers.

Colliculus inferior (CIR)

Macroscopically, the colliculus inferior (CIR), like the colliculus superior, consists of two prominently elevated hemispherical structures (Figs. 108, 109, 115–120, 124–126). The CIR is separated from the colliculus superior by the transverse collicular sulcus, and the hemispheres are separated from each other by a long, midline sulcus. The nucleus of CIR is densely populated with medium-sized cells mixed with some small and large neurons of the oval and fusiform types. The nerve cells show strong LDH, moderately strong SE and AC, moderate SDH and ATPase, mild AChE and MAO, and negligible AK activity. In MAO, the peripheral part of the nucleus of CIR shows stronger positive activity than the central part. Also, the neurons of larger size give more positive enzyme reaction than the small neurons. In AChE, the positive activity appears to be localized only in the cell membrane, leaving the cytoplasm negative. The neuropil gives moderately strong LDH and ATPase; moderate SDH, MAO, and SE; and mild AC, AChE, and AK activity. The glial cells in the CIR are more numerous than in the colliculus superior, and they show moderately strong LDH; moderate ATPase, SDH, and SE; mild AC and MAO; and negligible AK and AChE activity. The CIR is a highly vascular structure, and the blood vessels show AChE-, AK- and ATPase-positive activity.

PONS

Nuclei griseum pontis (GPO)

The GPO occupies the basilar part of the pons (Figs. 108, 109, 114, 119–123), split into irregular cell masses by longitudinal and transverse fibers. All these different cell masses are cytoarchitectonically and histo-

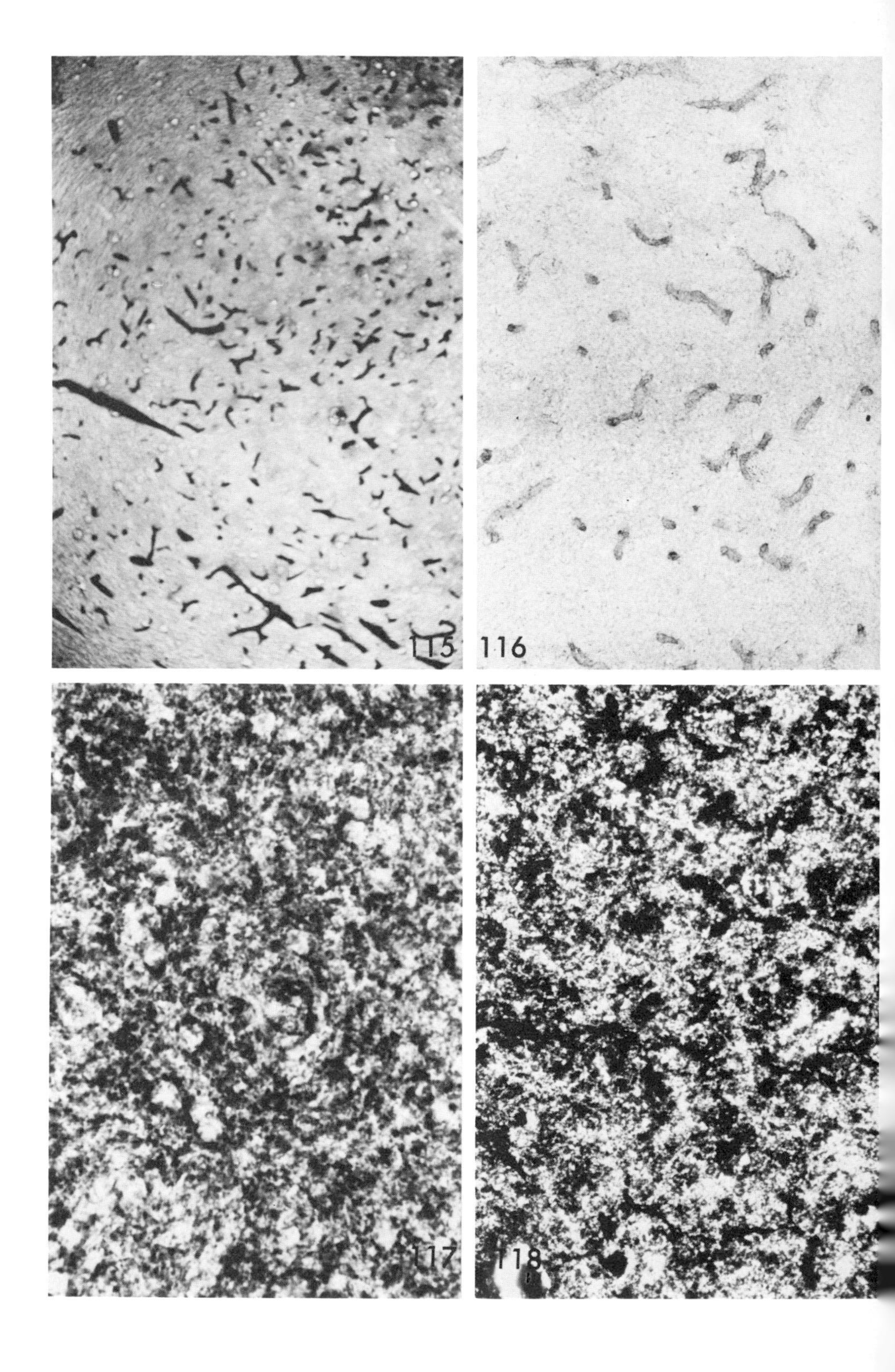
115
116
117
118

FIGURES 115–118. A comparative view of the nucleus colliculus inferior showing AK, AChE, LDH, and ATPase activity. The blood vessels show strong reactions in AK, AChE, and ATPase preparations. ×37, ×148, ×592, ×370.

chemically similar. The GPO is made up of a few large, mostly medium-sized, and some small neurons which are triangular, star-shaped, oval, or fusiform. They show very strong LDH, strong SE, moderately strong SDH and AC, moderate ATPase, mild to moderate AChE, mild MAO,

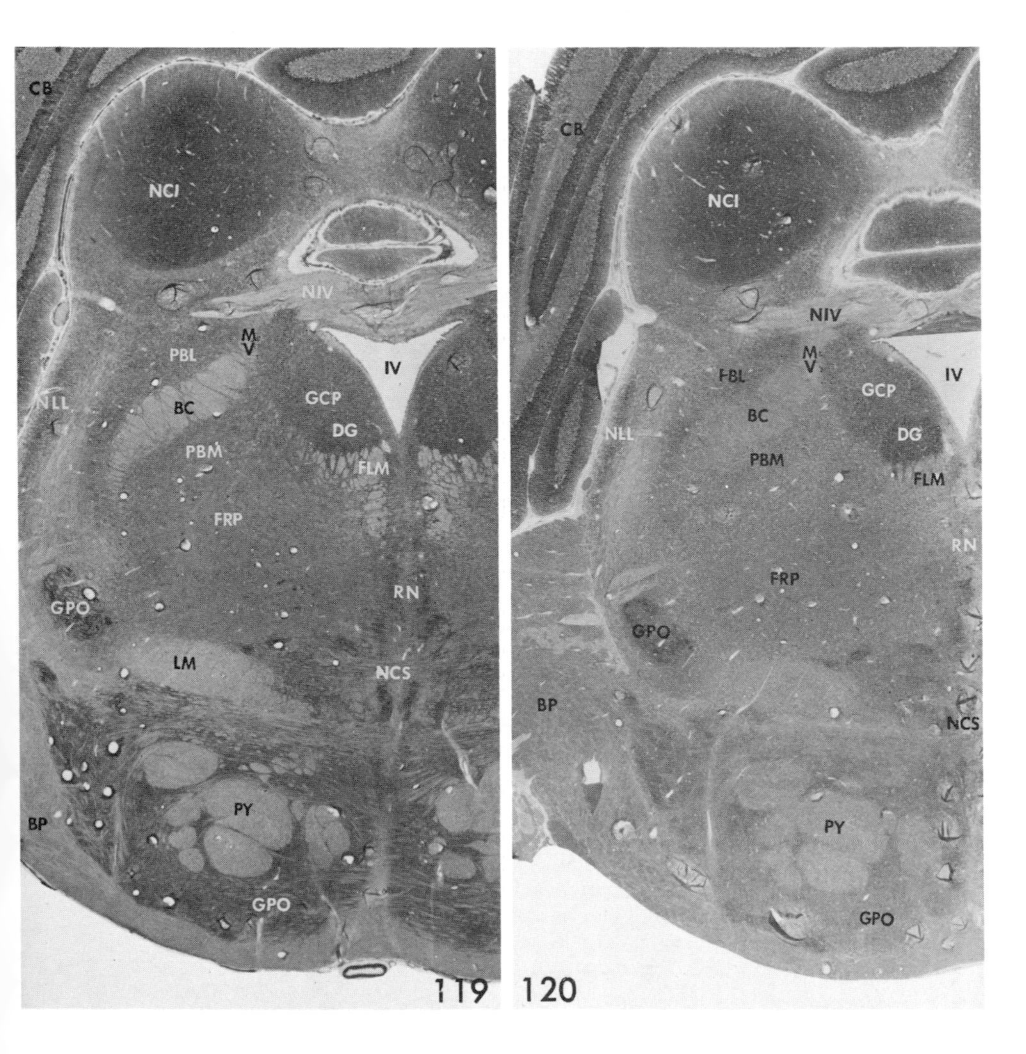

FIGURES 119, 120. LDH and SDH preparations at the level of the crossing of the nervus trochlearis in the midbrain. ×7.

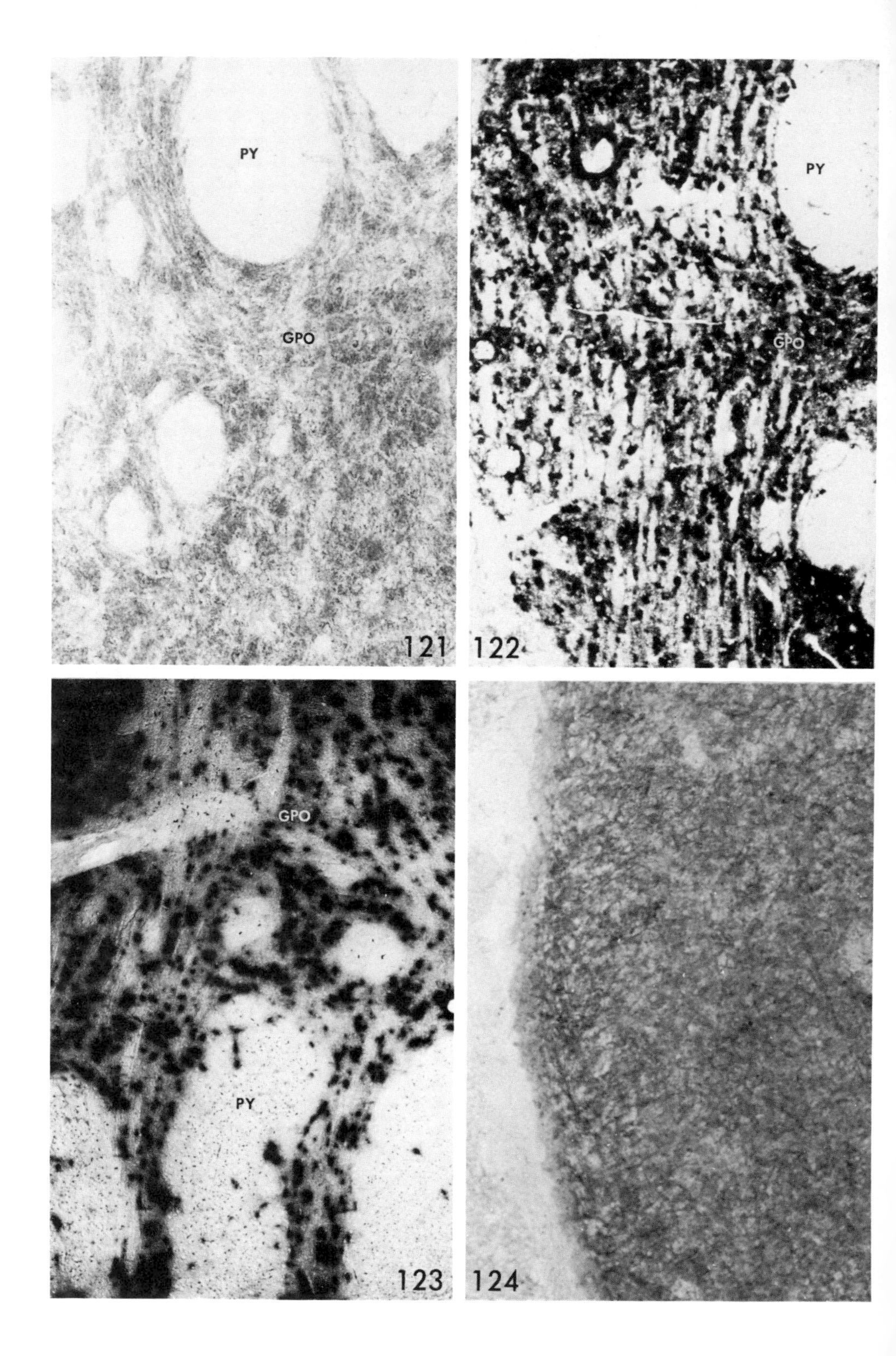
PY
GPO
121
PY
GPO
122
GPO
PY
123
124

FIGURES 121–123. A comparative view of the reactions given by griseum pontis for AChE, SE, and AC. The neurons give the strongest reaction in AC and the weakest in AChE preparations. ×148, ×37, ×37.

FIGURE 124. The peripheral part of the nucleus colliculus inferior showing strong MAO activity in the neuropil. ×148.

and negative AK activity. The neuropil shows moderately strong LDH and ATPase; moderate SDH, AChE, MAO, and SE; and mild AC activity. The fibrae transversae pontis (FTP) show mild to moderate SE, ATPase, LDH, and MAO; mild SDH, AC, and AChE; and negative AK activity. The blood vessels are positive in AK, ATPase, and AChE. The gray areas of the GPO are more vascular than the adjoining nuclei.

Nucleus locus coeruleus (LCR)

The LCR is found between the central gray of the pons and the nucleus parabrachialis medialis and brachium conjunctivum (Figs. 125, 126). The caudal part of the LCR is better defined than the rostral part. Its dorsal aspect is in intimate contact with the nucleus mesencephalicus nervi trigemini (MV), with which its neurons intermingle. The nerve cells of the LCR are stellate, triangular, and oval shaped and contain melanin and are mostly medium-sized and large, with prominent nucleoli. The neurons show strong LDH and AC; moderately strong SE; moderate SDH, ATPase, and MAO; and mild AChE activity. The neuropil shows moderately strong LDH and ATPase and moderate SE, AC, SDH, AChE, and MAO activity. The LCR contains many more AK-positive blood vessels than the adjoining central gray.

Nucleus olivaris superior (OS)

The OS is located in the caudal and lateral part of the pontine tegmentum (Figs. 125–138, 140). As in man and the chimpanzee, it is composed of two distinct groups of cells: medial and lateral. The medial group of cells is more compactly arranged than the lateral group, with a few loosely arranged cells between the two. The neurons are triangular, fusiform, spindle-shaped, and oval with long, prominent cell processes. The OS contains a mixture of small, medium-sized, and large neurons. They show strong LDH, SDH, and SE; moderately strong AC; moderate AChE; and mild MAO activity. Most neurons show mild ATPase activity in the perinuclear part of the cytoplasm, whereas at the periphery they exhibit moderately strong activity. The positive cytoplasmic enzyme activity extends into the cell processes for some distance, particularly in

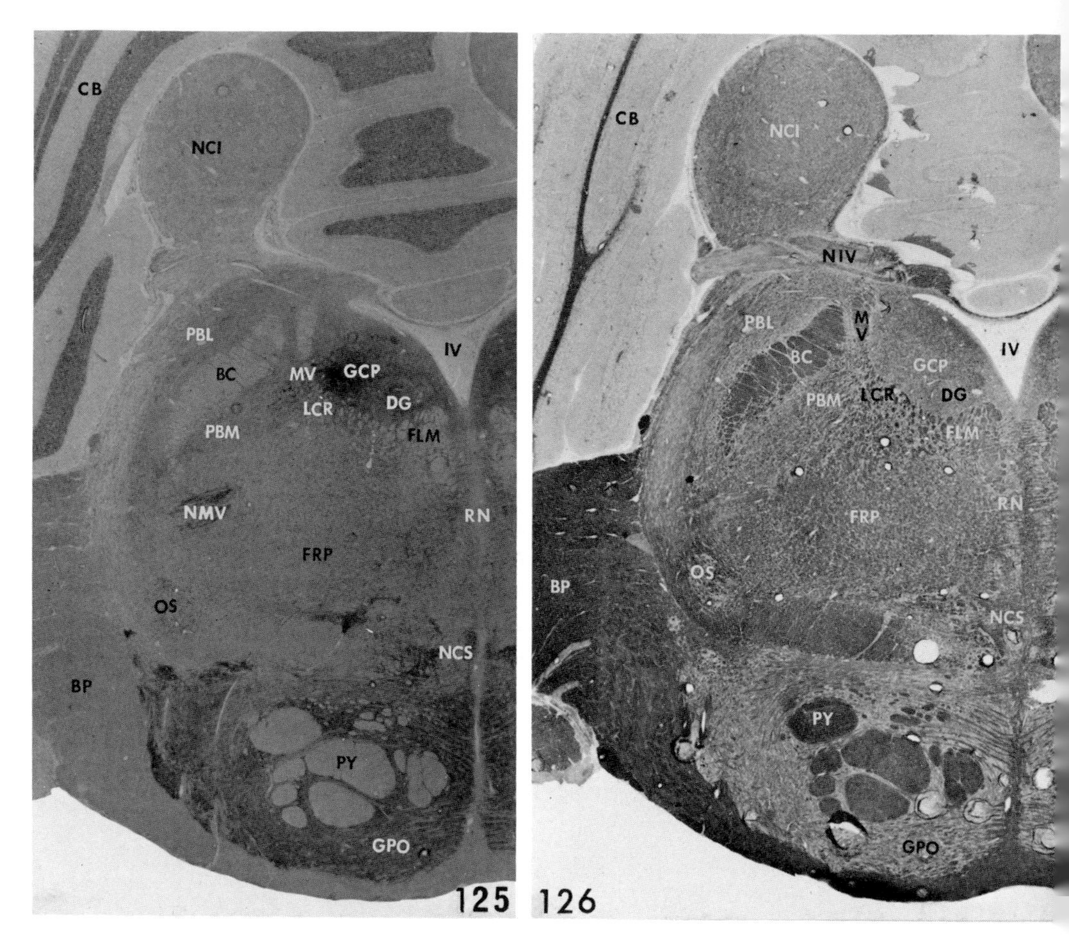

FIGURES 125, 126. AChE and MAO preparations at the level of the Gudden nucleus (nucleus dorsalis tegmenti) showing a comparative picture of these enzymes in various nuclei of the pons. ×7, ×7.

the cells of the medial part. The neuropil shows moderately strong LDH, SDH, and SE activity in the medial part and moderate activity in the lateral part. The neuropil of both lateral and medial parts show moderate AC, ATPase, and MAO and moderately strong AChE activity. The medial part has more AK-positive blood vessels than the lateral part.

Nucleus corpus trapezoidalis (NCT)

The NCT is formed by a loosely arranged mass of cells located on the ventromedial aspects of the OS (Figs. 131–138, 140). In contrast to the latter, it is composed of large cells, with a few small and medium-sized neurons. They are oval or triangular in shape with a few fusiform cells. The nerve cells show very strong LDH, strong SDH and SE, moderately strong AC, and mild to moderate AChE and MAO activity. The perinuclear part of the cytoplasm shows mild ATPase, whereas the peripheral part shows moderately strong ATPase activity. The neuropil shows less enzyme activity than the OS, probably because of decreased enzyme activity extending into cell processes. The NCT has few AK-positive blood vessels compared to the OS.

Nucleus lemniscus lateralis (NLL)

The NLL is formed by an elongated mass of cells found in the caudal part of the pons situated between the brachium pontis and the formatio reticularis pontis (Figs. 114, 119, 120). It is composed of both small and medium-sized fusiform, triangular, round, or oval neurons. The neurons show moderately strong LDH and SE; moderate SDH, ATPase, and AC; and mild MAO and AChE activity. The neuropil shows moderately strong ATPase; moderate LDH and MAO; and mild SDH, SE, AC, and AChE activity.

Nucleus praepositus (PP)

The PP, in the Snider and Lee (1961) atlas on the rhesus brain, has been described as the nucleus eminentiae teretis. It is a triangular nucleus situated at the rostral extremity of the nucleus hypoglossi. The subependymal layer separates this nucleus from the floor of the IVth ventricle. Rostrally, it extends almost to the level of the nucleus abducentis (Figs. 141–143, 148). The cells are small and medium-sized and are oval, fusiform, or triangular. The medium-sized neurons have prominent cell processes compared to the small cells. Most cells show moderately strong LDH, SE, and AC; moderate SDH; and mild ATPase, AChE, and MAO activity. The neuropil of the central part of this nucleus shows enzyme activity different from that of the peripheral part. The central part of this nucleus shows moderately strong LDH; moderate ATPase, MAO, and SE; whereas the peripheral part shows strong ATPase, moderately strong MAO, moderate LDH, and mild SE. Both central and peripheral parts show moderate AC, AChE, and SDH activity.

Nucleus interpositus (NIP)

This is a small, rather indistinct nucleus situated at the same level as the PP and composed of small elongate, fusiform, or oval neurons

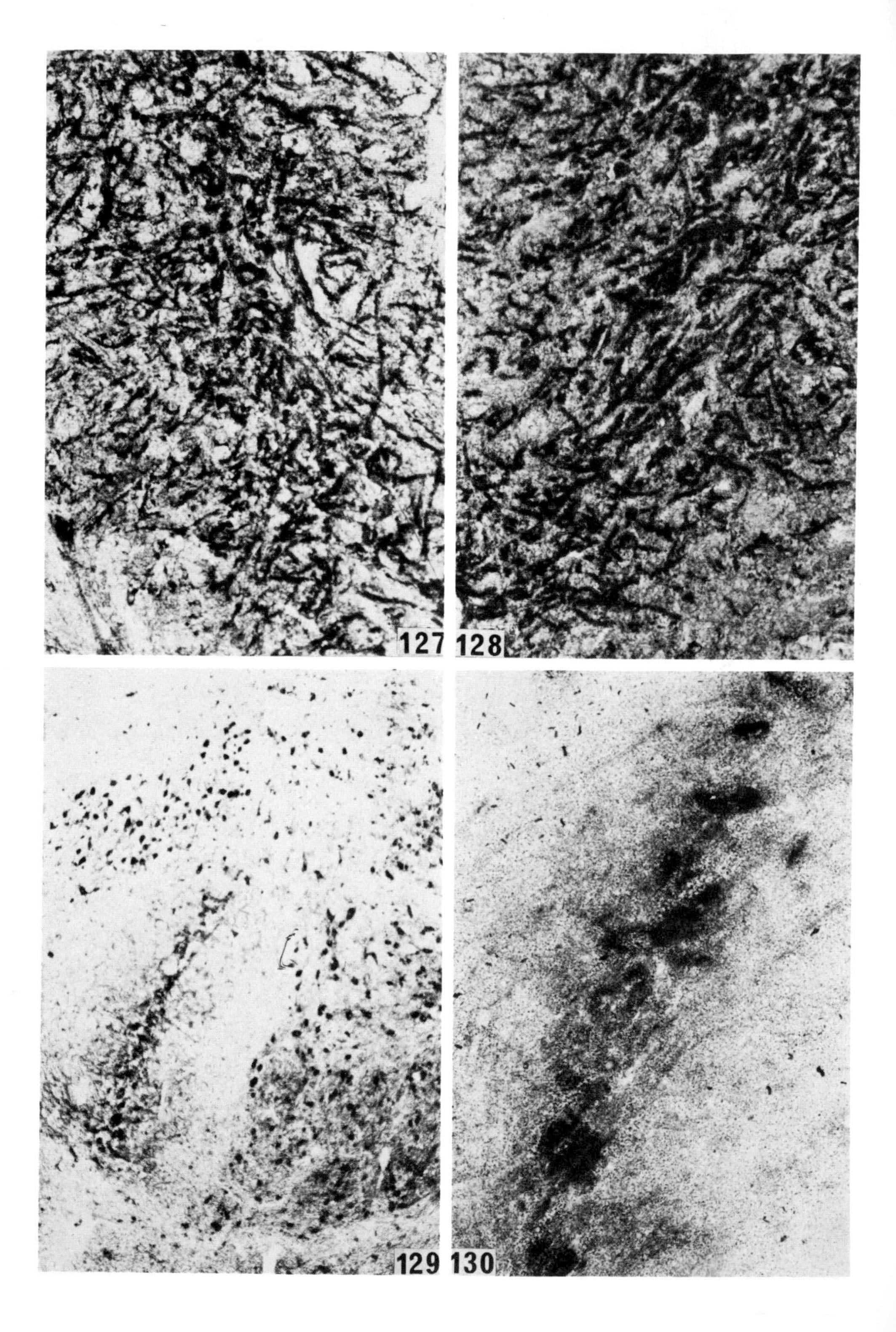
127
128
129
130

FIGURES 127–130. LDH, SDH, SE, and AC preparations of the nucleus olivaris superior. Although the neurons show strong reactions for all these enzymes, the activity does not pass far into the cell processes in the SE and AC preparations as was observed in those of LDH and SDH. ×148, ×148, ×37, ×148.

mixed with very few medium-sized cells. They show moderately strong LDH; moderate SE, SDH, and AC; and mild ATPase, AChE, and MAO activity. The neuropil is moderate in SDH, LDH, MAO, and ATPase and mild in SE, AC, and AChE preparations.

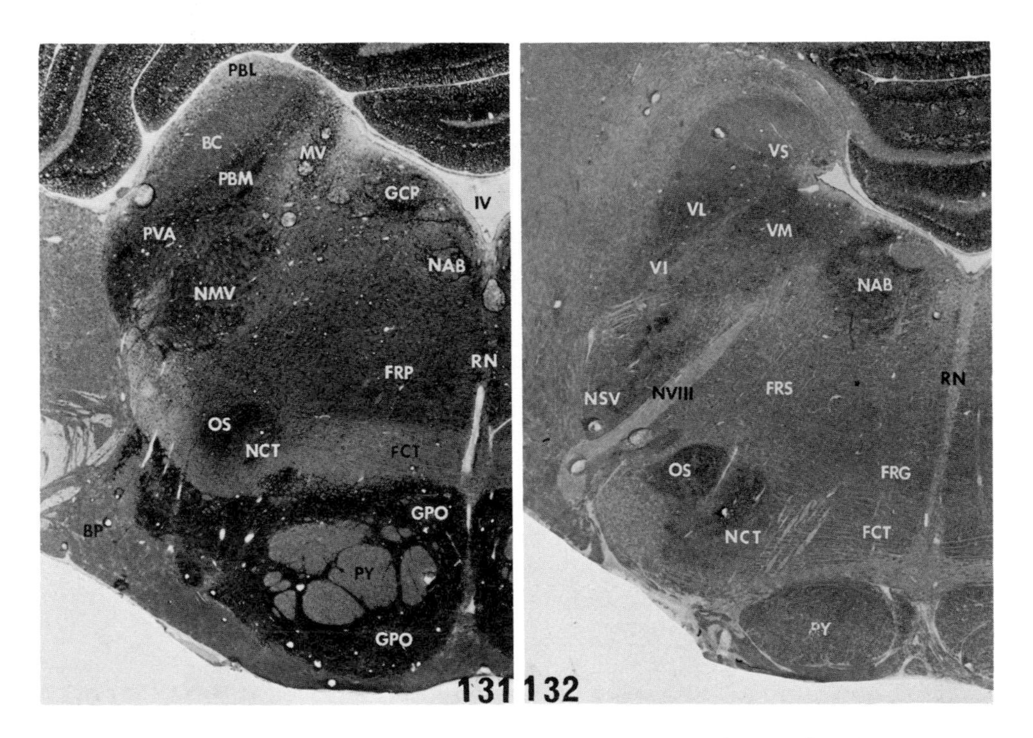

FIGURES 131, 132. A comparative picture of the SE and SDH distribution in the various nuclei at the level of the nucleus abducentis. ×7, ×7.

Nucleus centralis superior (NCS)

The NCS is irregular in shape and situated in the oral part of the pontine tegmentum (Figs. 108, 109, 119, 120, 125, 126). The irregular shape of the NCS is attributed to the various fiber tracts which surround this nucleus. Dorsally the main stem of the nucleus is separated by the fibers of decussation of brachium conjunctivum into a small, dorsal, cylindrical mass (called, by many workers, the nucleus annularis)

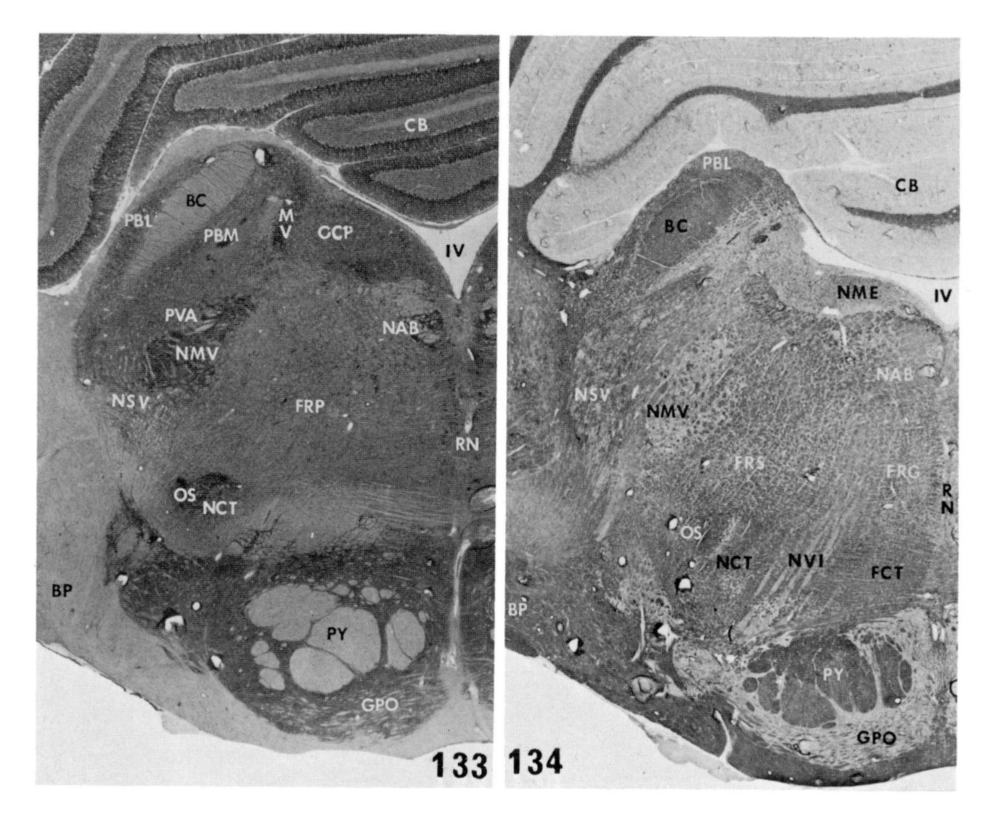

FIGURES 133, 134. A comparative picture of LDH and MAO preparations at the level of the nucleus abducentis. ×7, ×7.

and a large ventral (main) part. Olszewski and Baxter (1959) divided this nucleus in the human into dorsal, medial, lateral, and ventral subnuclei. Such clear-cut subdivisions, however, could not be recognized in the rhesus monkey. The NCS is composed of small and medium-sized cells with a few large ones which are oval, triangular, and spindle-shaped. The nerve cells show strong LDH; moderately strong AC and SE; moderate SDH; and mild AChE, MAO, and ATPase activity. The neuropil shows moderately strong LDH; mild to moderate ATPase, SDH, AChE, and MAO; and mild SE and AC activity.

Raphae nuclei (nucleus dorsalis [DR] and nucleus ventralis [VR] raphae)

These nuclei are situated in the middle of the caudal part of the pons and oral part of the medulla oblongata (Figs. 119, 120, 125, 126,

131–143, 148–150). The cells of these nuclei are linearly arranged in an anteroposterior direction, extending posteriorly to the posterior longitudinal fissure in the IVth ventricle, and are separated from the ependymal lining by subependymal gray. Anteriorly, the cell line extends to the lemniscus medialis. This cell mass has been divided somewhat artificially into ventral and dorsal raphae nuclei. The DR has more compactly

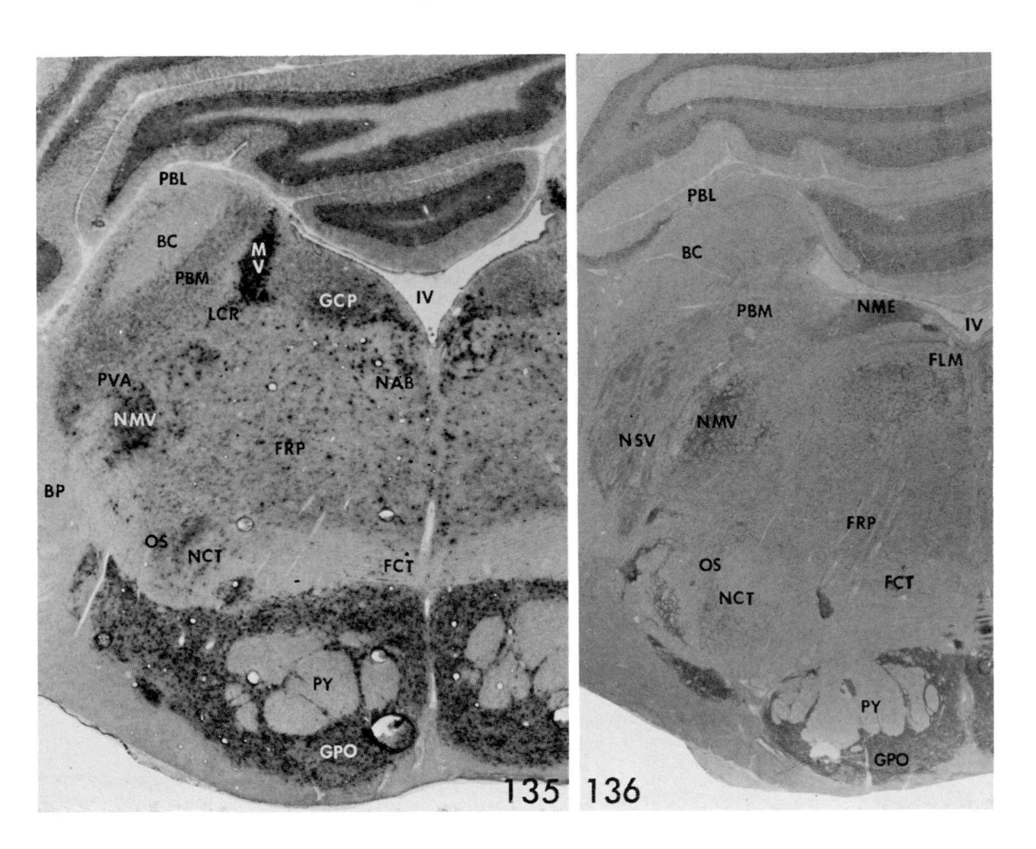

FIGURES 135, 136. AC and AChE preparations at the level of the nucleus olivaris superior. The difference in the enzyme content and in the distribution of these enzymes is evident. ×7, ×7.

arranged cells than the VR. The neurons are mostly medium-sized and are oval, fusiform, and triangular. They show strong LDH; moderate SDH, SE, AC, and ATPase; and mild AChE and MAO activity. The neuropil shows moderately strong LDH; moderate SE, SDH, ATPase, MAO, and AChE; and mild AC activity. These nuclei have fewer AK-positive blood vessels than the nucleus abducentis.

Nucleus of medial eminence (NME)

The NME is located dorsomedial to the fasciculus longitudinalis medialis and dorsal to the nucleus abducentis (Figs. 134, 136, 138, 140). This nucleus probably corresponds to the nucleus suprageniculatus of Olszewski and Baxter (1959) in the human and to the nucleus paramedianus dorsalis described in the stereotaxic atlas of the rhesus monkey brain by Snider and Lee (1961). This is a somewhat triangular nucleus situated beneath the ependyma of the IVth ventricle and composed mostly of small neurons, with a few medium-sized cells, which are oval and fusiform. The neurons show strong SDH; moderately strong SE, LDH, and AC; moderate ATPase; and mild MAO and AChE activity. Transversely cut nerve fibers show moderately strong MAO-positive activity compared to the moderate neuropil. The neuropil shows moderately strong LDH, SDH, and ATPase and moderate AC, SE, and AChE activity. This nucleus has fewer blood vessels than the neighboring nucleus abducentis.

Nuclei parabrachialis lateralis and medialis (PBL and PBM)

The PBL is located dorsolateral and the PBM ventromedial, to the brachium conjunctivum (Figs. 131, 133–136). The PBL is a narrow band consisting of comparatively lightly stained, sparingly distributed cells, whereas the PBM consists of moderately stained cells, aggregated into one mass. It is related ventromedially to the locus coeruleus and the nucleus mesencephalicus n. trigemini. The neurons in these nuclei are fusiform, oval, or triangular and are small with a few medium-sized cells. The nerve cells in the PBL show moderate LDH, SE, and AC and mild SDH, ATPase, AChE, and MAO activity. On the other hand, the neurons in the PBM show moderately strong SE and LDH; moderate AC, ATPase, and SDH; and mild AChE and MAO activity. The neuropil in the PBL shows moderate LDH and ATPase and mild SE, AC, AChE, MAO, and SDH activity. The neuropil of the PBM gives moderately strong LDH and ATPase; moderate SE and AChE; and mild SDH, MAO, and AC activity. The PBM is more vascular and shows a larger number of AK-positive blood vessels than the PBL.

MEDULLA OBLONGATA

Nucleus olivaris inferior (OI)

Located in the oral part of the medulla oblongata, lateral and dorsolateral to the pyramids, the OI is an irregular mass of cells taking a

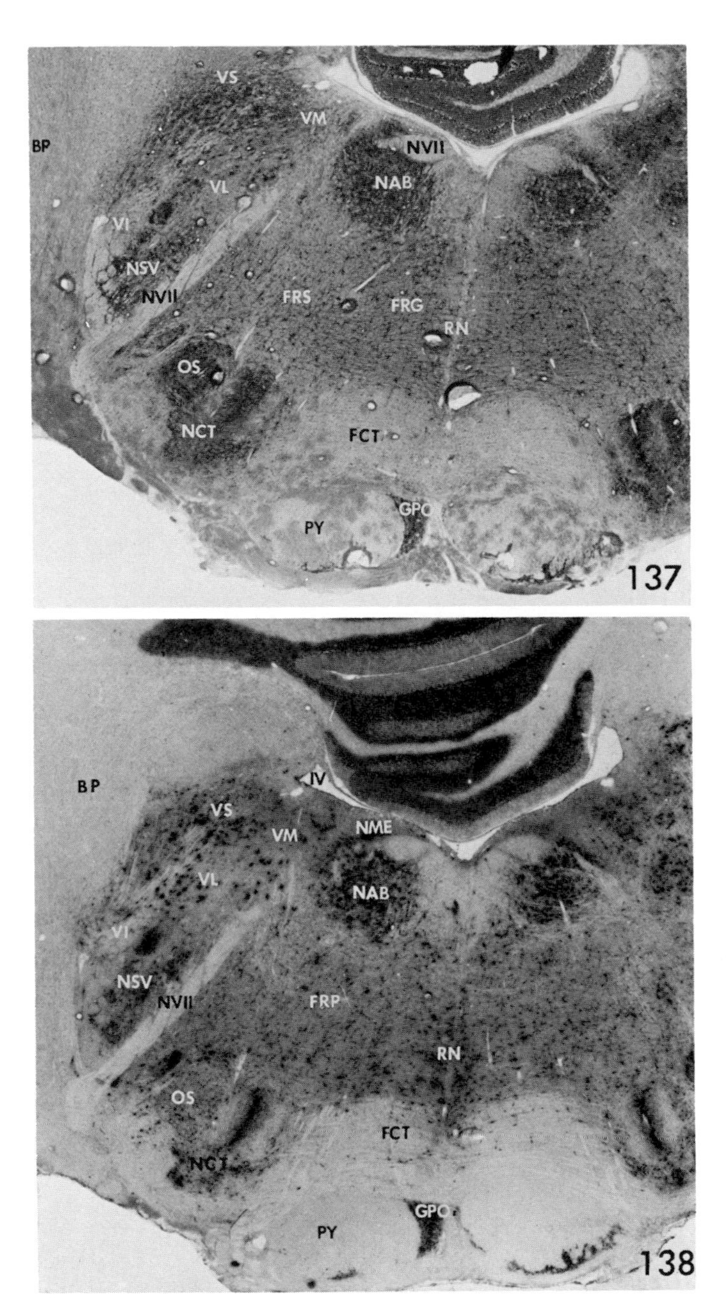

FIGURES 137, 138. Sections passing through the nucleus abducentis showing SE and AC activity, respectively, at this level. ×7, ×7.

number of zigzag courses, with its hilus directed medially (Figs. 139, 141, 143–150, 155–158). In addition to this main mass, there are smaller masses of neurons which are not as irregular in shape. The one located dorsally is called the dorsal olive (OID), and the one medially placed is medial olive (OIV). There are no noticeable histochemical differences in the neurons of the different parts of the olivary complex; hence, a

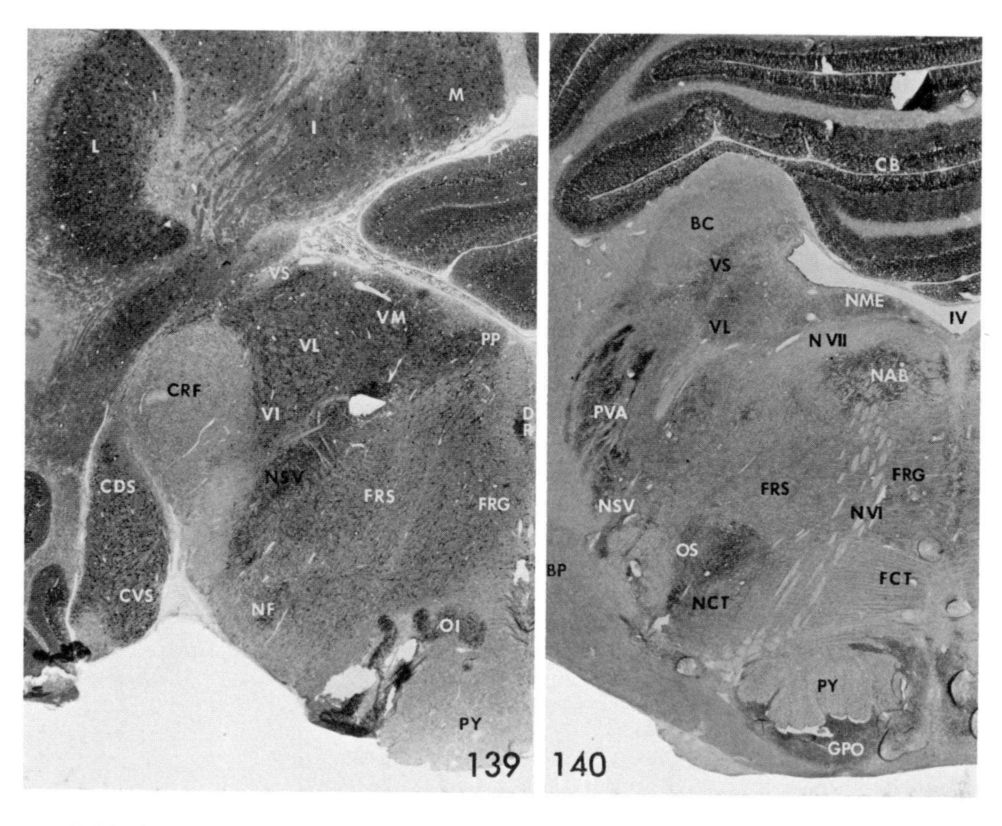

FIGURE 139. Simple esterase preparation showing enzyme reaction in the deep cerebellar, cochlear, and vestibular nuclei and in the medulla oblongata. ×7.

FIGURE 140. Section passing through the middle of the nucleus of the medial eminence also showing SDH activity in the nucleus principalis n. trigemini and vestibular nuclei. ×7.

common description will suffice. Most of the neurons are medium-sized, round or oval in shape, and compactly arranged. They give a strong reaction in LDH and AC; moderately strong in SE; moderate in SDH; and mild to moderate in ATPase, AChE, and MAO preparations. The neuropil is weaker than the neurons except in the ATPase. The glial cells show moderately strong LDH and moderate ATPase activity. A large number of closely arranged blood vessels are positive in AK, ATPase, and AChE.

Nucleus cuneatus

The cuneate nucleus consists of two main masses, the medial (CN) and the lateral (CNL) [Figs. 149, 150, 154, 156–160]. The CN is located

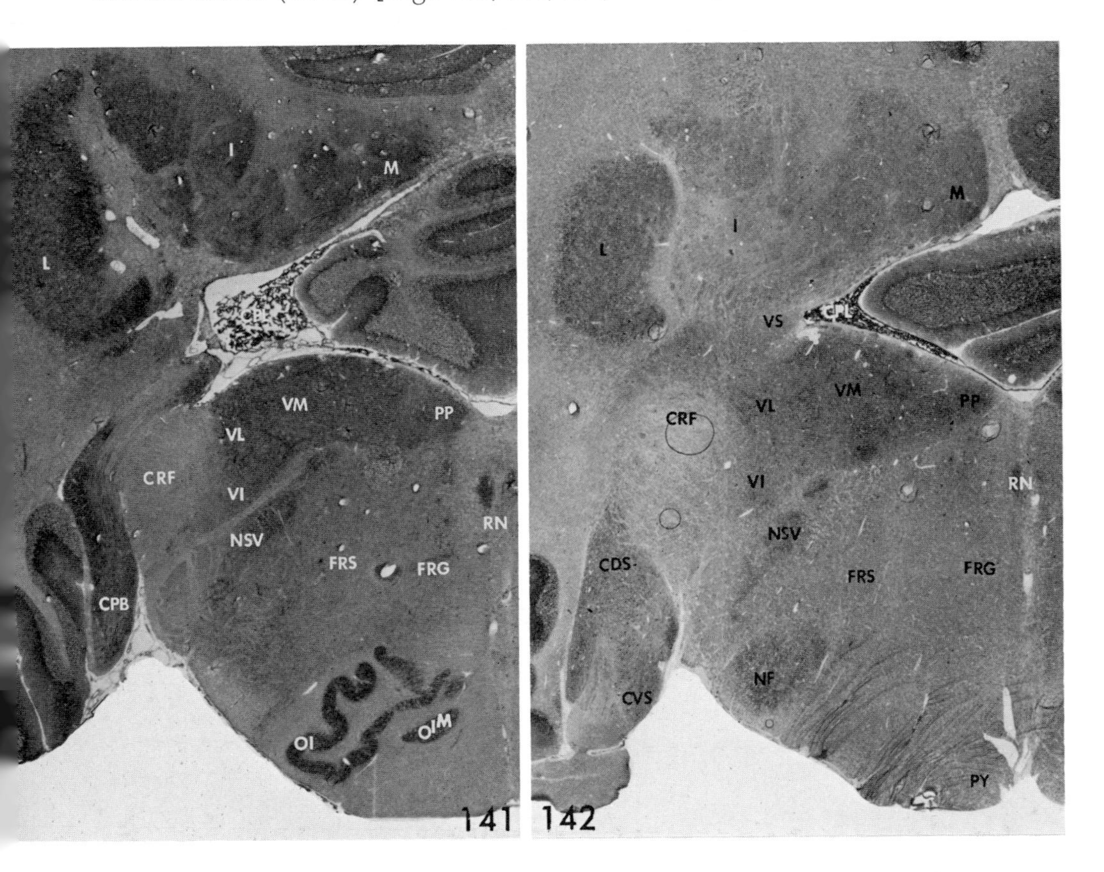

FIGURES 141, 142. LDH and SDH preparations of the medulla oblongata at the level of the vestibular and deep cerebellar nuclei showing differences in intensity of reaction in these preparations. ×7, ×7.

dorsolateral to the gracile nucleus and extends through the entire length of the latter. It consists of mostly medium-sized and some small, compactly arranged neurons, which are round, oval, or fusiform. The CNL is located dorsolateral to the CN and extends rostrally to about the caudal part of the nucleus vestibularis inferior and contains large neurons mixed with medium-sized cells which are not as compactly arranged as those of the CN. Most of the neurons are round, triangular, fusiform, or star-shaped with long cell processes. The CNL, in most histochemical preparations in the present study, shows a one-step-stronger reaction than the CN. For example, while there is a moderately strong neuronal activity

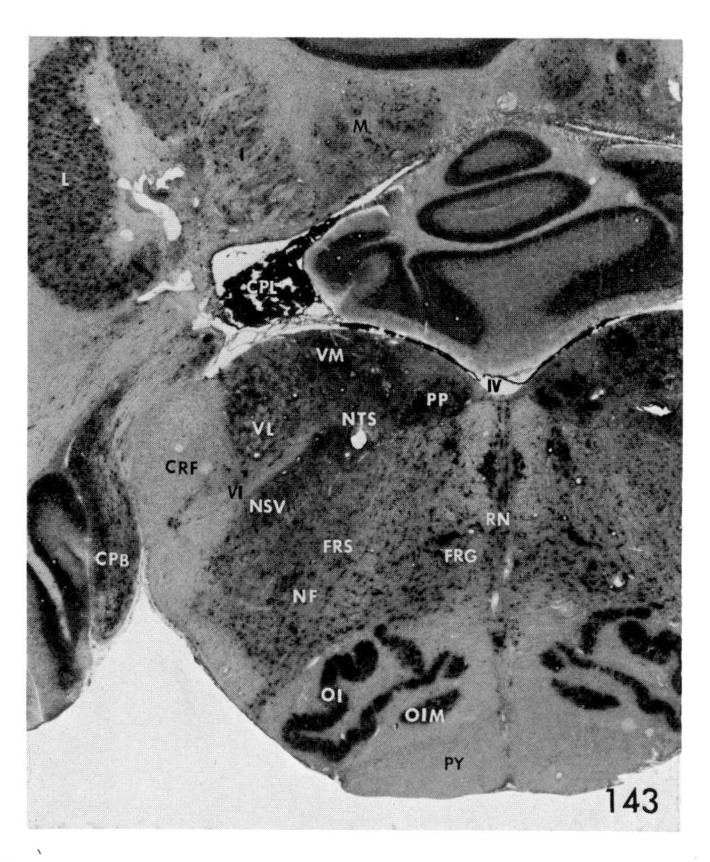

FIGURE 143. AC preparation of the medulla oblongata at the level of the deep cerebellar and vestibular nuclei. ×7.

of SE in the CN, there is a strong reaction in the CNL; additionally, while the neuropil in the CN is moderate, that in the CNL is moderately strong. In the AChE and MAO, however, the CN and CNL react identically with mild activity in the neurons and moderate activity in the neuropil. The neurons in the CN give a strong LDH, SDH, and AC; moderately strong SE; and moderate ATPase activity. The neuropil is moderately strong in LDH; moderate in SDH, SE, and ATPase; and mild in AC. The enzyme activity in the cell processes of neurons is weaker than in the perikarya and can be observed for a longer distance in the CNL than in the CN. The glial cells, as usual, give the strongest reaction in LDH. The blood vessels react positively in ATPase, AChE, and AK.

Nucleus gracilis (G)

The G is located medial to the CN in the lower part of the medulla oblongata (Figs. 156–160, 171, 172). The reticulated appearance of this nucleus is due to the passage of numerous fibers of the fasciculus gracilis and to the neurons dispersed between the fibers. Caudally, the cells are somewhat more compactly arranged than in the oral part. Most of the neurons are of medium size and are oval, fusiform, and triangular in outline. The histochemical reactions vary from strong in the LDH, SDH, AC, and SE preparations to moderate in ATPase, AChE, and MAO in the cells. The enzyme activity in the neuropil is strong in LDH and SDH, moderately strong in SE and ATPase, moderate in AChE and MAO, and mild in AC tests.

Nucleus Roller (NR)

The NR represents a group of large and medium-sized, compactly arranged neurons ventromedial to the hypoglossal nucleus (Figs. 149, 150). Although the cells of the latter are somewhat larger in size and less compactly arranged than those of the NR, the histochemical activity of the neurons in both nuclei is identical.

Nucleus intercalatus (NI)

The nucleus intercalatus is located between the nucleus dorsalis vagi dorsolaterally and the hypoglossal nucleus ventromedially (Figs. 103, 149, 150, 155, 156). It consists of small neurons, which are oval or elongate with long cell processes. The density of cell population is less than either of the adjoining nuclei. Histochemically, these cells show weaker enzyme activity. The neurons are moderate in LDH and AC; mild to moderate in SDH, SE, and ATPase; and mild in AChE and MAO activity. The neuropil is moderately strong in ATPase; moderate in LDH and MAO; and mild in SDH, SE, AC, and AChE. The glial cells are strongest in LDH and show moderate activity. There are fewer blood vessels in the NI than in the other cranial nerve nuclei of the medulla oblongata at this level.

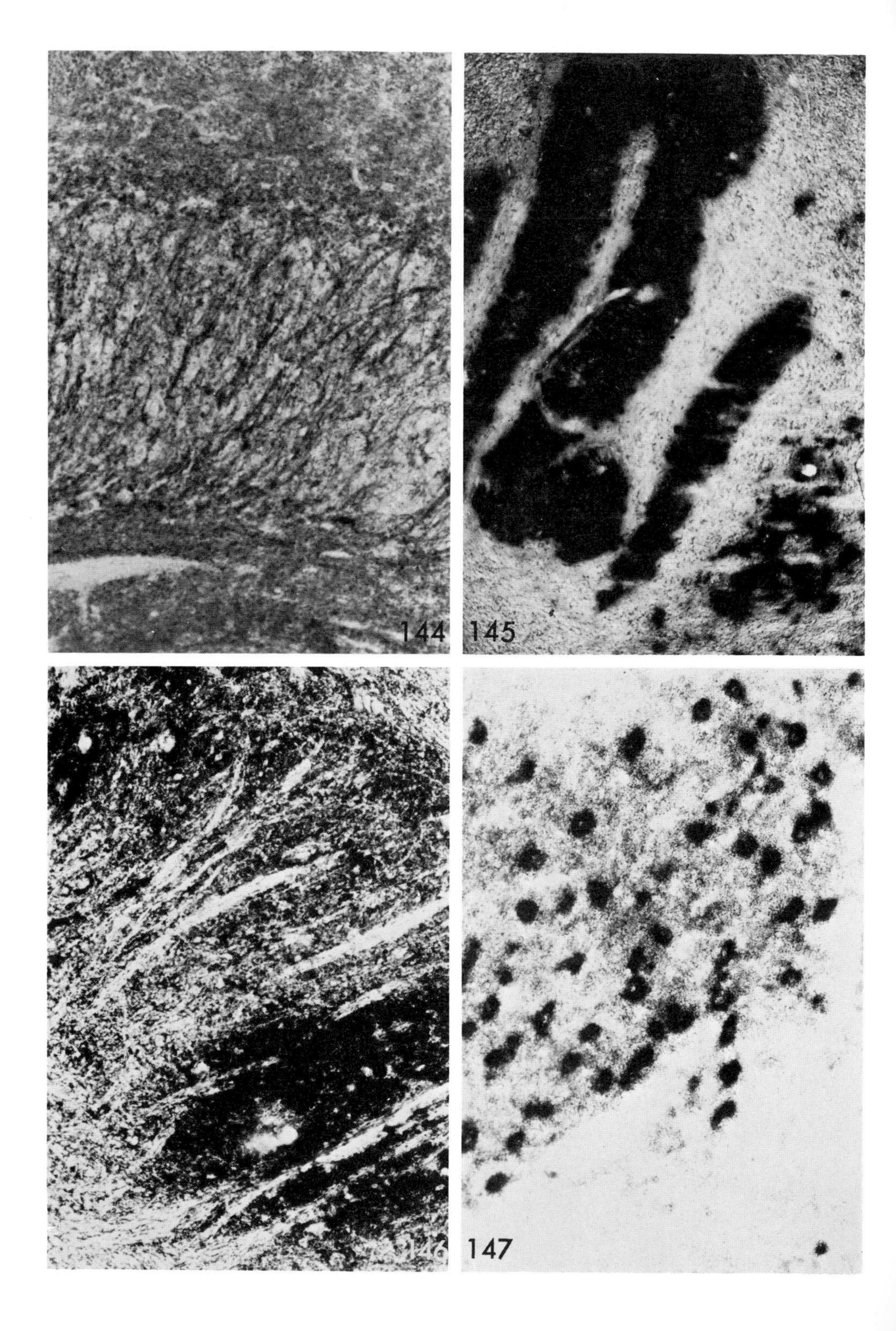
144
145
146
147

FIGURES 144–147. Photomicrographs of the nucleus olivaris inferior showing MAO, AC, ATPase, and SE activity. Whereas the neuropil is stronger in the ATPase, MAO activity predominates in the nerve fibers. ×148, ×37, ×148, ×148.

Area postrema (AP)

The area postrema is a V-shaped structure located in the lower lateral margin of the floor of the IVth ventricle (Figs. 103, 149, 151–153) and is separated from the fovea vagi by a glistening band of neuroglial processes known as the funiculus separans. The AP is highly vascular, as is indicated in the AK preparations. There has been considerable disagreement concerning the cellular morphology and neural structure of the AP in subhuman primates. This controversy has been discussed in a previous article on the chimpanzee brain (Shantha and Manocha, 1969). In our

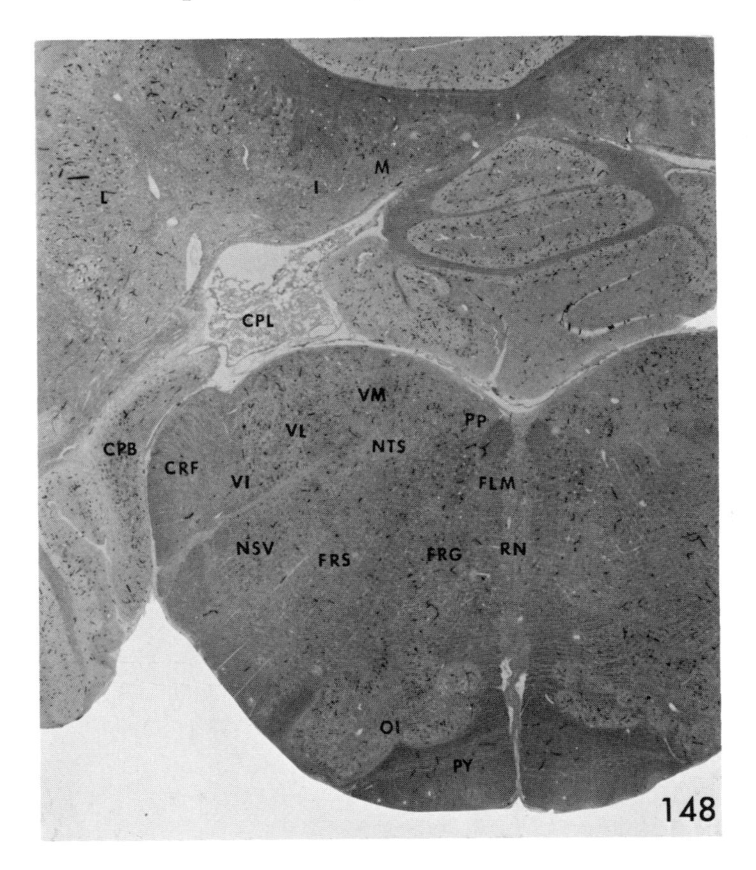

FIGURE 148. AK preparation showing activity in the blood vessels in the various nuclei of the medulla oblongata at the level of the deep cerebellar nuclei. Note the smaller number of blood vessels in the white matter compared to the gray areas. ×7.

rhesus preparations, we have observed a small number of medium-sized oval, rounded, fusiform, or triangular neurons, with numerous glialoid and glial cells. A large number of blood sinusoids and blood vessels are also observed. The histochemical reactions vary in the different components. The neurons give strong LDH and AC; moderate SDH, SE, and MAO; mild AChE and ATPase; and negative AK reactions. The neuropil is strong in LDH; moderate in SE, SDH, ATPase, AChE, and MAO; and mild in AC preparations. The glialoid cells are strong in LDH;

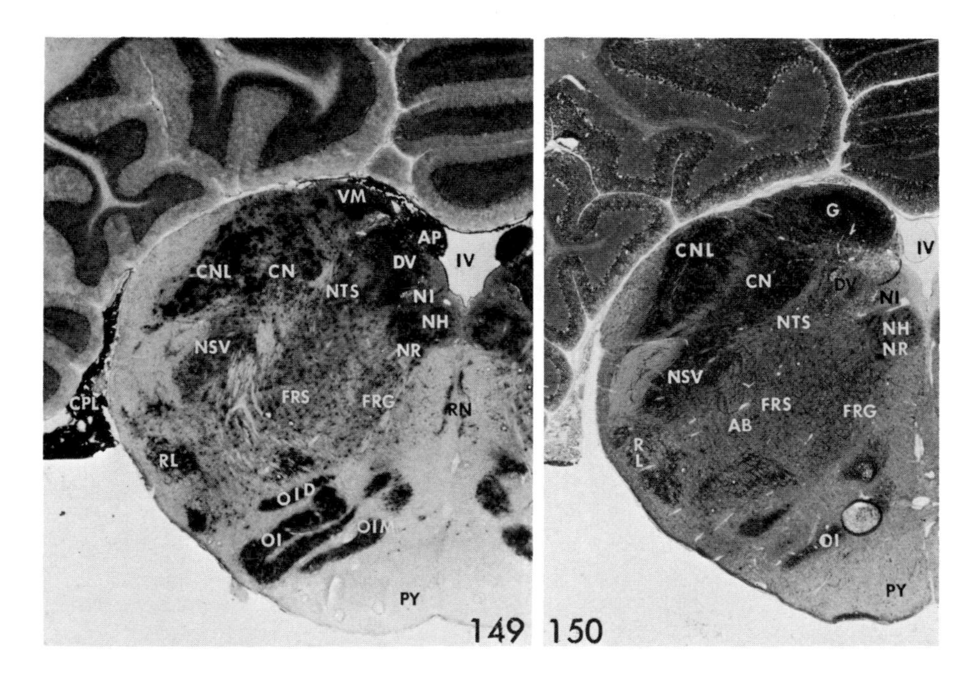

FIGURES 149, 150. AC and SE preparations at the level of hypoglossal and dorsal vagus nuclei showing the differences in the intensity of these enzymes in various areas. ×7, ×7.

moderately strong in SE and AC; moderate in SDH, ATPase, and MAO; and mild in AChE, whereas the glial cells show weaker enzyme activity. The sinusoids show mild AK activity, and the blood vessels are positive in AK and ATPase.

CENTRAL GRAY

This is the gray area around the aqueductus Sylvii (AS). Since it extends through the pons and the midbrain, it is generally divided into griseum centrale pontis (GCP) and griseum centrale mesencephali (GCM) [Figs. 108, 109, 114, 119, 120, 125, 126, 131, 133, 135, 161, 162]. The former consists mostly of smaller cells, whereas the GCM has a mixture of small and medium-sized cells. The histochemical activity is identical in the smaller cells of the GCP and GCM: the medium-sized cells show somewhat stronger enzyme activity. In Nissl preparations of the chimpanzee brain, Shantha and Manocha (1969) subdivided the GCM into subnuclei dorsalis, lateralis, and medialis on the basis of cell density and distribution of small and medium-sized cells. The dorsal subnucleus occupies an area dorsal to the aqueductus Sylvii, whereas the lateral subnucleus occupies an area lateral to the dorsal subnucleus, and the medial subnucleus is located ventromedial to the lateral subnucleus. The cell population of the lateral subnucleus is denser and more closely packed than that of other subnuclei. Since there are no histochemical differences in the various areas, a common description is given. Most neurons show moderately strong LDH; moderate SDH, AC, ATPase, SE, and MAO; and mild AChE enzyme activity. The neuropil is moderately strong in ATPase; moderate in LDH, AChE, and MAO; and mild in SE, AC, and SDH preparations.

RETICULAR FORMATION

The reticular formation occupies a large area in the medulla oblongata, pons, and midbrain; hence, is designated the formatio reticularis myelencephali (FRM), formatio reticularis pontis (FRP), nucleus reticularis tegmenti pontis (RTP), and formatio reticularis tegmenti mesencephali (FRTM). In the medulla oblongata, the reticular formation also includes a reticular complex situated dorsolateral to the inferior olivary nucleus and is called the nucleus reticularis lateralis (RL). In order to give a clear picture of the histochemical differences in various areas of the reticular formation in the medulla oblongata, the term FRM in this study refers to the caudal part of that region of the reticular formation which is centrally located in the medulla oblongata. Rostrally, however, extending from the level of the upper part of the olive to the lower part of the pons, the reticular formation is further differentiated into the medially located formatio reticularis gigantocellularis (FRG) and the laterally placed formatio reticularis parvocellularis (FRS). This subdivision is based on the predominance of large and small cells in these two respective areas.

The reticularis lateralis (RL) consists of mostly medium-sized and large cells (Figs. 149, 150, 156–158, 169, 170). They are oval, fusiform,

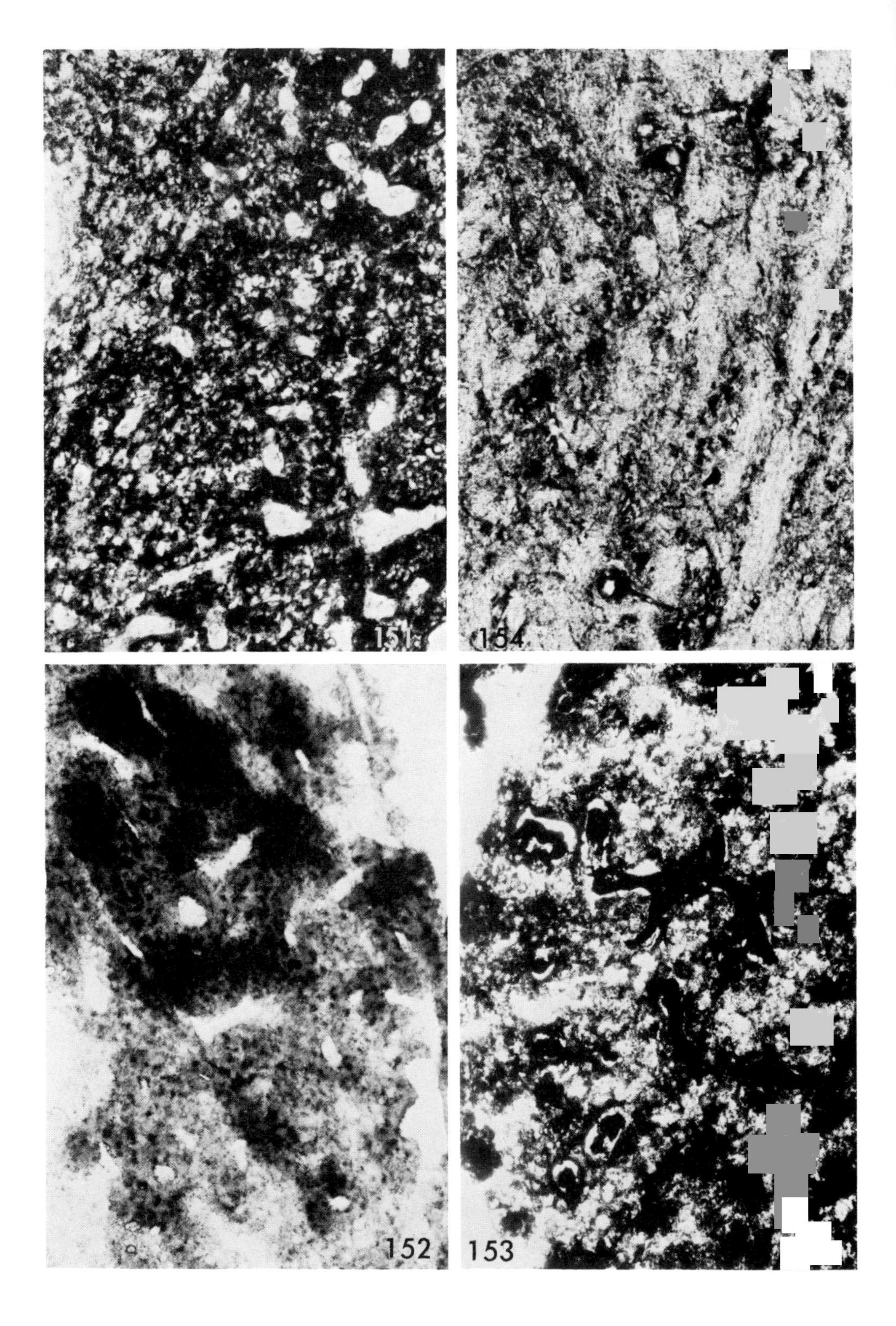
151
154
152
153

FIGURES 151–153. Preparations of the area postrema showing the varying intensity of LDH, AChE, and ATPase activity in the various components of this area. ×148, ×370, ×148.

FIGURE 154. LDH preparation of the nucleus cuneatus lateralis showing strong activity in the neurons and moderate activity in the neuropil. ×148.

or triangular in outline, and most of them are multipolar with elongated, thick cell processes. The histochemical reaction in the neurons and the neuropil is given in Table II. In the AChE, the positive activity in the neurons is restricted to the peripheral part of the perikarya and is not distributed throughout the cytoplasm as is the case in other enzyme preparations. The glial cells show strong LDH and moderate ATPase activity. The nerve fibers, cut longitudinally or transversely, give a moderate to moderately strong MAO reaction. The blood vessels are strong in ATPase and moderately strong in AChE and AK. A count of the latter indicates that the RL contains more blood vessels per section than the FRM, described below.

The formatio reticularis myelencephali (FRM) consists of a mixture of small and medium-sized cells, although the dorsal part of the FRM contains a greater number of small cells than the ventral part (Figs. 156–158). The histochemical reactions are given in Table II. The blood vessels are positive in ATPase, AChE, and AK. The larger blood vessels, particularly, show very strong ATPase activity. The nerve fibers, cut longitudinally and horizontally, give a moderate to moderately strong MAO reaction.

The histochemical differences in the formatio reticularis gigantocellularis (FRG) and parvocellularis (FRS) are based mostly on the cell size (Figs. 132, 134, 137–143, 148–150, 155, 166–168). The larger cells show stronger activity than the smaller ones (Table II). The FRG consists mostly of large and medium-sized cells. The larger cells are star-shaped, elongated, or triangular, whereas the medium cells are fusiform, oval, or triangular. The elongated cell processes of the FRG show enzyme activity in somewhat weaker form than the perikarya for a considerable distance. The glial cells in both areas show moderately strong LDH and moderate ATPase activity. As revealed in AK, the FRG has more blood vessels than the FRS.

The formatio reticularis pontis (FRP) occupies the major part of the tegmentum of the pons and consists of small and medium-sized cells which are oval, elongate, fusiform, or triangular with elongated cell processes. Most of the neurons are scattered within a network of cell processes (Figs. 119, 120, 125, 126, 131, 133, 135, 136).

The nucleus reticularis tegmenti pontis (RTP) is situated in the ante-

TABLE II*

HISTOCHEMICAL REACTIONS GIVEN BY THE VARIOUS COMPONENTS OF THE RETICULAR FORMATION IN THE RHESUS MONKEY BRAIN

	RL		FRM		FRG		FRS		FRP		FRTM		RTP	
	N	NP	N	NP	N	NP	N	NP	N	NP	N	NP	N	NP
Lactic dehydrogenase (LDH)	++++	++±	+++	++	+++	++	++±	++	+++	++±	++±	++	+++	++±
Succinic dehydrogenase (SDH)	+++	++±	++±	++	++	+	++	+	++±	+	++	++	++±	++±
Monoamine oxidase (MAO)	+	++	±	+	+	++	+	++	+	++	+	+	±	++
Simple esterases (SE)	+++	++±	++±	+	++±	+	++±	+	++±	+	++±	+	+++	++±
Acetyl cholinesterase (AChE)	±	+	+	+	+	++	+	++	+	++	+	++	+	++
Adenosine triphosphatase (ATPase)	++	++±	+	++	++	++	+	++	++	++	+	++	+	++±
Alkaline phosphatase (AK)	–	–	–	–	–	–	–	–	–	–	–	–	–	–
Acid phosphatase (AC)	++++	+	+++	±	++++	+	++±	+	++±	+	++	±	++	+

* RL: Nucleus reticularis lateralis; FRM: Formatio reticularis myelencephali; FRG: Formatio reticularis gigantocellularis; FRS: Formatio reticularis parvocellularis; FRP: Formatio reticularis pontis; FRTM: Formatio reticularis tegmenti mesencephali; RTP: Nucleus reticularis tegmenti pontis; N: Neurons; NP: Neuropil

+ = mild enzyme activity; ++ = moderate activity; ++± = moderately strong activity; +++ = strong activity; ++++ = very strong activity; ± = negligible activity; – = negative activity

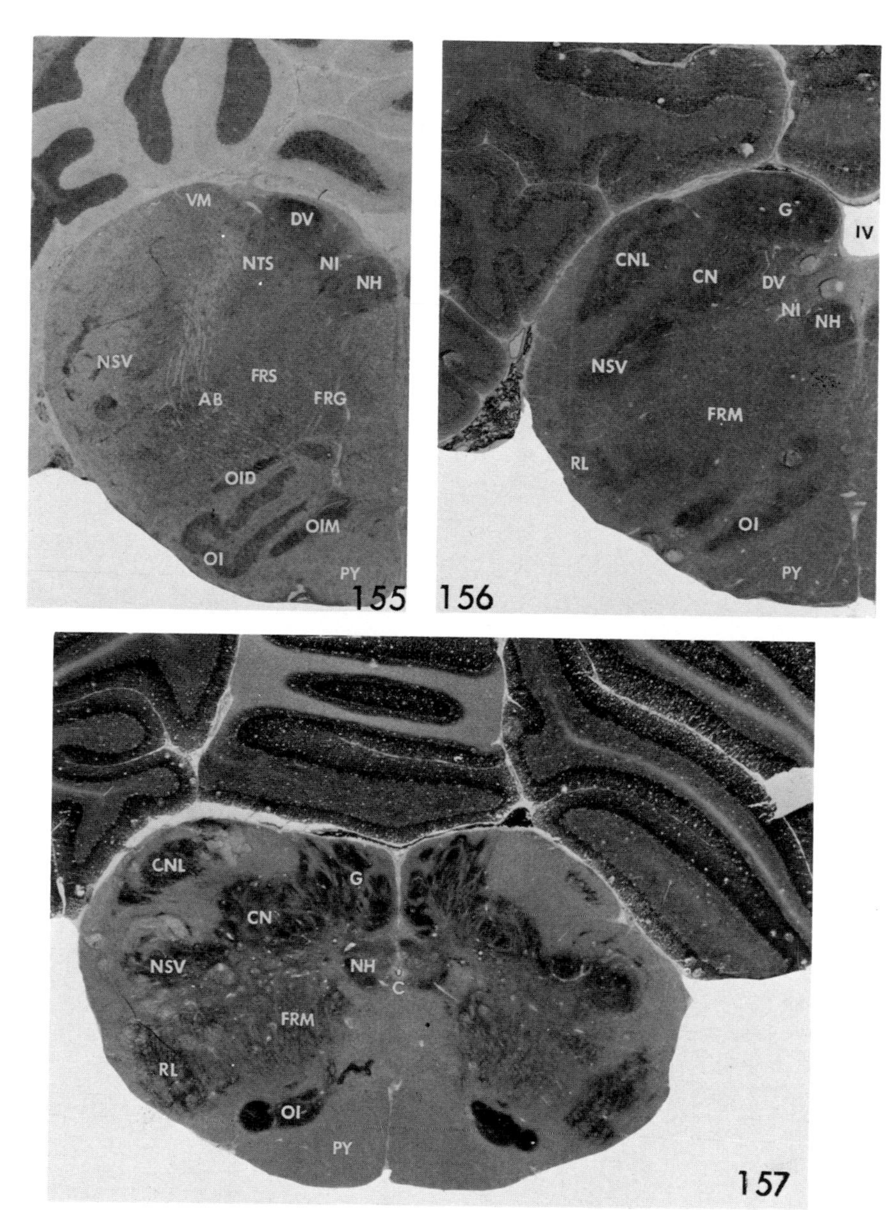

FIGURES 155–157. AChE, SDH, and LDH preparations through different levels of the medulla oblongata showing variations in the distribution of these enzymes. ×7, ×7, ×7.

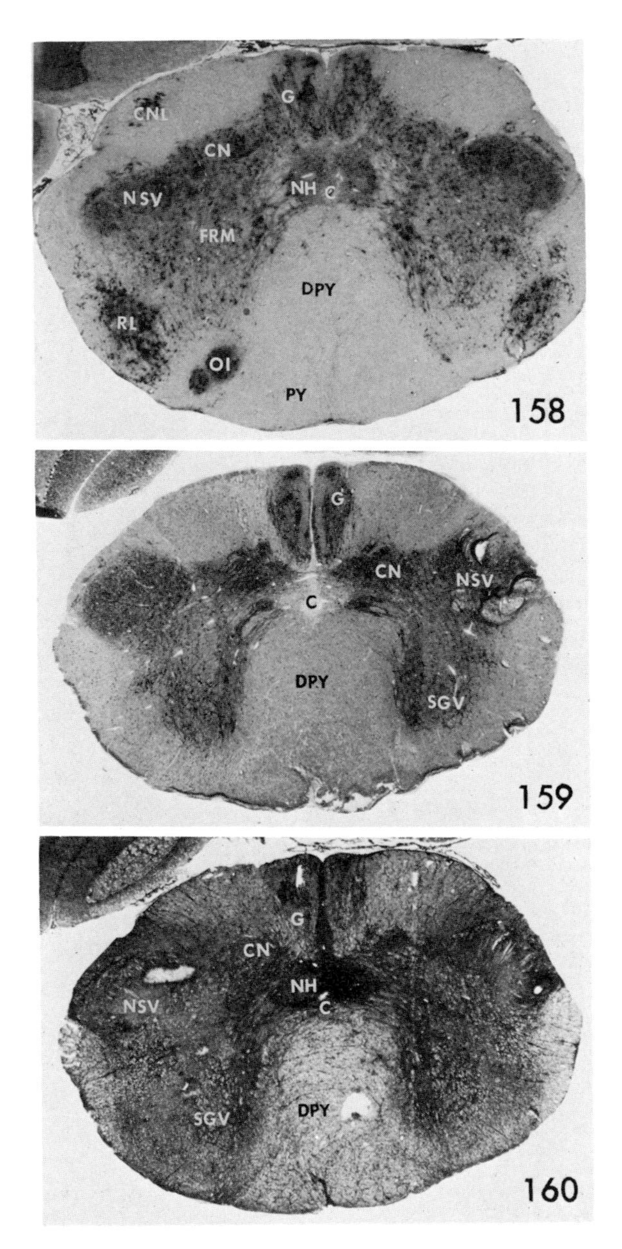

FIGURES 158–160. Sections passing through the lower part of the medulla oblongata and the upper part of the cervical spinal cord showing AC, SE, and ATPase activity in the various components at these levels. ×7, ×7, ×7.

rior part of the tegmentum of the pons and is surrounded by fibers of the nucleus lemniscus medialis (posteriorly and on the sides) as well as the fibrae pontis transversae (anteriorly) [Figs. 108, 109, 163, 165]. This nucleus is made up mostly of elongated, oval, or triangular, usually medium-sized and some small cells, which are not closely packed. The histochemical reactions are given in Table II.

The formatio reticularis tegmenti mesencephali (FRTM) is the oral extension of the FRM into the midbrain and occupies a considerable area of the tegmentum ventral to the colliculi (Figs. 59, 108, 109, 114). Rostrally, the FRTM is greatly reduced because of the appearance of the nucleus ruber in the tegmentum. The cell population of the FRTM consists of small, with some medium-sized and a few large, neurons. The larger cells show somewhat stronger activity than the smaller ones (Table II).

GENERAL DISCUSSION

OXIDATIVE ENZYMES AND MONOAMINE OXIDASE

The complex nuclear pattern of the brain stem and its histochemistry have been studied in serial caudocranial sections through the brain of the guinea pig (Friede, 1959a, 1959b), cat (Friede, 1961), and squirrel monkey (Manocha and Bourne, 1966a, 1966b, 1966c, 1967a, 1967b, 1968; Manocha *et al.,* 1967; Manocha and Shantha, 1969). The selective localization of the various enzymes in these investigations exhibited beautiful patterns in different areas of the brain. The reactions given by the brain stem of the rhesus monkey are essentially similar. The motor cranial nuclei, nucleus gracilis, cuneatus, olivaris inferior, cochlearis, and vestibularis show strong SDH, CYO, and LDH and weak MAO reactions. The nucleus intercalatus and tractus solitarius show weaker SDH and CYO and stronger LDH activity. Marked MAO activity is observed in the nuclei interpeduncularis, centralis superior, annularis, etc., while SDH, CYO, and LDH have been more prominent in the lateral and medial geniculate nuclei; superior and inferior colliculi; substantia grisea centralis; nuclei oculomotorius, trochlearis; and so on. In the superior colliculus, the observations on the squirrel and rhesus monkey brains (stronger activity in the upper layers, except for the upper two, progressively decreasing in the deeper layers) are identical to those on the guinea pig (Friede, 1959b), cat (Friede, 1961), and human being (Friede and Fleming, 1962). Wawrzyniak (1963), in the rabbit and

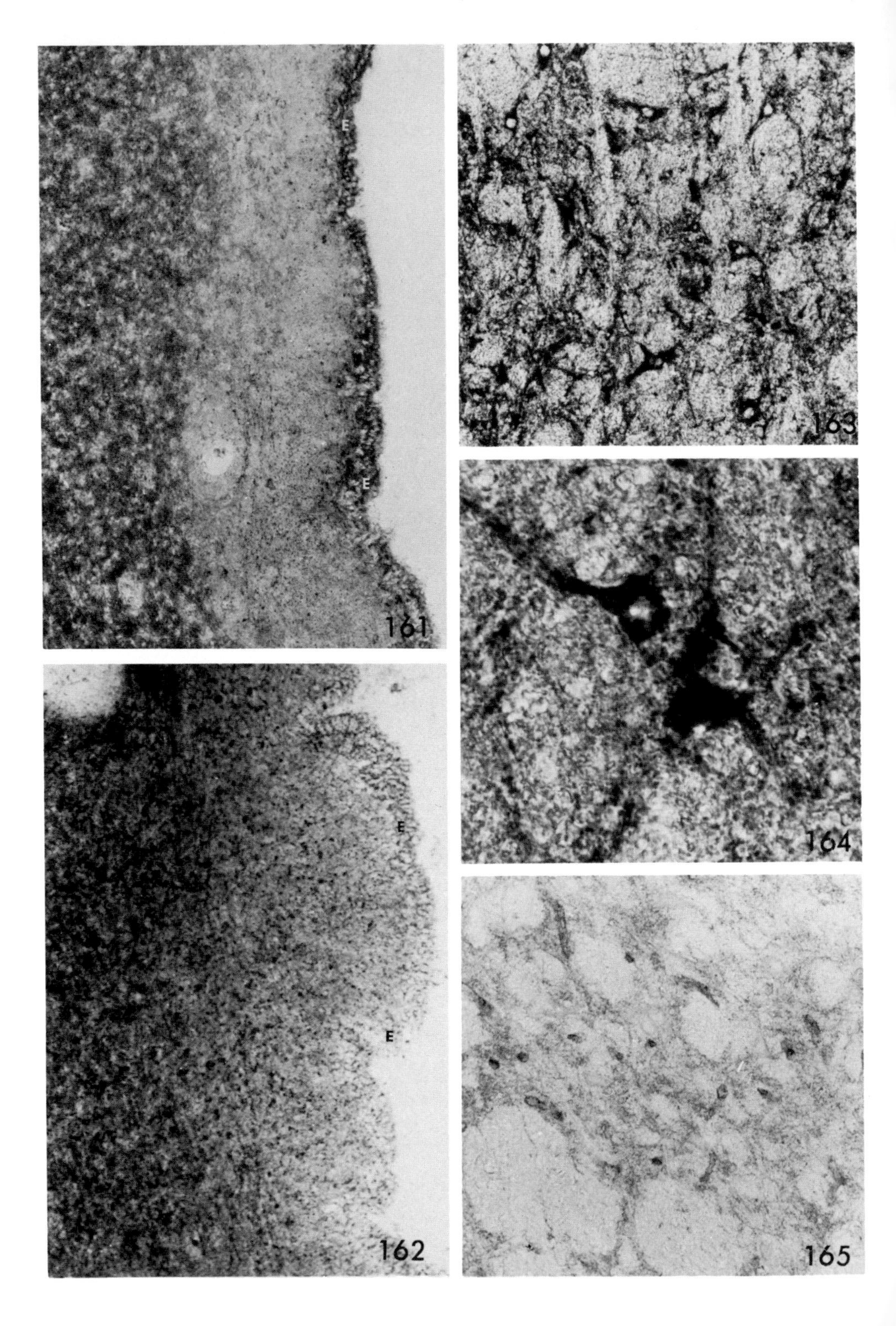
E
E
161
E
E
162
163
164
165

FIGURES 161, 162. SDH and MAO preparations of the mesencephalic central gray. Note the stronger activity in the central gray compared to the subependymal layer. ×592, ×142.

FIGURES 163–165. LDH, SDH, and AChE preparations, respectively, of the nucleus reticularis tegmenti pontis giving a comparative view of the enzyme activity. Note the negligible AChE activity in the neurons in Fig. 165. ×148, ×592, ×148.

guinea pig brain, has observed, however, intense SDH activity in the deeper layers compared to the two peripheral gray layers. The neuroglial elements in the deeper layers show moderate SDH and CYO reactions.

In general, the distribution of SDH and CYO in the various brain stem nuclei indicates their selective localization in the cells of the gray matter, dendritic branches, synaptic terminals and other proven sites of mitochondria in the neuropil, and that they play an active role in oxidative phosphorylation (Hogeboom *et al.*, 1948; Abood *et al.*, 1952; Shimizu *et al.*, 1957; Friede, 1959a, 1959b, 1961, 1966). SDH and CYO show similar distribution, appearing to be proportional to oxygen consumption and metabolic activity and may be taken as an index of the functional activity of the latter (Himwich, 1951; Schiebler, 1955; Shimizu *et al.*, 1957; Friede, 1959a, 1961; Manocha and Bourne, 1966b, 1966c). The CYO activity in the different regions of the brain has been studied in the monkey (Tolani and Talwar, 1963) and rabbit (Ridge, 1967). The results obtained for monkey and rabbit brain agreed in relative activities of the brain regions (Tipton and Dawson, 1968). In the pig brain, Tipton and Dawson showed that the distribution of α-glycerophosphate dehydrogenase (αGPD) paralleled that of SDH and CYO. They believed that these three enzymes are part of the mitochondrial enzyme content of the brain. The distribution of the enzymes concerned with the glycolytic pathway showed a great similarity to that of the citric acid cycle enzymes. However, a few areas such as the area postrema may depend more on glycolytic metabolism than on the citric acid cycle. It may be generalized that the perikarya of the neurons and the oligodendroglia demonstrate glycolytic metabolism more than do the axonal endings and the neuropil; the latter have a greater amount of citric acid cycle enzymes (Friede and Fleming, 1963; Manocha and Bourne, 1966a, 1966b, 1966c, 1968). It is evident that the neuropil is rich in synaptic terminals, and the mitochondria are concentrated at these synaptic endings. The mitochondria contain all the succinoxidase present in the cytoplasm and play an important role in the Krebs cycle series of reactions and in oxidative phosphorylation (Shimizu and Morikawa, 1957).

The DPN diaphorase activity is more pronounced in the neurons, whereas the activity of TPN diaphorase is prominent in the glial cells,

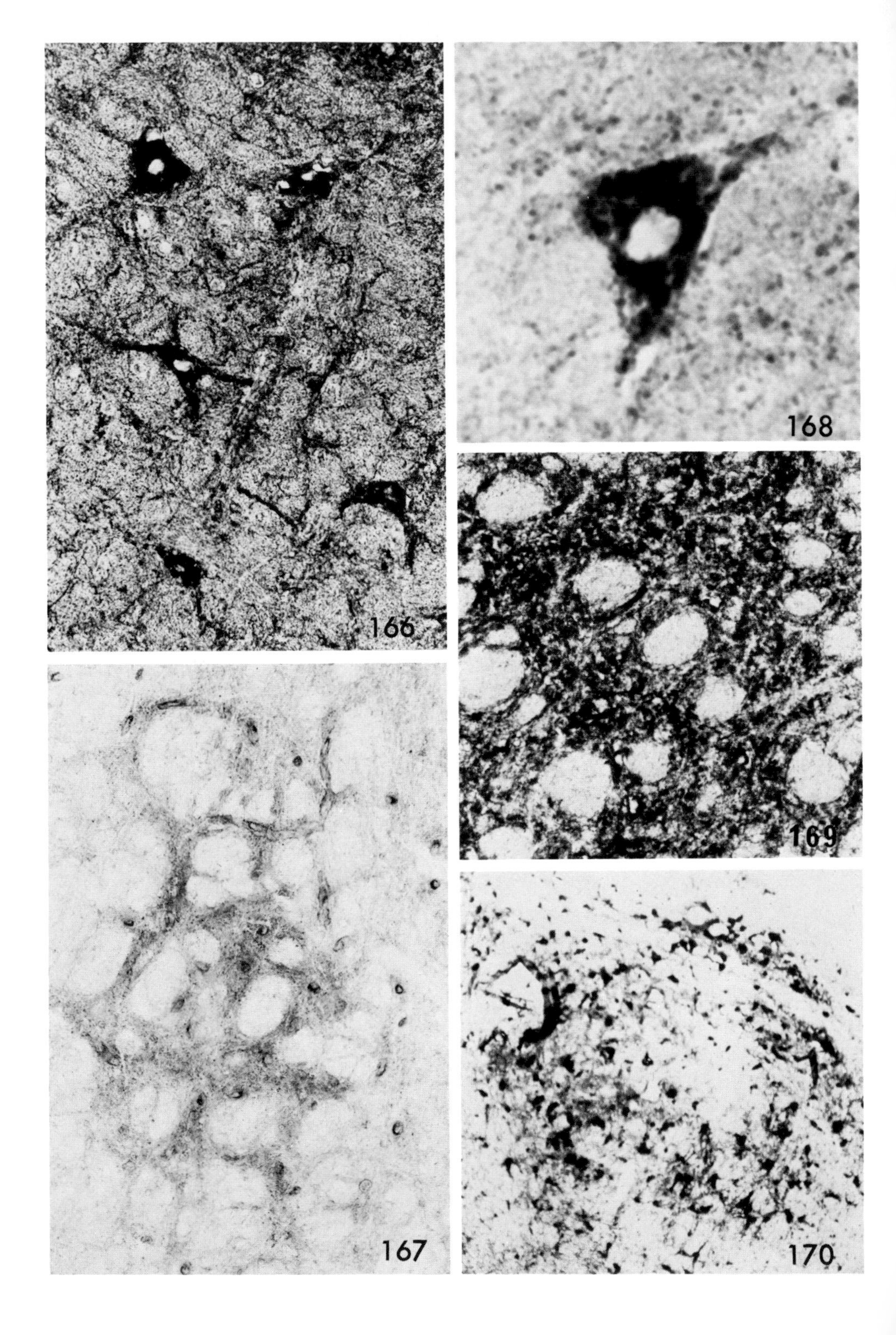
166
167
168
169
170

FIGURES 166–168. Formatio reticularis gigantocellularis showing LDH, AChE, and SDH activity. The neurons are strong in LDH and SDH and mild in AChE activity. ×148, ×37, ×370.

FIGURES 169, 170. LDH and SE preparations of the nucleus reticularis lateralis of medulla oblongata showing strongly positive neurons. The enzyme activity passes on to the cell processes, particularly in the LDH preparation. ×148, ×37.

Schwann cells, and amphicytes. Romanul and Cohen (1960) showed that with respect to enzyme activity, the dehydrogenase system (succinic, malic, lactic, glutamic, glucose-6-phosphate, and isocitric dehydrogenase) is arranged in order of decreasing activity in the neurons and increasing activity in the satellite cells. These results are somewhat different from those of Hamburger (1961) who along with Hydén (1959, 1960) has advocated the intimate functional relationship between the nerve cells and glia at the biochemical level. According to Hamburger, the nerve cell bodies mainly utilize glutamate, whereas succinate and pyruvate are essentially oxidized by the glia, and α-oxoglutamate is used equally by the two components. He ascribed less enzyme activity to the glia than to the nerve cells, whereas Lowry *et al.* (1954) and Elliott (1959) reported the glia to be more active metabolically.

The neuroglia outnumber the neuronal cells by approximately ten to one (Wolfgram and Rose, 1959). Histochemically, the activity of cytochrome oxidase and succinic, malic, lactic, and α-glycerophosphate dehydrogenase were demonstrated in the perinuclear cytoplasm of the three types of neuroglia in the adult rat and cat (Wolfgram and Rose, 1959). In our preparations of the rhesus monkey, only the oligodendroglia show high LDH activity. Since glucose-6-phosphate dehydrogenase is higher in white matter containing numerous oligodendroglia compared to the gray matter, Pope (1960) believed that the hexosemonophosphate shunt is predominant in these glia. Friede (1962) showed that normal astrocytes have the least activity of oxidative enzymes among all the cells of the central nervous system, indicating very little oxidative metabolism in them. Hamburger (1961), however, believed that some of the steps of the TCA cycle are most effectively performed in the surrounding glia, while others predominate in the neurons.

In a comparison of oligodendrocytes and astrocytes, Pope and Hess (1957) and Hydén and Pigon (1960) showed that the former have a much higher oxidative metabolism than the latter and may even approach the neuronal level. A very intimate link between the oligodendroglia and the neuron is also indicated when RNA, as well as respiratory enzyme activity, is decreased on stimulation (Hydén and Pigon, 1960). These workers concluded, therefore, that at ordinary levels of activity, the oligodendrocytes supply the nerve cell with energy-rich compounds, probably by pinocytosis. At higher levels of activity, however, the nerve cells get priority to easily available energy for protein production and the glial cells resort to some other source of energy. This explains the high lipid content of the glia which may utilize lipoprotein and allow nerve cells to utilize glucose in periods of higher functional demands.

Some areas in the brain stem need special mention because their histochemical reactions differ from those of other nuclei. These are the nucleus interpeduncularis (IP), locus coeruleus (LCR), solitary active cells, and the area postrema. The IP and LCR show strong LDH and moderate SDH, as well as strong MAO. This is in contrast to the general pattern that nuclei with strong LDH, SDH, and CYO activity exhibit weak MAO reaction and vice versa (Manocha and Bourne, 1966a, 1966b, 1966c, 1968). Furthermore, a strong LDH and weaker SDH and CYO activity suggest more of a dependence upon anaerobic glycolytic metabolism. Due to prominent glycolytic metabolism, the LCR and some other nuclei such as the DV, SN, and subcommissural organs appear to be peculiarly resistant to anoxia (Adams, 1965). It is interesting that the LCR, in spite of low concentrations of respiratory enzymes (SDH and CYO), has a very rich capillary density which is exceeded only by the magnocellular hypothalamic nuclei. The capillaries in the LCR are unusually close to the nerve cells (Smith, 1963). Also the presence of strong MAO activity makes these two areas all the more important and may indicate that some aromatic monoamines such as catecholamines and serotonin are concentrated in this area (Maeda *et al.*, 1960). They concluded that the LCR contains aromatic monoamines acting as a parasympathetic transmitter and the electrical stimulation of the LCR displayed the characteristic parasympathetic reactions. Shimizu *et al.* (1959) attributed MAO activity in the IP and LCR in the brain of rodents as an indication of MAO being related "to the general metabolism of the autonomic neurons rather than to such a specialized function as the adrenaline or noradrenaline-MAO system." Shimizu (1961) suggested the presence of an active aromatic amine metabolism and that these nuclei display a mechanism similar to the autonomic centers such as the hypothalamus. Kusunoki *et al.* (1966) postulated that the IP controls visceral coordination and that MAO may play a role in the specialized oxidation in these areas. Additionally, in the IP and LCR, where MAO activity is strong, the AChE reaction is moderate. This is in contrast to the general pattern that areas rich in MAO are poor in AChE and vice versa. Wawrzyniak (1965) showed that the ventral part of the LCR is less cholinergic than the dorsal part, but the former is intensely stained with MAO, while the latter shows only slight MAO activity.

The solitary active cells of Thomas and Pearse (1961, 1964) and Duckett and Pearse (1964) showing activity of TPN-diaphorase are interesting in the context of the special metabolic role ascribed to some nuclei. These authors believe in the regulating function of these active cells in the various areas of the brain because they are highly resistant to noxious agents such as carbon monoxide. The locus coeruleus particularly showed high TPN-diaphorase (Thomas and Pearse, 1964) and high G6PD (Felgenhauer and Stammler, 1962). The latter enzyme is known to initiate the first oxidative step in the pentose phosphate cycle. Some of the specialized studies also indicate the great role of the solitary active cells in the locus coeruleus. The latter has a relay function in the central nervous system's cortical and subcortical mediation of general vegetative functions (Russell, 1955). These solitary active cells are the main part of the pneumotaxic center. Their stimulation leads to inspiration, and it is believed that the respiratory bulbar center survives for 30 minutes after a complete arrest of blood circulation (Baxter and Olszewski, 1955; Friede, 1959a).

Certain other nuclei which show exceptional metabolic activity may be mentioned (Friede *et al.*, 1963; Friede, 1966): for example, the nucleus dorsalis vagi, scattered cell groups near the floor of the fourth ventricle at the level of the pons, the nucleus supratrochlearis, the nucleus paraventricularis and supraopticus in the hypothalamus, a cell group at the ventral aspect of the pallidum with a small extension ventrally between putamen and pallidum, scattered cells in the ventral septal area, scattered cell groups dorsal and ventral to the anterior commissure. These cell groups show strong reaction for LDH and G6PD but give a weak reaction for SDH and CYO.

The area postrema is a controversial region with regard to its structure as well as function. The major differences among the various workers concern the neuronal or nonneuronal nature of the parenchymal cells. We have discussed this controversy in our earlier article on the chimpanzee brain (Shantha and Manocha, 1969). The histochemical study of Iijima *et al.* (1967) on the squirrel monkey showed that the nerve, glial, and ependymal cells can be individually recognized in the area postrema, and that they derive their energy mainly through the TCA cycle. The perivascular sheath shows a high anaerobic metabolism. The ependymal cells of the area are metabolically similar to those in other areas such as the spinal cord (Manocha *et al.*, 1967), showing glycogen (Shimizu and Kumamoto, 1952), amylophosphorylase (Shimizu and Okada, 1957; Amakawa, 1959; Iijima *et al.*, 1967), and SDH (Shimizu *et al.*, 1957).

Monoamine oxidase is present more in the gray matter than in the white matter, although the activity in some of the fiber tracts (e.g., tractus retroflexus Meynert) is as strong as in the habenular nuclei. Birkhäuser (1940), in a parallel study of MAO and AChE in dog and cow brains, showed that whereas AChE activity among different areas varied 35-fold, the MAO varied less than threefold. Tyrer *et al.* (1968) showed that areas with the strongest MAO reaction have about four times the activity of areas with the weakest response. Radioautographic studies of Ishii and Friede (1967) using tritiated norepinephrine showed excessively strong binding of norepinephrine at the surface membranes of pigmented nerve cells in the substantia nigra, locus coeruleus, dorsalis vagi, etc., which was not found in the nonpigmented nuclei. The melanin pigmentation of the nerve cells may, therefore, be related to the catecholamine-containing synapses at the surface of the neurons.

The activity of MAO and the levels of serotonin, dopamine, and norepinephrine from birth to adult in the rat and mouse have been studied by Robinson (1968) and Agrawal *et al.* (1966, 1968). The latter observed that the accumulation of these amines occurs concurrently with the manifestation of functional and behavioral maturation of the central nervous system. The locus coeruleus and nucleus ambiguus showed prominent MAO activity at birth, and other nuclei were very weak. The brain stem nuclei of the rat show more MAO at 15 days than the cerebrum (Robinson, 1967). Wawrzyniak (1965) found increased activity of MAO in the developing rodent brain, with an increase in the number of fibers along with the progress of myelination. In the cultures of CNS, MAO is present in small and medium-sized cells, but not in the large neurons (O'Steen and Callas, 1964).

Rodriguez (1967), as well as the present authors, observed MAO and AChE activity simultaneously in the same nervous structure in several instances. The concentration of MAO in the noncholinergic nerve endings could give it a role similar to that of AChE in cholinergic ones (Arnaiz and De Robertis, 1962). But, as Koelle and Valk (1954) noted that there is no selective association of MAO with the adrenergic nerves analogous to that of AChE with cholinergic nerves, the MAO activity may be found in adrenergic afferent or cholinergic neurons. Woohsmann (1963) believed there is no evidence of the metabolism of adrenaline and noradrenaline by MAO in the central nervous system in the form of serotonin oxidase and dopamine oxidase. Woohsmann's views seem to be farfetched. Tyrer *et al.* (1968) noted that the "presence of monoamine oxidase in blood vessel walls and around the cerebral ventricles, and in certain fiber tracts, implies that monoamines probably have function in the brain apart from a possible role in mediating transmission at certain synapses."

At the cytological level, the results on the distribution of MAO have clearly shown that most of the MAO activity is located in the mitochondria (Whittaker, 1953; Weiner, 1960; Arnaiz and De Robertis, 1962; Aprison *et al.*, 1962, 1964; Tipton and Dawson, 1968; Squires, 1968). The purification of MAO, CYO, and G6PD paralleled each other when mitochondria were prepared, suggesting that the three enzymes are bound to the same particles. Weiner (1960) observed that although both MAO and SDH were located in crude mitochondrial fractions, their ratio varied in different anatomical regions of the brain, which may indicate a biochemical heterogeneity of these particles (Arnaiz and De Robertis, 1962). Along with the hypothesis about the heterogeneity of mitochondrial particles, it must be remembered that thus far there is no identity of views on whether MAO is a single enzyme or consists of multiple enzymes (Blaschko, 1952). The work of Squires (1968) in the mouse indicates the existence of at least three forms of mitochondrial MAO, which may vary from one anatomical area to another. For example, the neurons containing serotonin, noradrenaline, and dopamine, respectively, may contain different forms of MAO, which are more or less specific for the deamination of these amines, containing specific enzymes for every substance, such as tryptamine, serotonin, tyramine, and catecholamines (Zeller, 1951; Pletscher *et al.*, 1960). By selective inhibition of MAO by *N*-methyl-*N*-propargyl-3 (2,4-dichlorophenoxy) propylamine hydrochloride, Johnston (1968) concluded that MAO is a binary system of enzymes, each of which has a detectably different sensitivity to this particular inhibitor.

CHOLINESTERASES

The overall distribution of cholinesterases, particularly the specific type (AChE), has ben studied in detail in a number of animals (Sinder and Scharrer, 1949; Scharrer and Sinder, 1949; Hard and Peterson, 1950; Koelle, 1951, 1952, 1954, 1955, 1963; Koelle and Koelle, 1959; Gerebtzoff, 1959; Lewis and Shute, 1959; Shute and Lewis, 1960a; Snell, 1961; Abrahams, 1963). In the brain stem, the cells of the cranial nerve nuclei show considerable AChE activity, although the intensity of reaction varies from one nucleus to another. Positive activity has also been observed in the nuclei gracilis, cuneatus, solitarius, and intercalatus; olivary nucleus; pontine nucleus; substantia nigra; superior and inferior colliculus; and reticular formation. In the superior colliculus more AChE activity is seen in gray layers, whereas the nucleus of the inferior colliculus is weak except for a stronger activity in its ventrolateral part. High intracellular AChE activity has also been described in the nucleus dorsalis vagi, locus coeruleus, and Edinger–Westphal nucleus (Lewis and Shute, 1959; Shute and Lewis, 1960a, 1960b; Friede 1966). Abrahams (1963) observed high AChE activity in the neurons of most motor nuclei, certain correlation centers, and several tertiary afferent neurons; moderate activity in the second neurons of several sensory pathways and in other correlation centers; and little or no activity in the primary sensory neurons, certain neurons synapsing directly with motor neurons, and in other correlation centers. In the cat brain stem AChE was present not only in the normal cytoplasm and in the neuronal membranes, but also in some glia, hippocampal astrocytes, protoplasmic astrocytes, and subependymal cell plates.

A study of the literature reveals that the AChE activity in the rat is higher than in the primates. Gerebtzoff (1959) pointed out that, in general, the smaller the brain, the higher the activity of AChE per gram. In the rat, very strong activity is observed in the nuclei hypoglossus, vagus, ambiguus, facial, abducens, trigeminal, trochlear, oculomotor, and the preganglionic cells of the nucleus Edinger–Westphal. The level of AChE activity in the rhesus monkey is comparatively weak. In addition, the nuclei belonging to the cerebellar system are weak, whereas in the vestibular nuclei, the lateral nuclei show strong activity compared to decreasing activity in superior, inferior, and medial nuclei. This observation is in contrast to the general observations that those nuclei where the sensory fibers terminate are poor in AChE. Gerebtzoff suggested that in the vestibular nuclei, there are differences in the number of fibers from associative centers, which end at the surface of the neurons in these nuclei.

In the central nervous system the intracellular sites of AChE reaction vary from one group of neurons to another. In some neurons the activity is cytoplasmic in distribution (for example, lateral horn cells in spinal cord, basal ganglia, etc.), whereas in others the maximum activity is observed in the cell membrane, particularly at the sites of synaptic junctions (for example, anterior horn cells of spinal cord, motor nuclei of cranial nerves, cerebral and cortical neurons, and Ammon's horn). AChE, in addition to its postsynaptic site of action, may have a presynaptic role, triggering the release of the actual transmitter both at the cholinergic and at the noncholinergic synapses (Koelle, 1961, 1962). Shute and Lewis (1963) believed that cholinesterase may act constantly on the sites of acetylcholine production in order to prevent the concentration of the latter from passing beyond a certain level. Gerebtzoff (1959) suggested that the possibility always exists that cholinergic and noncholinergic terminals may be present in the same cell, and a negative or a very weak reaction may only mean that acetylcholine is not playing the principal role. As rightly pointed out by Silver (1967), the data of AChE localization must be interpreted with caution, and attention must be paid to species differences and specificity of the methods used. This may be particularly true of the area postrema and some hypothalamic nuclei, where noradrenaline is more active (Vogt, 1957).

The cerebral cortex does not show any appreciable AChE activity, but since cortical cells are excited by acetylcholine and come under cholinergic influence, it may be that the AChE-containing pathways in the forebrain are the rostral extensions of the reticular formation of the midbrain (Shute and Lewis, 1963). Krnjevic and Silver (1966) also described a number of links between the forebrain and AChE-rich foci: e.g., nucleus interpeduncularis, substantia nigra, nucleus subthalamicus, area preoptica, and corpus striatum.

It is well known that different enzyme systems in the developing brain appear at different times during development. Aprison and Himwich (1954) showed in the rabbit that the cholinesterase activity in general was low during the gestation period and rose to maximum values at different periods after birth: the medulla oblongata at 15 days, colliculus superior at about 8–9 months, and the caudate nucleus and frontal cortex at approximately 18 months. In the adult, the AChE activity increased in the same order, with the exception of the frontal cortex. Youngstrom (1941), working with the different parts of the CNS of the human fetus, also found a correlation between the periods of the most rapid changes in the cholinesterase activity and the phyletic position of the neuroaxis.

Capillaries in most parts of the body do not seem to give a positive reaction for AChE (Crook, 1963), but in our histochemical preparations of the rhesus monkey brain a large number of capillaries in the brain stem, particularly in the gray areas, give a positive AChE reaction. AChE-positive capillaries are more concentrated in areas which have a dense capillary network; for example, the colliculus inferior. A corresponding section stained with AK, however, shows a larger number of blood vessels, which may indicate that not all the capillaries are AChE-positive. Gerebtzoff (1959) believed that a positive AChE reaction in the capillary walls is due to a common diffusion artifact. Cholinesterase activity in the capillary walls has been shown in the spinal cord and brain stem of the chicken (Cavanagh and Holland, 1961).

A high pseudocholinesterase (BChE) activity has been recorded in some nuclear groups, particularly in the lateral nucleus of the substantia nigra, the third nerve nucleus, red nucleus, and the deep layers of the superior colliculus (Abrahams, 1963).

Most of the brain stem neurons, neuropil, and glial cells show SE activity. The white matter shows comparatively less SE activity. According to Meyer (1963), SE is less marked in the small neurons and not present in glia. On the contrary, the presence of SE in glial cells has been reported by Shantha *et al.* (1967).

PHOSPHATASES

Among the phosphatases, the activity of alkaline phosphatase in the brain stem may be generalized to be present in the blood vessels, pia-arachnoid, and the choroid plexus. The intensity of reaction in the different nuclei varies not only within a species, but also within the same nucleus of a species. Rabbits (Shimizu, 1950; Rogers, 1963) and chickens (Landow *et al.*, 1942; Shimizu, 1950) have been reported to have higher activity of alkaline phosphatase in the neuropil than other mammalian species. The AK activity is mainly concentrated in the blood vessels (i.e., in arterioles and capillaries), whereas the large venous sinuses are negative, indicating physiologic difference between arteriolar and venous blood (Hard and Hawkins, 1950). Additionally, high concentrations of AK are observed in the arterioles (Bannister and Romanul, 1963). The locus coeruleus and nucleus colliculus inferior are richly supplied with capillaries, hence showing more AK activity. There is also a difference in the enzyme content of the gray and white matter. Fewer vessels of the white matter contain AK activity when compared to those in the gray matter. It is hard to comment on the physiological significance of the distribution of AK. Alkaline phosphatase in blood vessels in the brain appears to play a significant role in the transport of various chemicals and nutrients through the membranes of blood vessels and glial cells. It is interesting to note that the glycogen was histochemically demonstrated at the site of high AK activity (Meyer, 1963). This enzyme has also been implicated in playing a role in the blood barrier mechanism (Samorajski and McCloud, 1961). Surprisingly, AK activity is not present in the pulmonary capillaries and certain sinusoids (Landers *et al.*, 1962). Since nucleoli of cells appear to have high dephosphorylating activity, suggestions concerning the role of phosphatases in the synthesis of proteins have been made (Bourne, 1958).

The acid phosphatase (AC) activity in the brain stem nuclei of the rhesus monkey is variable. High activity of this enzyme has been observed in the cranial nerve nuclei. The AC activity is highest in the perikarya and is comparatively weak in the glial tissue and the neuropil. Simple esterases are also found in approximately the same sites as AC. The larger neurons contain many lysosomes and are, therefore, more deeply stained than the smaller ones. The cytological distribution of AC differs greatly among the different species and has been described in rodents, cats, and birds by Shimizu (1950), in the squirrel monkey by Iijima *et al.* (1969), and in the rhesus monkey by Friede (1966). The reaction in the neurons is mostly granular, but is of a variable intensity and distribution and is localized in the cytoplasm and the cell processes. While some neurons show a diffuse and mild reaction in their cytoplasm, others show a strongly positive dense mass in one area and a granular reaction in the other. In other neurons strongly reactive bodies are distributed unevenly in a lightly reacting cytoplasm. Throughout the midbrain, pons, and medulla, the neurons of the cranial nerve nuclei stain deeply and show those details exhibited by the cortical neurons (Naidoo and Pratt, 1951). The present study, as well as those of Kumamoto and Bourne (1963) and Friede and Knoller (1965), has shown that the AC reaction in the neurons is particularly strong at the site of lipofuscin pigments. This activity, however, decreases when the lipofuscin deposits fill the entire perikarya (Anderson and Song, 1962; Silva–Pinto and Coimbra, 1963). The AC activity in the neurons and fibers of senile brain is stronger (Josephy, 1949), probably due to the presence of more aging pigment. Coarse particles of moderate reaction also occurred along the cell processes and around the cell borders. These particles corresponded closely to the synaptic regions as demonstrated by Bodian's preparations. Friede and Knoller (1965) found that the distribution of AC activity in the various regions of the gray matter of the brain showed less variation compared to the oxidative enzymes, and therefore, the respective distribution in the brain of lysosomal and mitochondrial enzymes appears to be unrelated. They speculated that the biological significance of AC is more closely related to the maintenance and metabolism of cells than to the energy metabolism. Neuroglia also give AC activity, though the degree of positive activity varies from one type of neuroglia to another and from one nucleus to another. In the white matter, moderate AC activity has been observed in the glial cells. The capillaries do not show any AC activity, but prolonged incubation may produce some reaction (Kaluza and Burstone, 1964).

The presence in the brain of a number of isoenzymes of acid phosphatases which may represent different enzymes has been shown by a number of workers (e.g., Anderson *et al.*, 1961; Barron *et al.*, 1964; Bernsohn and Barron, 1964; Shuttleworth and Allen, 1966; Kalina and Bubis, 1968). Barron *et al.* (1964) showed three to four bands of neural AC in the human brain. These enzymes have different substrate specificities and sensitivities to inhibition by fluoride or Cu^{++} (Kalina and Bubis, 1968). One of the bands was active with β-glycerophosphate as well as α-naphthol phosphate substrates; whereas the other bands were active only with α-naphthol phosphate and were inhibited by Cu^{++} but not by fluoride. Since these phosphatases are active at a pH optimum of approximately 5.0, Shuttleworth and Allen (1966) believed these enzyme activities to be of lysosomal origin.

The ATPase is widely present in the central nervous system, and its association with the mitochondria has been demonstrated by biochemists (Schneider and Hogeboom, 1951). In the brain stem, the cranial nerve nuclei, colliculus superior, and nucleus ruber, there is prominent ATPase activity. There is some variation in ATPase localization in fresh frozen and formalin-fixed material. In the latter case, the inhibition of the enzyme may be due to inactivation of the mitochondrial ATPase (Novikoff *et al.*, 1958). Recently Fahn and Côté (1968) studied the activity of Na-K-activated ATPase in 21 regions of the rhesus monkey brain. The highest activity was observed in the cerebral cortex, cerebellar cortex, thalamus, and the colliculi, whereas the white matter showed the lowest activity. In the fresh frozen sections the nucleolus shows considerable activity. It is hard to confirm the intranuclear localization of ATPase activity, and more careful scrutiny is needed concerning the authenticity of this reaction, which may be due to nonspecific absorption of Pb^{++} and Ca^{++} as also explained by Adams (1965). Sandler and Bourne (1962) described, however, that the nuclear activity was banished by using $MgSO_4$ instead of $MgNO_3$ as an activator in either the Wachstein–Meisel (Pb^{++}) or Padykulla and Herman (Ca^{++}) substrates. ATPase is distinctly localized on neural cell membranes and cell processes. This enzyme at this site plays a role in supplying energy for the ion transport concerned with impulse conduction and synaptic transmission. Blood vessels all over the brain stem show ATPase-positive activity.

REFERENCES

Abood, L. G., Gerard, R. W., Banks, J., and Tschirgi, R. D. Substrate and enzyme distribution in cells and cell fractions of the nervous system. *Am. J. Physiol.* **168**:728–738 (1952).

Abrahams, V. C. Histochemical localization of cholinesterases in some brain stem regions of the cat. *J. Physiol.* (*London*) **165**:55P (1963).

Adams, C. W. M. (ed.) Histochemistry of the cells in the nervous system. In: *Neurohistochemistry,* pp. 253–331. Elsevier, Amsterdam, 1965.

Agrawal, H. C., Glisson, S. N., and Himwich, W. A. Changes in monoamines of rat brain during postnatal ontogeny. *Biochim. Biophys. Acta* **130**:511–513 (1966).

Agrawal, H. C., Glisson, S. N., and Himwich, W. A. Developmental changes in monoamines of mouse brain. *Intern. J. Neuropharmacol.* **7**:97–103 (1968).

Amakawa, F. Studies on the polysaccharide synthesis of the central nervous tissue with special reference to the phosphorylase reaction and several enzymes (in Japanese). *Kumamoto Igaku Zassi* **33**: Suppl. 8, 1–37 (1959).

Anderson, P. J., and Song, S. K. Acid phosphatase in the nervous system. *J. Neuropathol. Exptl. Neurol.* **21**:263–283 (1962).

Anderson, P. J., Song, S. K., and Christoff, N. The cytochemistry of acid phosphatase in neural tissue: separation, validation, and localization. In: *Fourth International Congress of Neuropathology,* Vol. I, p. 75. Thieme, Stuttgart, 1961.

Aprison, M. H., and Himwich, H. E. Relationship between age and cholinesterase activity in several rabbit brain areas. *Am. J. Physiol.* **179**:502–506 (1954).

Aprison, M. H., Wolf, M. A., Poulos, G. L., and Folkerth, T. L. Neurochemical correlates of behavior. III. Variation of serotonin content in several brain areas and peripheral tissues of the pigeon. *J. Neurochem.* **9**:575–584 (1962).

Aprison, M. H., Takahashi, R., and Folkerth, T. L. Biochemistry of the avian central nervous system. I. The 5-hydroxytryptophan decarboxylase-monoamine oxidase and cholineacetylase-acetylcholinesterase systems in several discrete areas of the pigeon brain. *J. Neurochem.* **11**:341–350 (1964).

Arnaiz, G. R., and De Robertis, E. D. P. Cholinergic and non-cholinergic nerve endings in the rat brain. II. Subcellular localization of monoamine oxidase and succinate dehydrogenase. *J. Neurochem.* **9**:503–508 (1962).

Bannister, R. G., and Romanul, F. C. A. The localization of alkaline phosphatase activity in cerebral blood vessels. *J. Neurol. Neurosurg. Psychiatry* **26**:333–341 (1963).

Barron, F., Bernsohn, J., and Hess, A. Zymograms of neural acid phosphatases. Implications for slide histochemistry. *J. Histochem. Cytochem.* **12**:42–44 (1964).

Baxter, D. W., and Olszewski, J. Respiratory responses evoked by electrical stimulation of pons and mesencephalon. *J. Neurophysiol.* **18**:276–287 (1955).

Bernsohn, J., and Barron, K. Multiple forms of brain hydrolases. *Intern. Rev. Neurobiol.* **7**:297–344 (1964).

Birkhäuser, H. Fermente im Gehirn geistig normaler Menschen (Cholinesterase, Mono-und Diamin-oxydase, Cholin-oxydase) *Helv. Chim. Acta* **23**:1071–1086 (1940).

Blaschko, H. Amine oxidase and amine metabolism. *Pharmacol. Rev.* **4**:415–458 (1952).

Bourne, G. H. Histochemical demonstration of phosphatases in the central nervous system of the rat. *Exptl. Cell. Res. Suppl.* **5**:101–117 (1958).

Cavanagh, J. B., and Holland, P. Cholinesterase in the chicken nervous system. *Nature* **190**:735–736 (1961).

Crook, J. C. Acetylcholinesterase activity of capillary blood vessels in the central nervous system of the rabbit. *Nature* **199**:41–43 (1963).

Duckett, S., and Pearse, A. G. E. The nature of the solitary active cells of the central nervous system. *Experientia* **20**:259–260 (1964).

Elliot, K. A. C. Progress in neurobiology. IV. *The Biology of Myelin* (S. R. Korey, ed.). Harper (Hoeber), New York, 1959.

Fahn, S., and Côté, L. J. Regional distribution of sodium-potassium activated adenosine triphosphatase in the brain of the rhesus monkey. *J. Neurochem.* **15**:433–436 (1968).

Felgenhauer, K., and Stammler, A. Das Verteilungsmuster der Dehydrogenasen und Diaphorasen im Zentralnervensystem des Meerschweinchens. *Z. Zellforsch.* **58**:219–233 (1962).

Friede, R. L. Histochemical investigations on succinic dehydrogenase in the central nervous system. II. Atlas of medulla oblongata of the guinea pig. *J. Neurochem.* **4**:111–123 (1959a).

Friede, R. L. Histochemical investigations on succinic dehydrogenase in the central nervous system. III. Atlas of the midbrain of the guinea pig including pons and cerebellum. *J. Neurochem.* **4**:290–303 (1959b).

Friede, R. L. *A Histochemical Atlas of Tissue Oxidation in the Brain Stem of the Cat.* S. Karger, Basel and New York, 1961.

Friede, R. L. The relation of the formation of lipofuscin to the distribution of oxidative enzymes in the human brain. *Acta Neuropathol.* **2**:113–125 (1962).

Friede, R. L. A quantitative mapping of alkaline phosphatase in the brain of the rhesus monkey. *J. Neurochem.* **13**:197–203 (1966).

Friede, R. L., and Fleming, L. M. A mapping of oxidative enzymes in the human brain. *J. Neurochem.* **9**:179–198 (1962).

Friede, R. L., and Fleming, L. M. A mapping of the distribution of lactic dehydrogenase in the brain of the rhesus monkey. *Am. J. Anat.* **113**:215–234 (1963).

Friede, R. L., and Knoller, M. A quantitative mapping of acid phosphatase in the brain of the rhesus monkey. *J. Neurochem.* **12**:441–450 (1965).

Friede, R. L., Fleming, L. M., and Knoller, M. A comparative mapping of enzymes involved in hexosemonophosphate shunt and citric acid cycle in the brain. *J. Neurochem.* **10**:263–277 (1963).

Gerebtzoff, M. A. *Cholinesterases.* International Series of Monographs on Pure and Applied Biology (Division: Modern Trends in Physiological Sciences), (P. Alexander and Z. M. Bacq, eds.). Pergamon Press, New York, 1959.

Hamburger, V. Experimental analysis of the dual origin of the trigeminal ganglion in the chick embryo. *J. Exptl. Zool.* **148**:91–124 (1961).

Hard, W. L., and Hawkins, R. K. Histochemical studies on the area postrema. *Anat. Rec.* **108**:577 (1950).

Hard, W. L., and Peterson, A. C. The distribution of choline esterase in nerve tissue of the dog. *Anat. Rec.* **108**:57–64 (1950).

Himwich, H. E. *Brain Metabolism and Cerebral Disorders.* Williams and Wilkins, Baltimore, 1951.

Hogeboom, G. H., Schneider, W. C., and Palade, G. E. Cytochemical studies of mammalian tissues. I. Isolation of intact mitochondria from rat liver; some biochemical properties of mitochondria and submicroscopic particulate material. *J. Biol. Chem.* **172**:619–635 (1948).

Hydén, H. Quantitative assay of compounds in isolated fresh nerve cells and glial cells from control and stimulated animals. *Nature* **184**:433–435 (1959).

Hydén, H. The Neuron. In: *The Cell* (J. Brachet and A. C. Mirsky, eds.), Vol. IV, Ch. 5. Academic Press, New York and London, 1960.

Hydén, H., and Pigon, A. A cytophysiological study of the functional relationship between oligodendroglial cells and nerve cells of Deiter's nucleus. *J. Neurochem.* **6**:57–72 (1960).

Iijima, K., Bourne, G. H., and Shantha, T. R. Histochemical studies on the distribution of enzymes of glycolytic pathways in the area postrema of the squirrel monkey. *Acta Histochem.* **27**:42–54 (1967).

Iijima, K., Shantha, T. R., and Bourne, G. H. Histochemical studies on the nucleus basalis of Meynert of the squirrel monkey (*Saimiri sciureus*). *Acta Histochem.* **30**:96–108 (1968).

Iijima, K., Shantha, T. R., and Bourne, G. H. Histochemical study on the pigments and distribution of various enzymes in the dorsal vagal nucleus and hypoglossal nucleus of the squirrel monkey. *Acta Histochem.* **32**:18–36 (1969).

Ishii, T., and Friede, R. L. Distribution of a catecholamine-binding mechanism in rat brain. *Histochemie* **9**:126–135 (1967).

Johnston, J. P. Some observations upon a new inhibitor of monoamine oxidase in brain tissue. *Biochem. Pharmacol.* **17**:1285–1297 (1968).

Josephy, H. Acid phosphatase in the senile brain. *Arch. Neurol. Psych.* **61**:164–169 (1949).

Kalina, M., and Bubis, J. J. Histochemical studies on the distribution of acid phosphatase in neurones of sensory ganglia; light and electron microscopy. *Histochemie* **14**:103–113 (1968).

Kaluza, J. S., and Burstone, M. S. Staining patterns of phosphatases of the central nervous system with azo dye methods. *J. Neuropathol. Exptl. Neurol.* **23**:477–486 (1964).

Koelle, G. B. The elimination of enzymatic diffusion artifacts in the histochemical localization of cholinesterases and a survey of their cellular distributions. *J. Pharmacol. Exptl. Therap.* **103**:153–171 (1951).

Koelle, G. B. Histochemical localization of cholinesterases in the central nervous system of the rat. *J. Pharmacol. Exptl. Therap.* **106**:401 (1952).

Koelle, G. B. Histochemical localization of cholinesterases in the central nervous system of the rat. *J. Comp. Neurol.* **100**:211–228 (1954).

Koelle, G. B. The histochemical identification of acetylcholinesterases in cholinergic adrenergic and sensory neurons. *J. Pharmacol. Exptl. Therap.* **114**:167–184 (1955).

Koelle, G. B. A proposed dual neurohumoral role of acetylcholine: it functions at the pre- and post-synaptic sites. *Nature* **190**:208–211 (1961).

Koelle, G. B. A general concept of the neurohumoral functions of acetylcholinesterase. *J. Pharm. Pharmacol.* (*London*) **14**:65–90 (1962).

Koelle, G. B. (ed.) Cytological distribution and physiological functions of cholinesterases. In: *Handbuch der Experimente Pharmakologie,* pp. 187–298. Springer-Verlag, Berlin, 1963.

Koelle, W. A., and Koelle, G. B. The localization of external or functional acetylcholinesterase at the synapses of autonomic ganglia. *J. Pharmacol. Exptl. Therap.* **126**:1–8 (1959).

Koelle, G. B., and Valk, A. Physiological implications of the histochemical localization of monoamine oxidase. *J. Physiol.* (*London*) **126**:434–447 (1954).

Krnjevic, K., and Silver, A. Acetylcholinesterase in the developing forebrain. *J. Anat.* (*London*) **100**:63–89 (1966).

Kumamoto, T., and Bourne, G. H. Experimental studies on the oxidative enzymes and hydrolytic enzymes in spinal neurons. I. A histochemical and biochemical investigation of the spinal ganglion and the spinal cord following sciatic neurotomy in the guinea pig. *Acta Anat.* **55**:255–277 (1963).

Kusunoki, T., Ishibashi, H., and Masai, H. The distribution of monoamine oxidase and melanin pigment in the central nervous system of amphibia. *J. Fur Hirnforschug.* **9**:63–70 (1966).

Landers, J. W., Chason, J. L., Gonzalez, J. E., and Palutke, W. Morphology and enzymatic activity of rat and cerebral capillaries. *Lab. Invest.* **11**:1253–1259 (1962).

Landow, H., Kabat, E. A., and Newman, W. Distribution of alkaline phosphatase in normal and neoplastic tissues of the nervous system. *Arch. Neurol. Psychiat.* **48**:518 (1942).

Lewis, P. R., and Shute, C. C. D. Selective staining of visceral efferents in the rat brain stem by a modified Koelle technique. *Nature* **183**:1743–1744 (1959).

Lowry, O. H., Roberts, N. R., Leiner, K. Y., Wu, M. Y., Farr, A. L., and Albers, R. W. The quantitative histochemistry of brain. III. Ammon's horn. *J. Biol. Chem.* **207**:39–49 (1954).

Maeda, T., Abe, T., and Shimizu, N. Histochemical demonstration of aromatic monoamine in the locus coeruleus of the mammalian brain. *Nature* **188**:326–327 (1960).

Manocha, S. L. Cranial nerve nuclei of the chimpanzee brain (*Pan troglodytes*). In: *Proceedings Second International Congress of Primatology,* Vol. III, pp. 79–83. S. Karger, Basel and New York, 1969.

Manocha, S. L., and Bourne, G. H. Histochemical mapping of monoamine oxidase and lactic dehydrogenase in the pons and mesencephalon of squirrel monkey (*Saimiri sciureus*). *J. Neurochem.* **13**:1047–1056. (1966a).

Manocha, S. L., and Bourne, G. H. Histochemical mapping of succinic dehydrogenase and cytochrome oxidase in the spinal cord, medulla oblongata and cerebellum of squirrel monkey (*Saimiri sciureus*). *Exptl. Brain Res.* **2**:216–229 (1966b).

Manocha, S. L., and Bourne, G. H. Histochemical mapping of succinic dehydrogenase and cytochrome oxidase in the pons and mesencephalon of squirrel monkey (*Saimiri sciureus*). *Exp. Brain Res.* **2**:230–246 (1966c).

Manocha, S. L., and Bourne, G. H. Histochemical mapping of succinic dehydrogenase and cytochrome oxidase in the diencephalon and basal telencephalic centers of the brain of squirrel monkey (*Saimiri sciureus*). *Histochemie* **9**:300–319 (1967a).

Manocha, S. L., and Bourne, G. H. Histochemical mapping of oxidative enzymes in the brain of *Saimiri sciureus. Anat. Rec.* **157**:287 (1967b).

Manocha, S. L., and Bourne, G. H. Histochemical mapping of lactate dehydrogenase and monoamine oxidase in the medulla oblongata and cerebellum of squirrel monkey (*Saimiri sciureus*). *J. Neurochem.* **15**:1033–1040 (1968).

Manocha, S. L., and Shantha, T. R. Enzyme histochemistry of the nervous system. In: *Structure and Function of Nervous Tissue* (G. H. Bourne, ed.), Vol. II, pp. 137–140. Academic Press, New York and London, 1969.

Manocha, S. L., Shantha, T. R., and Bourne, G. H. Histochemical mapping of the distribution of monoamine oxidase in the diencephalon and basal telencephalic centers of the brain of squirrel monkey (*Saimiri sciureus*). *Brain Res.* **6**:570–586 (1967).

Meyer, P. Histochemistry of the developing human brain. *Acta Neurol. Scand.* **39**:123–138 (1963).

Naidoo, D., and Pratt, O. E. The localization of some acid phosphatases in brain tissue. *J. Neurol. Neurosurg. Psychiatry.* **14**:287–294 (1951).

Novikoff, A. B., Hausman, D. H., and Podher, E. The localization of adenosine triphosphatase in liver: *in situ* staining and cell fractionization studies. *J. Histochem. Cytochem.* **6**:61–71 (1958).

O'Steen, W. K., and Callas, G. Histochemical study of monoamine oxidase in central nervous system cultures. *Anat. Rec.* **150**:257–264 (1964).

Olszewski, J., and Baxter, D. *Cytoarchitecture of the Human Brain Stem.* Lippincott, Philadelphia, 1959.

Pearson, A. A. Facial motor nuclei. *J. Comp. Neurol.* **85**:461–476 (1946).

Pletscher, A., Gey, K. F., and Zeller, P. Monoamine oxidase inhibitors. *Fortschr. Arzneimittelforsch.* **2**:417–490 (1960).

Pope, A. Quantitative histochemistry of the cerebral cortex. *J. Histochem. Cytochem.* **8**:425–430 (1960).

Pope, A., and Hess, H. H. Cytochemistry of neurones and neuroglia. In: *Metabolism of the Nervous System* (D. Richter, ed.). Pergamon Press, London, 1957.

Ridge, J. W. The distribution of cytochrome oxidase activity in rabbit brain. *Biochem. J.* **102**:612–617 (1967).

Robinson, N. Histochemistry of monoamine oxidase in the developing rat brain. *J. Neurochem.* **14**:1083–1089 (1967).

Robinson, N. Histochemistry of rat brain stem monoamine oxidase during maturation. *J. Neurochem.* **15**:1151–1159 (1968).

Rodriquez, S. Über die transneuronale Degeneration der optischen Zeutren der Ratte, besonders des corpus geniculatum laterale. *Anat. Anz.* **120**:187–197 (1967).

Rogers, K. T. Studies on chick brain differentiation. IV. Comparative biochemical alkaline phosphatase studies on an attricial bird brain and chick kidney and intestine. V. Comparative histochemical alkaline phosphatase studies on chick retina and blackbird, mouse, rabbit, cat and human brains. *J. Exptl. Zool.* **153**:15–35 (1963).

Romanul, F. C. A., and Cohen, R. B. A histochemical study of dehydrogenases in the central and peripheral nervous systems. *J. Neuropathol. Exptl. Neurol.* **19**:135–136 (1960).

Russell, G. V. Nucleus locus coeruleus (dorsolateralis tegmenti). *Texas Rep. Biol. Med.* **13**:939–988 (1955).

Samorajski, T., and McCloud, J. Alkaline phosphomonoesterase and blood brain permeability. *Lab. Invest.* **10**:492–501 (1961).

Sandler, M., and Bourne, G. H. Intranuclear histochemical localization of adenosine triphosphatase. *J. Histochem. Cytochem.* **10**:636 (1962).

Scharrer, E., and Sinder, J. A. A contribution to the "chemoarchitectonics" of the optic tectum of the brain of the pigeon. *J. Comp. Neurol.* **91**:331–336 (1949).

Schiebler, T. H. Herzstudie II. Mitteilung. Histologische, histochemische und experimentelle Untersuchungen am Atrioventrikularsystem. von Huf-und Nagetieren. *Z. Zellforsch.* **43**:243–306 (1955).

Schneider, W. C., and Hogeboom, G. H. Cytochemical studies of mammalian tissues: the isolation of cell components by differential centrifugation: a review. *Cancer Res.* **11**:1–22 (1951).

Shantha, T. R., Iijima, K., and Bourne, G. H. Histochemical studies on the cerebellum of squirrel monkey (*Saimiri sciureus*). *Acta Histochem.* **27**:129–162 (1967).

Shantha, T. R., and Manocha, S. L. The brain of chimpanzee (*Pan troglodytes*). III. Midbrain, pons, medulla oblongata and cerebellum. In: *Handbook of Chimpanzee* (G. H. Bourne, ed.), Vol. I, pp. 318–330. S. Karger, Basel and New York, 1969.

Shimizu, N. Histochemical studies on the phosphatase of the nervous system. *J. Comp. Neurol.* **93**:201–218 (1950).

Shimizu, N. Histochemical properties of the locus coeruleus. *J. Histochem. Cytochem.* **9**:617–618 (1961).

Shimizu, N., and Kumamoto, T. Histochemical studies on the glycogen of the mammalian brain. *Anat. Rec.* **114**:479–498 (1952).

Shimizu, N., and Morikawa, N. Histochemical studies of succinic dehydrogenase of the brain of mice, rats, guinea pigs, and rabbits. *J. Histochem. Cytochem.* **5**: 334–345 (1957).

Shimizu, N., and Okada, M. Histochemical distribution of phosphorylase in the rodent brain from newborn to adults. *J. Histochem. Cytochem.* **5**:459–471 (1957).

Shimizu, N., Morikawa, N., and Ishii, Y. Histochemical studies of succinic dehydrogenase and cytochrome oxidase of the rabbit brain, with special reference to the results in paraventricular structures. *J. Comp. Neurol.* **108**:1–21 (1957).

Shimizu, N., Morikawa, N., and Okada, M. Histochemical studies of monoamine oxidase of the brain of rodents. *Z. Zellforsch.* **49**:389–400 (1959).

Shute, C. C. D., and Lewis, P. R. The salivatory center in the rat. *J. Anat.* (*London*) **94**:59–73 (1960a).

Shute, C. C. D., and Lewis, P. R. The use of cholinesterase techniques combined with operative procedures to follow nervous pathways in the brain. *Bibliotheca. Anat.* (*Basel*) **2**:34–49 (1960b).

Shute, C. C. D., and Lewis, P. R. Cholinesterase-containing systems of the brain of the rat. *Nature* **199**:1160–1164 (1963).

Shuttleworth, E. C., Jr., and Allen, N. Acid hydrolases in pia-arachnoid and ependyma of rat brain. *Neurology* **16**:979–985 (1966).

Silva-Pinto, M., and Coimbra, A. Comparative studies of the central nervous system phosphatases employing the Gomori and azo-dye methods. *Acta Anat.* **52**:157–173 (1963).

Silver, A. Cholinesterases of the central nervous system with special reference to the cerebellum. *Intern. Rev. Neurobiol.* **10**:57–109 (1967).

Sinder, J. A., and Scharrer, E. Distribution of certain enzymes in the brain of the pigeon. *Proc. Soc. Exptl. Biol. Med.* **72**:60–62 (1949).

Smith, B. Monoamine oxidase in pineal gland, neurohypophysis and brain of the albino rat. *J. Anat.* (*London*) **97**:81–86 (1963).

Snell, R. S. The histochemical localization of cholinesterase in the central nervous system. *Bibliotheca. Anat.* **2**:50–58 (1961).

Snider, R. S., and Lee, J. C. *A Stereotaxic Atlas of the Monkey Brain* (*Macaca mulatta*). The Univ. of Chicago Press, Chicago, 1961.

Squires, R. F. Additional evidence for the existence of several forms of mitochondrial oxidase in the mouse. *Biochem. Pharmacol.* **17**:1401–1411 (1968).

Thomas, E., and Pearse, A. G. E. The fine localization of dehydrogenases in the nervous system. *Histochemie* **2**:266–282 (1961).

Thomas, E., and Pearse, A. G. E. The solitary cells, histochemical demonstration of damage resistant nerve cells with a TPN-diaphorase reaction. *Acta Neuropathol.* **3**:238–249 (1964).

Tipton, K. F., and Dawson, A. P. The distribution of monoamine oxidase and α-glycerophosphate dehydrogenase in pig brain. *Biochem. J.* **108**:95–101 (1968).

Tolani, A. J., and Talwar, G. P. Differential metabolism of various brain regions. Biochemical heterogeneity of mitochondria. *Biochem. J.* **88**:357–362 (1963).

Tyrer, J. H., Eadie, M. J., and Kukums, J. R. Histochemical measurements of relative concentrations of monoamine oxidase in various regions of the rabbit brain. *Brain* **91**:507–519 (1968).

Vogt, M. Distribution of adrenalin and noradrenalin in the central nervous system and its modification by drugs. In: *Metabolism of the Nervous System* (D. Richter, ed.), pp. 553–565. Pergamon Press, Oxford, 1957.

Wawrzyniak, M. Histochemical activity of some enzymes in the mesencephalon during the ontogenetic development of the rabbit and guinea-pig. I. Colliculus superior. *1st Intern. Symp. Histochem, May 13–16, 1963.*

Wawrzyniak, M. The histochemical activity of some enzymes in the mesencephalon during the ontogenetic development of the rabbit and guinea pig. II. Histochemical development of acetylcholinesterase and monoamine oxidase in the nontectal portion of the midbrain of the guinea pig. *Z. Zellforsch.* **72**:261–305 (1965).

Weiner, N. The distribution of monoamine oxidase and succinic oxidase in brain. *J. Neurochem.* **6**:79–86 (1960).

Whittaker, V. D. The specificity of pigeon brain acetylcholinesterase. *Biochem. J.* **54**:660–664 (1953).

Wolfgram, F., and Rose, A. S. The histochemical demonstration of dehydrogenases in neuroglia. *Exptl. Cell. Res.* **17**:526–530 (1959).

Woohsmann, H. Ein Beitrag zur Monoaminoxydase des Zentralnervensystems. (A contribution to the monoaminoxidase of the central nervous system.) *Acta Biol. Med. Ger.* **11**:164–168 (1963).

Youngstrom, K. A. Acetylcholinesterase concentration during the development of the human fetus. *J. Neurophysiol.* **4**:473–477 (1941).

Zeller, E. A. Oxidation of amines. In: *The Enzymes, Chemistry and Mechanism of Action* (J. B. Summer and K. Myrback, eds.), Vol. II, Part 1. Academic Press, New York, 1951.

VII

SPINAL CORD

OXIDATIVE ENZYMES AND MONOAMINE OXIDASE

The activity of lactic dehydrogenase in the various components of the spinal cord is generally stronger than that of SDH (Figs. 171, 174–177, 181), and there is more enzyme activity in the gray matter than in the white matter. In the gray areas of the ventral, dorsal, and lateral horns, the neurons give a stronger enzyme reaction than do the neuropil and the substantia gelatinosa. The latter shows moderately strong LDH and moderate SDH activity, whereas the white matter shows mild LDH and negligible to mild SDH activity. The glial cells in the white matter show a strong LDH and a moderate SDH reaction. Among the neurons of the ventral, dorsal, and lateral horns, the cytoplasm shows strong to very strong activity for both SDH and LDH, although the degree of intensity of LDH reaction is generally higher. In the neurons of the lateral horn, some lightly stained (mild or moderate) cells can also be observed in the SDH preparations. The cytoplasmic reaction extends into the cell processes for a considerable distance. A majority of the neurons show homogeneous distribution of enzyme activity in the cytoplasm and the processes. A very small number of neurons, particularly those of the dorsal horn cells, are observed in which the enzyme activity shows either a perinuclear or peripheral aggregation in the cytoplasm. The synapses, generally observed as darkly stained beads, are present in large numbers and show moderately strong and strong activity for SDH and LDH. The internuncial neurons give moderate SDH and moderately strong LDH reactions. In the developing

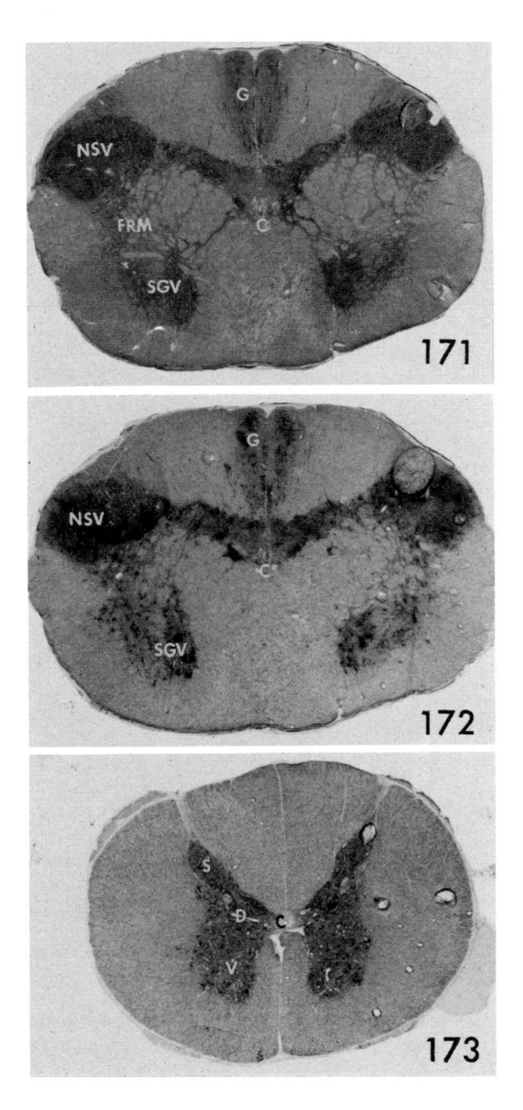

FIGURES 171–173. Sections passing through the various levels of the cervical spinal cord showing LDH, AC, and SE activity in the gray and white matter. ×7, ×7, ×7.

human fetus, the anterior horn cells show enzyme activity a few weeks earlier than those of the posterior horn (Duckett and Pearse, 1969).

In the monoamine oxidase preparations, the cytoplasmic reaction of the neurons of various nuclei is negligible or mild to moderate, whereas the nucleus and nucleolus are negative to negligible. The internuncial neurons give a negligible MAO reaction. The substantia gelatinosa is mild to moderate and the nerve fibers moderately reactive. The white matter reacts negligibly. The cells of the intermediolateral and intermediomedial columns show strong MAO activity (Hashimoto *et al.*, 1962).

Oxidative enzymes in the spinal cord have been studied in a number of animals, particularly the rat and the squirrel monkey. Kumamoto and Bourne (1963) and Nandy and Bourne (1964a, 1964b) studied the distribution of SDH, cytochrome oxidase, and DPN diaphorase in rat spinal cord; Manocha *et al.* (1967) studied the localization of SDH, LDH, glucose-6-phosphate dehydrogenase, phosphorylases (phosphorylase a, a+b, and total phosphorylase), and glycogen transferase in the squirrel monkey spinal cord. In the squirrel and rhesus monkeys, the present authors observed variation in the degree of positive activity from neuron to neuron of a given nucleus; this, in our opinion, depends upon the functional state of the cells and number of mitochondria found within the perikarya. The correlation between the oxidative enzyme activity and the number of mitochondria has been obtained from studies of Scarpelli and Pearse (1958), who showed the presence of SDH, cytochrome oxidase, and DPN diaphorase in the mitochondria at light and electron microscope levels. Using quantitative methods Howe and Flexner (1947) reported a 12% reduction of SDH activity in anterior horn cells within 31 to 54 days after unilateral nerve root sectioning. The nucleolus, in our studies, also shows varying locations; i.e., central to close to the nuclear membrane.

In the spinal cord, Nandy and Bourne (1964a, 1964b) have described various types of neurons based solely upon the relationship of nucleolus to nuclear membrane and degree of the enzyme reaction products in the cell cytoplasm, especially in SDH and ATPase preparations of dorsal horn cells of the rat spinal cord. So far there is no strong evidence indicating that all the neurons in the spinal cord of the squirrel monkey (Manocha *et al.*, 1967) and the rhesus monkey show such a relation. In interpreting the relationship between the nucleolus and cytoplasm, one should consider the possibility that when a large nucleolus is found within a small nucleus (as in dorsal horn cells), its precise location in a sectioned cell can be easily mistaken, and the nucleolar relationship to perinuclear and cytoplasmic enzyme activity misinterpreted. For further details on this topic, the reader is referred to the chapters on the dorsal root ganglion and the cerebellum.

The glycogen body generally found in the birds is a gelatinous structure composed of vesicular, glycogen-storing cells located between the dorsal tracts in the lumbosacral region of the spinal cord. No such body has been reported in the primates (Friede and Vossler, 1964). In a study of the histochemistry of the glycogen body of the turkey spinal cord, Friede and Vossler showed high levels of glycolytic and pentose shunt enzymes, suggesting a predominantly glycolytic metabolism. A large quantity of glucose-6-phosphate dehydrogenase also indicates that the glycogen body could utilize the hexosemonophosphate shunt, and they suggested that the absence of cholinesterases indicates that there is no cholinergic control of the glycogen-storing tissue. Low levels of citric acid cycle enzymes in the glycogen body have been reported earlier by Masai and Matano (1961). The alkaline phosphatase reaction in the glycogen body is very weak, and this rules out the possibility of its being derived from the meninges, since the latter always give a strong AK reaction (Friede and Vossler, 1964).

The present authors have studied in detail the distribution of phosphorylases in the squirrel monkey spinal cord and have shown that, whereas the cytoplasm exhibits greater activity of SDH and LDH than the surrounding gray matter, the phosphorylases show strong activity in the neuropil and mild activity in the cytoplasm of neurons. The moderate reaction of phosphorylase a and the strong to very strong reaction of phosphorylase a+b and total phosphorylase in the neuropil probably indicate the presence of large quantities of inactive phosphorylases in the gray matter of the spinal cord. The preponderance of phosphorylase b over a has been recorded earlier in various tissues (Godlewski, 1961, 1962; Manocha *et al.*, 1967; Shantha *et al.*, 1967) and sharply contrasts to the opposite picture shown by skeletal muscles (Fisher and Krebs, 1955; Leonard, 1957) which are actively engaged in glycogen breakdown at all times. The phosphorylase reaction is strong in the nucleoli in contrast to the low activity of other enzymes. The significance of the presence of phosphorylases in the nucleolus is difficult to explain at this time, but it is probable that they are involved in the breakdown of glycogen to glucose-1-phosphate, a source of energy for nucleolar activities.

Enzyme localization in the spinal cord tracts has been studied by Friede (1966). He noted an inverse relationship between the presence of oxidative enzymes in the axons and the number of oligodendroglial cells with enzyme content. For example, he found that the pyramidal tract and some of the ascending tracts in the spinal cord have marked enzyme activity in their axons with relatively few oligodendroglia exhibiting significant enzyme reaction. In other regions of the white matter, such as the cerebral and cerebellar hemispheres, the oxidative enzyme activity was barely recognizable, but these regions contained numerous oligodendroglia with marked enzyme activity. Additionally, tracts with stronger

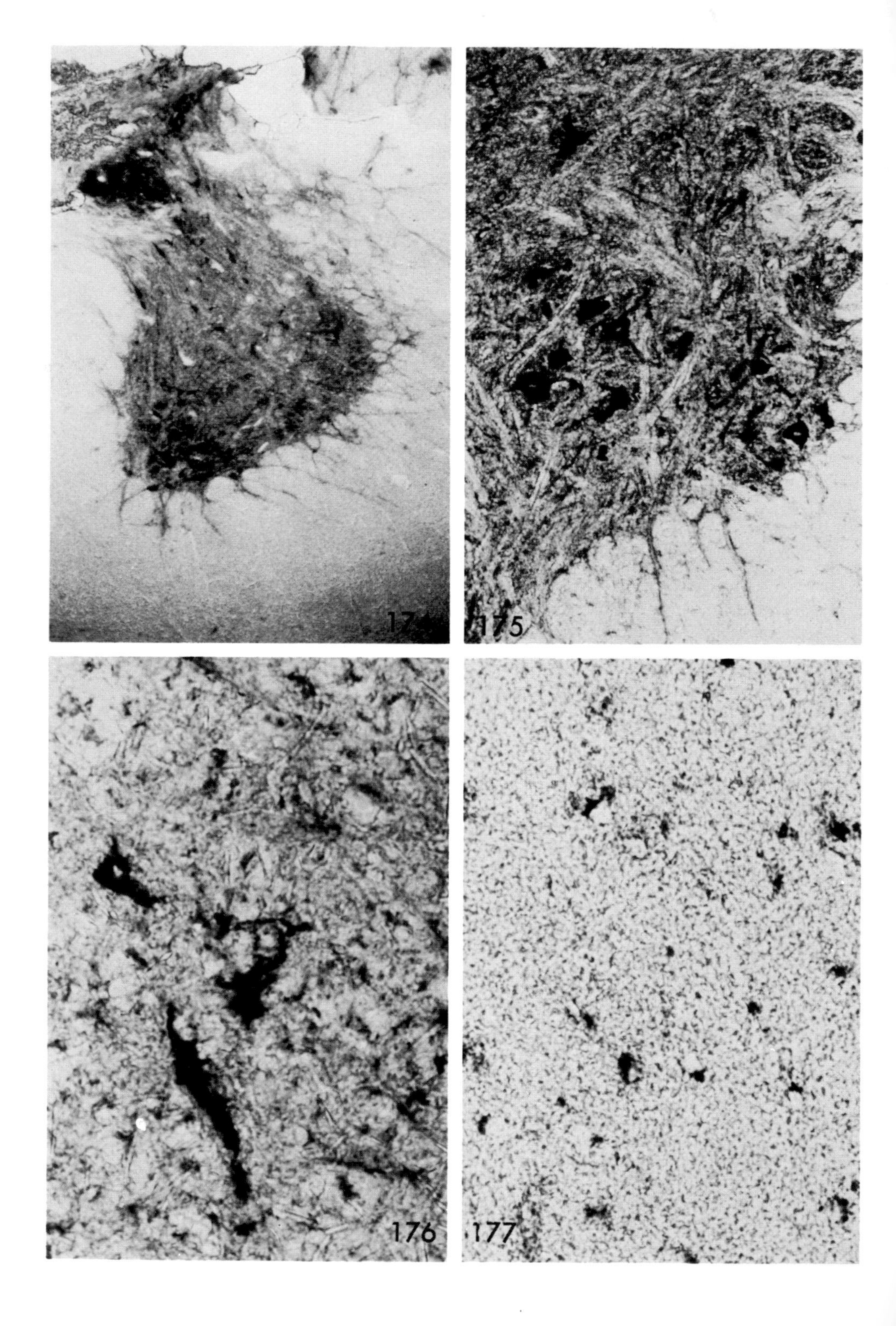
174
175
176
177

FIGURES 174, 175. SDH and LDH preparations of the ventral horn of the spinal cord showing strong and very strong activity in the neurons and moderately strong activity in the neuropil. Note the extension of the enzyme activity into the cell processes for long distances. ×37, ×148.

FIGURES 176, 177. White matter of the spinal cord showing strong LDH and moderate SDH positive glial cells. ×592, ×592.

enzyme reaction showed only a few oligodendroglia with good enzyme activity, and vice versa. Friede, in his studies on the tractus gracilis and cuneatus, has found that the LDH, SDH, malic dehydrogenase, glucose-6-phosphate dehydrogenase, and cytochrome oxidase gradually increased along the terminal portion of the tract. This slight gradient was more pronounced in the tractus gracilis than in cuneatus. Most of the oxidative enzymes encountered in the white matter were found in the axons and glial cells, while the myelin sheath contained little or no activity. The white matter showed glucose-6-phosphate dehydrogenase in the rat and the dog (Thomas and Pearse, 1961), although Adams *et al.* (1963) could not confirm this in the rat.

The areas of synaptic contact, both at the cell surface and in cell processes, show moderate to strong reactions for various oxidative enzymes, phosphorylases, esterases, and phosphatases, which clearly demonstrates that these areas are highly active metabolically. Similar observations have been made by Kumamoto and Bourne (1963) and Nandy and Bourne (1964a, 1964b). The synaptic areas show variations in the extent of the contact, ranging from small granules to large vesicles. The electron microscopic studies of Palay (1956) showed vesicles of 200–650 Å in diameter near the synapse. The strong activity of oxidative enzymes and ATPase suggests the presence of a large number of mitochondria near these vesicles as reported earlier (Bodian, 1940; De Robertis, 1958, 1959; Boycott *et al.*, 1960; Gray and Guillery, 1961). Palay (1956) suggested that these vesicles may provide the morphological representative of the prejunctional subcellular units of neurohumoral discharge at the synapse demanded by physiological evidence. The presence of phosphorylases in the synaptic areas as shown by Manocha *et al.* (1967) further suggests the potential to break down glycogen to supply energy. It has been difficult to quantitate the synaptic areas in the present study in terms of cellular and dendritic surface, but the present authors are inclined to believe that in the spinal cord the synapses clothe the entire cells and dendritic surfaces like a mosaic; as Illis (1964) suggested, 50 to 70% of the surface of the cells and their processes are occupied by synaptic contacts.

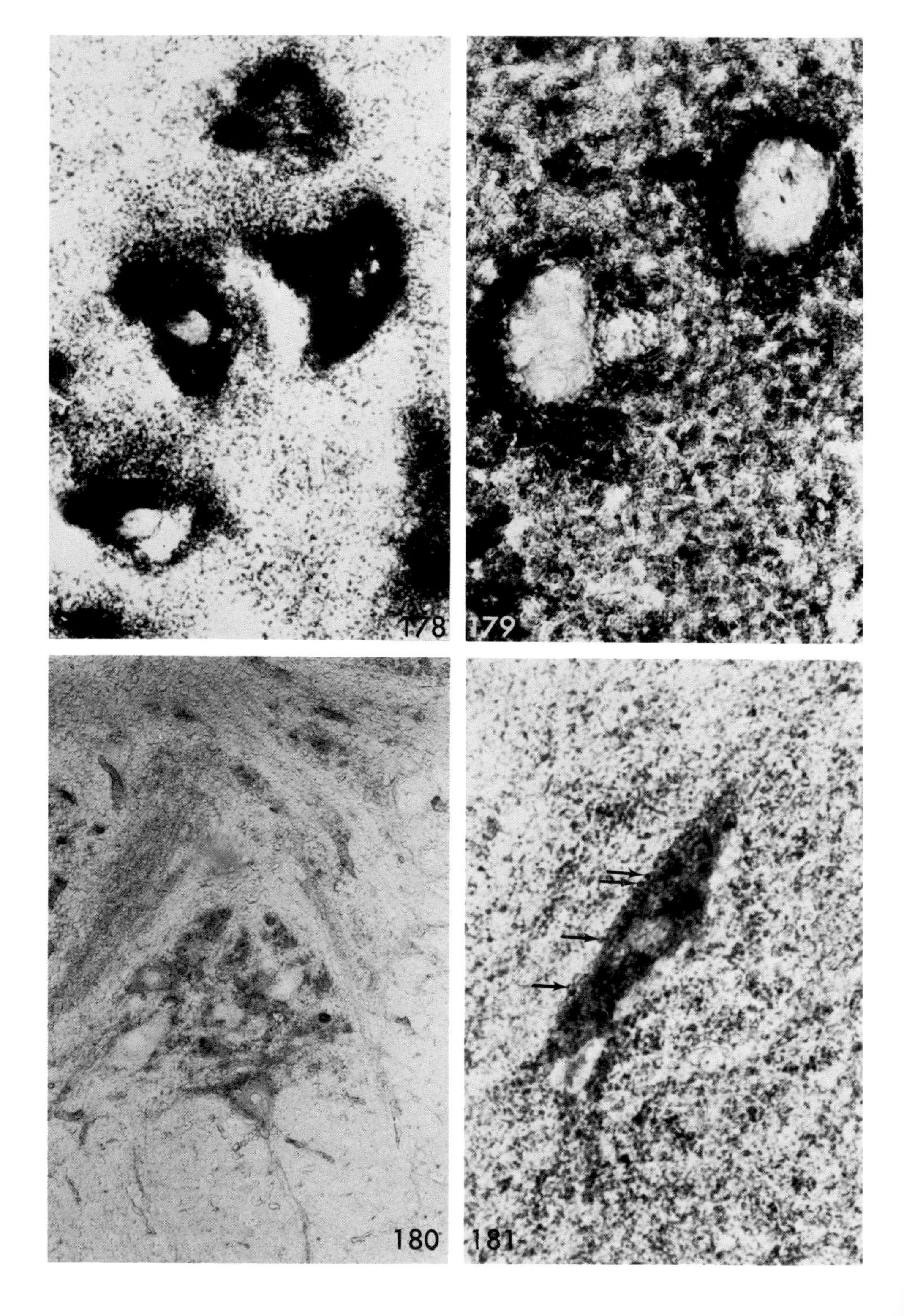
178
179
180
181

FIGURES 178–181. Neurons from the ventral horn of the spinal cord showing SE, ATPase, AChE, and SDH activity in the neurons and neuropil. It is interesting to note the differences in the AChE content in the different neurons of the ventral horn (Fig. 180). The arrows indicate the synapses. ×148, ×592, ×148, ×592.

ESTERASES

The simple esterases (SE) show maximum activity in the cytoplasm of all neurons as well as in the internuncial neurons, in contrast to a negligible reaction in the nucleus and nucleolus (Figs. 173, 178). In the cytoplasmic reaction, the activity varies from very strong in the neurons of the ventral and dorsal nuclei to strong in the lateral nucleus and moderate in the internuncial neurons. The neuropil and the substantia gelatinosa give a moderate reaction, whereas the white matter shows mild SE activity.

AChE activity in the neurons of the various nuclei ranges from mild in some neurons to moderate and strong in the others, extending into the cell processes (Fig. 180). The ventral horn cells show the strongest AChE reaction of the neurons. In contrast, it is the lateral horn cells in the squirrel monkey which show stronger AChE content than other nuclei. The nuclei and nucleoli give a negative and the internuncial neurons a mild to moderate AChE reaction. The neuropil shows mild activity and the substantia gelatinosa mild to moderate activity. The white matter activity is negligible. Of the blood vessels, only those located in the gray matter give a moderately strong AChE reaction. This is in agreement with the observations of Ishii and Friede (1967) who found that in the gray matter, maximum activity is observed in the motor neurons of the anterior horn, autonomic centers in the intermediolateral and intermediomedial columns, and in the substantia gelatinosa. The remaining gray matter shows very weak AChE activity. Earlier studies by Koelle (1952, 1954) in the rat brain support our conclusions. The AChE activity in the rat is concentrated mainly at the perikaryon membranes and along the axons and dendrites of a number of neurons. The motor neurons show the highest AChE activity in the perikarya as well as in their dendritic processes and axons (Silver, 1967), with very weak or negative reaction in the primary sensory neurons. Giacobini and Holmstedt (1958) measured the AChE content of the neurons and described them under two classes in which the AChE activity differed by a factor of four, the stronger AChE reaction being in the motor neurons. McIsaac and Koelle (1959) classified the AChE activity as functional and reserve AChE. The functional, or "external," AChE is located at the synapses; this form of the enzyme is primarily inhibited *in vivo* by anticholinesterase compounds. The reserve, or "internal," AChE is stored in the perikaryon of cholinergic nerve cells. According to these authors, the latter is probably formed within the endoplasmic reticulum of Nissl substance.

The distribution of AChE in the entire spinal cord has not been exhaustively studied, with the exception of data on selected segments of the upper cervical cord in the rat and other species. Giacobini and Holmstedt (1958) and Giacobini (1959) described the activity of AChE in the gray matter of the rat, with emphasis on the anterolateral cell column and its cell processes. Koelle (1954) and Gerebtzoff (1959) described a strong AChE reaction in the anterior column cells and surrounding neuropil. The motor neurons of the ventral gray column always showed strong AChE and NAD-diaphorase activity in the perikarya, although the AShE activity in the neuropil was less than the NAD-diaphorase activity (Ishii and Friede, 1967). In the chicken spinal cord, AChE activity was found mainly in the cytoplasm of neurons, except in the dorsal column (Cavanaugh and Holland, 1961), in which the AChE reaction was noted mainly in the neuropil. Low AChE staining in Clarke's column and high activity in the lateral column have also been described by Roessmann and Friede (1967). In the rat spinal cord, Nandy and Bourne (1964a) reported that AChE activity is present in the synaptic areas of the cell surface but is absent in the cytoplasm of cells in the ventral and dorsal horns. On the contrary, in the squirrel monkey, Manocha *et al.* (1967) found the presence of this enzyme in the neurons of this area. Gerebtzoff (1959) summarized the AChE content of the spinal cord in this manner: the highest activity is found in the anterior horn cells, with synaptic reaction particularly prominent on the surface of the perikaryon and the proximal part of the dendrites. The AChE activity decreases in the following order: preganglionic neurons of the intermediolateral region, substantia gelatinosa, Clarke's column, small interneurons of other regions, etc. In the lumbar portion of the spinal cord, AChE was studied in 14 intact rabbits and 31 animals after the administration of tetanus toxin (Zhebroskava, 1965). This experiment revealed that AChE disappears from the majority of the motor neurons of the anterior horn after the administration of tetanus toxin, which may contribute to the disturbed reflex function of the spinal cord. Galabov *et al.* (1968) showed that AChE and acetylocholine increase their activity after transection of the spinal cord and believed that there is no correlation between enhanced AChE activity and increased content of acetylcholine after spinal cord transection.

In the present study as well as in that of the squirrel monkey (Manocha *et al.,* 1967), we have observed AChE activity in the cytoplasm of some neurons of the ventral and dorsal horn, as well as in the synaptic areas. Other neurons show AChE activity in the synaptic areas and not in the cytoplasm. Similar observations are made in the MAO preparations, and some anterior horn cells show stronger MAO activity than AChE activity. This may suggest that the neurons which are rich in AChE are poor in MAO and vice versa and may be either cholinergic or adrenergic. Shanthaveerappa and Bourne (1965a, 1965b) and Shantha *et al.* (1967) in the olfactory bulb and cerebellum came to a similar conclusion. In a study of the effects of LSD-25 on cholinesterases and monoamine oxidase in the rat spinal cord (Nandy and Bourne, 1964a) and cerebral cortex (Shanthaveerappa *et al.*, 1963), inhibition of these enzymes was demonstrated, and it was suggested that LSD-25 may act as a hallucinogenic agent by inhibiting the enzyme mechanism which controls the passage of excess impulses across the synapses and also by raising the levels of acetylcholine, adrenaline, and adrenochrome.

In our preparations, the qualitative difference in the degree of AChE positive activity in the neurons of the anterior horn may indicate a different functional state of the individual cells. For example, more functionally active cells may contain less AChE, indicating that they may have used most of the enzyme to hydrolyze acetylcholine, whereas the less active cells may contain more enzyme, indicating that these cells are not using as much of the enzyme or are storing it. An alternative explanation is that the cells containing less enzyme may be less active and those with more enzyme may be more active. It has been thought that the concentration of AChE (as indicated by the intensity of AChE activity for the histochemical test) in a given neuron reflects the extent of participation of acetylcholine in synaptic transmission, both pre- and postsynaptically. In a study of the developing human spinal cord, Duckett and Pearse (1969) showed that the appearance of AChE activity in neuronal cytoplasm in the lumbar segment of the spinal cord coincides with the earliest detectable movement in the lower limbs. The AChE activity is generally weak in all the fiber tracts of the spinal cord, except in the ventral roots, which show a marked reaction in the axons (Ishii and Friede, 1967).

PHOSPHATASES

The neurons of the various nuclei do not show any considerable alkaline phosphatase activity in the cytoplasm of the nucleus. Most neurons are negligible, although some show mild AK activity in the peripheral part of the cytoplasm. The neuropil, substantia gelatinosa, and white matter show a negligible reaction in the AK preparations. This is in sharp contrast to the acid phosphatase preparations where the cytoplasm of the neurons reacts quite strongly. This activity does not continue into the cell processes for long distances. The nucleus and the nucleoli exhibit negative to negligible activity. The internuncial neurons give a moderate AC reaction in the cytoplasm (Fig. 172). The neuropil gives mild, substantia gelatinosa moderate, and white matter negligible activity for acid phosphatase. It is estimated that the gray matter has AC activity about 2.5 times that of the white matter, whereas the AK activity in the gray matter is approximately 10 times that of the white (Samorajski and Fitz, 1961). The synapses of the neurons give an enzyme reaction which is as strong as that of the cytoplasm. The intracellular distribution of AC activity is presumed to be located in the lysosomes, which are randomly distributed in the cytoplasm (Novikoff, 1957; DeDuve, 1959; Becker *et al.*, 1960; Nandy and Bourne, 1965). Becker *et al.* (1960) also observed the enzyme reaction in the dendritic processes. AC activity in the neurons varies directly with the amount of Nissl substance (LaVelle *et al.*, 1954), and it has been shown by a number of workers (Wolf *et al.*, 1943; Pope *et al.*, 1949; Shimizu, 1950; Leduc and Wislocki, 1953; Manocha *et al.*, 1967) that the larger neurons show greater phosphatase reaction than the smaller ones. Under stress conditions in chromatolytic cells, for example, there is an increased cytoplasmic reaction (Roizin, 1951), and until normal conditions are restored, the breakdown would overbalance replacement and the involved enzyme activity would be high (LaVelle *et al.*, 1954). There is also a considerable increase in the AC content of the spinal cord after the transection of the nerve (Samorajski and Fitz, 1961). Similarly, Bodian and Mellors (1943), Bueker *et al.* (1949), and Barron and Tuncbay (1962) showed increased AC activity in the cytoplasm of chromatolytic neurons. It has been suggested that the AC in neurons may be part of an enzyme system acting upon a ribonucleoprotein reserve to release phosphate for metabolic processes in cell maintenance and function (LaVelle *et al.*, 1954).

Nandy and Bourne (1963), using the Burstone method and the Gomori method for AK in the spinal cord, showed a great discrepancy in results with these two techniques. With the Burstone method the neurons were negative and the synaptic regions positive; on the other hand, the Gomori method showed strong nucleolar activity with variable activity in the nucleus, cytoplasm, cell processes, and areas of synapses. The present authors had a similar experience and decided in favor of recording observations from the slides prepared by the Burstone technique. Gomori's method gives some nonspecific deposition when incubation is prolonged even slightly beyond the optimum time, and this differs from tissue to tissue. The quantitative histochemical studies of Samorajski and Fitz (1961) at various ventrodorsal levels of the rabbit spinal cord have shown that, whereas the alkaline phosphatase activity increases about tenfold from the ventral tract to the anterior column, it decreases in the region of the lateral column and increases again in the posterior column to the level of the anterior column or slightly higher. These authors have shown that the maximum AC activity is in the dorsal column, and the activity in the anterior column is approximately double that of the anterior tracts.

The ATPase reaction is most prominent in the cytoplasm of the neurons and the neuropil (Fig. 179). In the former, a number of cells show a strong reaction in the peripheral part of the cytoplasm and a moderate reaction in the inner part. The opposite situation is not found in our preparations, although Nandy and Bourne (1964b) found a stronger reaction in the perinuclear area and a lighter one in the peripheral area. The nuclei give a negligible to mild reaction and the nucleoli a moderate one. The substantia gelatinosa and white matter show moderate activity, whereas the neuropil is moderately strong. The areas of synaptic contact in the neurons show diffuse and fine granular reactions of moderate to strong intensity.

The nucleoli in the ATPase preparations give a moderate reaction but do not demonstrate any significant correlation between the size of the nucleoli and histochemical reaction as observed in the rat (Nandy and Bourne, 1964b). There are earlier reports of variations in size of the nucleolus and its metabolic role in protein synthesis (Brachet, 1942; Gates, 1942; Geiger, 1957; Stich, 1959; Wischmitzer, 1960). Increased nucleolar volume has been taken as an indication of increased functional activity. Grimm (1949) and Fukuda and Koelle (1959) indicated that there may be a direct relationship between the RNA of the nucleolus and adenosine triphosphatase activity.

In a study of the TPPase-positive material of the rat spinal cord, Shantha and Bourne (1966) described a complex configuration of the Golgi apparatus in various cellular components. For example, they observed dense, darkly staining Golgi networks in the anterior horn cells, which extended for a considerable distance into the cell processes, starting from a darkly staining "basal body." On the contrary, the neurons in the nucleus lateralis, nucleus dorsalis, and Clarke's nucleus showed less complex and less dense Golgi apparatus, and in the neurons of the substantia gelatinosa, the Golgi material occupied only one part of the cells.

Studies on the spinal cord of the squirrel and rhesus monkeys also show similar results. It is not known for certain why the Golgi complex is so rich in an enzyme which decomposes cocarboxylase.

REFERENCES

Adams, C. W. M., Davison, A. N., and Gregson, N. A. Enzyme inactivity of myelin: histochemical and biochemical evidence. *J. Neurochem.* **10**:383–395 (1963).

Barron, K. D., and Tuncbay, T. O. Histochemistry of acid phosphatase and thiamine pyrophosphatase during axon reaction. *Am. J. Pathol.* **40**:637–652 (1962).

Becker, N. H., Goldfischer, S., Shin, Woo-Yung, and Novikoff, A. B. The localization of enzyme activities in the rat brain. *J. Biophys. Biochem. Cytol.* **8**:649–663 (1960).

Bodian, D. Further notes on the vertebrate synapse. *J. Comp. Neurol.* **73**:323–337 (1940).

Bodian, D., and Mellors, R. C. Phosphatase activity in chromatolytic nerve cells. *Proc. Soc. Exptl. Biol. Med.* **55**:243–245 (1943).

Boycott, B. B., Gray, E. G., and Guillery, R. W. A theory to account for the absence of boutons in silver preparations of the cerebral cortex, based on a study of axon terminals by light and electron microscopy. *J. Physiol.* (*London*) **152**:3 (1960).

Brachet, J. La localisation des acides pentosenucleiques, dans les tissus animaux et les oeufs d'amphibiens en voie de developpement. *Arch. Biol.* (*Paris*) **53**: 207–257 (1942).

Bueker, E. D., Solnitzky, O., and Meyers, C. E. Acid phosphatase reaction during the regenerative cycle of motor neurons in cats and domestic fowl after interruption of the sciatic nerve. *Bull. Georgetown Univ. Med. Center* 3:59–68 (1949).

Cavanaugh, J. B., and Holland, P. Cholinesterase in the chicken nervous system. *Nature* **190**:735–736 (1961).

DeDuve, C. Lysosomes, a new group of cytoplasmic particles. In. *Subcellular Particles* (T. Hayashi, ed.), pp. 128–159. Ronald Press, New York, 1959.

De Robertis, E. Submicroscopic morphology and function of synapse. *Exptl. Cell Res., Suppl.* **5**:347 (1958).

De Robertis, E. Submicroscopic morphology of the synapse. *Intern. Rev. Cytol.* **8**:61–96 (1959).

Duckett, S., and Pearse, A. G. E. Histoenzymology of the developing human spinal cord. *Anat. Rec.* **163**:59–65 (1969).

Fisher, E. H., and Krebs, E. G. Conversion of phosphorylase b to phosphorylase a in muscle extracts. *J. Biol. Chem.* **216**:121–131 (1955).

Friede, R. L. *Topographic Brain Chemistry*. Academic Press, New York and London, 1966.

Friede, R. L., and Vossler, A. E. Histochemistry of the glycogen body of the turkey spinal cord. *Histochemie* **4**:330–335 (1964).

Fukuda, T., and Koelle, G. B. The cytological localization of intracellular neuronal acetylcholinesterase. *J. Biophys. Biochem. Cytol.* **5**:433–440 (1959).

Galabov, G., Manolov, S., Veakov, L., Nikolov, T., and Itchev, K. Studies on cholinesterase activity of anterior horn cells after transection of the spinal cord and plexus brachialis. *Acta Anat.* **68**:432–442 (1968).

Gates, R. R. Nucleoli and related nuclear structures. *Botan. Rev.* **8**:337–409 (1942).

Geiger, A. Chemical changes accompanying activity in the brain. In: *Metabolism of Nervous System* (D. Richter, ed.), pp. 245–256. Pergamon Press, New York, 1957.

Gerebtzoff, M. A. *Cholinesterases.* International Series of Monographs on Pure and Applied Biology (Division: Modern Trends in Physiology Sciences), (P. Alexander and Z. M. Bacq, eds.). Pergamon Press, New York, 1959.

Giacobini, E. Quantitative determination of cholinesterase in individual spinal ganglion cells. *Acta Physiol. Scand.* **45**:238–254 (1959).

Giacobini, E., and Holmstedt, B. Cholinesterase content of certain regions of the spinal cord as judged by histochemical and Cartesian diver technique. *Acta Physiol. Scand.* **42**:12–27 (1958).

Godlewski, H. G. The histochemically demonstrable phosphorylase and branching enzyme activity in some animals ascites tumor cells. *Acta Histochem.* **11**:58–67 (1961).

Godlewski, H. G. Histochemistry of the glycogen synthetase and phosphorylases in normal and pathologic tissues. *Acta Histochem.* Suppl. IV, 30–51 (1962).

Gray, E. G., and Guillery, R. W. The basis for silver staining of synapses of the mammalian spinal cord: a light and electron microscope study. *J. Physiol. (London)* **157**:581–588 (1961).

Grimm, V. Über die Gröbenbeziehung zwischen Kern und Nucleolus menschlicher Ganglionzellen, Ph.D. Dissertation (Bern, 1949).

Hashimoto, P. H., Maeda, T., Torii, K., and Shimizu, N. Histochemical demonstration of autonomic regions in the central nervous system of the rabbit by means of monoamine oxidase staining. *Med. J. Osaka Univ.* **12**:425–465 (1962).

Howe, H. A., and Flexner, J. B. Succinic dehydrogenase in regenerating neurons. *J. Biol. Chem.* **167**:633–671 (1947).

Illis, L. Spinal cord synapses in the cat: The normal appearances by the light microscope. *Brain* **87**:543–554 (1964).

Ishii, T., and Friede, R. L. A comparative histochemical mapping of the distribution of acetylcholinesterase and nicotinamide adenine-dinucleotide-diaphorase activities in the human brain. *Intern. Rev. Neurobiol.* **10**:231–275 (1967).

Koelle, G. B. Histochemical localization of cholinesterases in the central nervous system of the rat. *J. Pharmacol. Exptl. Therap.* **106**:401 (1952).

Koelle, G. B. The histochemical localization of cholinesterases in the central nervous system of the rat. *J. Comp. Neurol.* **100**:211–228 (1954).

Kumamoto, T., and Bourne, G. H. Experimental studies on the oxidative enzymes and hydrolytic enzymes in spinal neurons. I. A histochemical and biochemical investigation of the spinal ganglion and the spinal cord following sciatic neurotomy in the guinea pig. *Acta Anat.* **55**:255–277 (1963).

LaVelle, A., Lui, C. N., and LaVelle, F. W. Acid phosphatase activity as related to nucleic acid sites in the nerve cell. *Anat. Rec.* **119**:305–324 (1954).

Leduc, E. H., and Wislocki, G. B. The histochemical localization of acid and alkaline phosphatases, non-specific esterase, and succinic dehydrogenase in the structures comprising the hemato-encephalic barrier of the rat. *J. Comp. Neurol.* **97**:241–279 (1953).

Leonard, S. L. The effect of hormones on phosphorylase activity in skeletal muscle. *Endocrinology* **60**:619–629 (1957).

Manocha, S. L., Shantha, T. R., and Bourne, G. H. Histochemical studies on the spinal cord of squirrel monkey (*Saimiri sciureus*). *Exptl. Brain Res.* **3**:25–39 (1967).

Masai, H., and Matano, S. Comparative neurological studies on respiratory enzymic activity in the central nervous system of submammals. I. Birds. *Yokohama Med. Bull.* **12**:265–270 (1961).

McIsaac, R. J., and Koelle, G. B. Comparison of the effects of inhibition of external, internal and total acetylcholinesterase upon ganglionic transmission. *J. Pharmacol. Exptl. Therap.* **126**:9–20 (1959).

Nandy, K., and Bourne, G. H. Alkaline phosphatases in brain and spinal cord. *Nature* **200**:1216–1217 (1963).

Nandy, K., and Bourne, G. H. A histochemical study of the localization of the oxidative enzymes in the neurones of the spinal cord in rat. *J. Anat. (London)* **98**:647–653 (1964a).

Nandy, K., and Bourne, G. H. A histochemical study on the localization of adenosine triphosphatase and 5-nucleotidase in the rat spinal cord. *Arch. Neurol.* **11**:547–553 (1964b).

Nandy, K., and Bourne, G. H. Histochemical studies on the distribution of acid naphthol AS-phosphatase in the spinal cord of the rat. *Acta Anat.* **61**:84–91 (1965).

Novikoff, A. B. Biochemical heterogeneity of the cytoplasmic particles. *Symp. Soc. Exptl. Biol.* **10**:92–109 (1957).

Palay, S. L. Synapses in the CNS. *J. Biophys. Biochem. Cytol.* Suppl. 2, 193–202 (1956).

Pope, A., Meath, J. A., Caveness, W. F., Livingston, K. E., and Thompson, R. H. Histochemical distribution of cholinesterase and acid phosphatase in the prefrontal cortex of psychotic and non-psychotic patients. *Trans. Amer. Neurol. Ass.* **74**:147–153 (1949).

Roessmann, U., and Friede, R. L. The segmental distribution of acetyl cholinesterase in the cat spinal cord. *J. Anat. (London)* **101**:27–32 (1967).

Roizin, L. Comparative morphologic and histometabolic studies of nerve cells in brain biopsies and topectomies. *J. Neuropath. Exptl. Neurol.* **10**:177–189 (1951).

Samorajski, T., and Fitz, G. R. Phosphomonoesterase changes associated with spinal cord chromatolysis. *Lab. Invest.* **10**:129–143 (1961).

Scarpelli, D. G., and Pearse, A. G. E. Cytochemical localization of succinic dehydrogenase in mitochondria. *Anat. Rec.* **132**:133–151 (1958).

Shantha, T. R., and Bourne, G. H. The thiamine pyrophosphatase technique as an indicator of morphology of the Golgi apparatus in the neurons. IV. Studies on the spinal cord, hippocampus and trigeminal ganglion. *Ann. Histochem.* **2**:337–351 (1966).

Shantha, T. R., Iijima, K., and Bourne, G. H. Histochemical studies on the cerebellum of the squirrel monkey (*Saimiri sciureus*). *Acta Histochem.* **27**:129–162 (1967).

Shanthaveerappa, T. R., Nandy, K., and Bourne, G. H. Histochemical studies on the mechanism of action of the hallucinogens *d*-lysergic acid diethylamide tartrate (LSD-25) and *d*-2-bromolysergic acid tartrate (BOL-148) in rat brain. *Acta Neuropathol.* **3**:29–39 (1963).

Shanthaveerappa, T. R., and Bourne, G. H. Histochemical studies on distribution of dephosphorylating and oxidative enzymes and esterases in olfactory bulb of the squirrel monkey. *J. Nat. Cancer Inst.* **35**:153–165 (1965a).

Shanthaveerappa, T. R., and Bourne, G. H. Histochemical studies on the olfactory glomeruli of squirrel monkey. *Histochemie* **5**:125–129 (1965b).

Shimizu, N. Histochemical studies on the phosphatase of the nervous system. *J. Comp. Neurol.* **93**:201–218 (1950).

Silver, A. Cholinesterases of the central nervous system with special reference to the cerebellum. *Intern. Rev. Biol.* **10**:57–109 (1967).

Stich, H. F. Changes in nucleoli related to alteration in cellular metabolism. In: *Developmental Cytology* (D. Rudnick, ed.), p. 105. Ronald Press, New York, 1959.

Thomas, E., and Pearse, A. G. E. The fine localization of dehydrogenases in the nervous system. *Histochemie* **2**:266–282 (1961).

Wischmitzer, S. Ultrastructure of nucleus and nucleocytoplasmic relations. *Intern. Rev. Cytol.* **10**:137–162 (1960).

Wolf, A., Kabat, E. A., and Newman, W. Histochemical studies in tissue enzymes III. A study of the distribution of acid phosphatases with special references to the nervous tissue. *Am. J. Pathol.* **19**:423–440 (1943).

Zhebrovskava, N. E. Histochemical study of the cholinesterase distribution in the spinal cord and spinal ganglia of rabbits in experimental tetanus intoxication. *Arkh. Patol.* **27**:71–72 (1965).

VIII

DORSAL ROOT GANGLION CELLS

OXIDATIVE ENZYMES AND MONOAMINE OXIDASE

In general, the activity of oxidative enzymes in the dorsal root ganglion is more prominent in the neurons than in the neuropil. The satellite cells show moderate and strong LDH, moderate SDH, and mild MAO reactions (Figs. 182, 184, 187). The cytoplasm of the neurons shows moderately strong, strong, and very strong LDH and SDH activity and mild to moderate MAO activity. The enzyme activity is both diffuse and granular in character. The nucleus and nucleolus show little or no activity. The variability of enzyme reactions in different cells of the same ganglion, as seen in Figs. 182–189, seems to be a common phenomenon and also has been observed in the squirrel monkey (Shantha *et al.*, 1967). While studying the oxidative enzymes such as SDH, LDH, DPN and TPN diaphorase, malic dehydrogenase, glucose-6-phosphate dehydrogenase, and glycerophosphate dehydrogenase in the ganglia (spinal, mesenteric, etc.), most workers have reported variability of the enzyme reaction in the individual neurons (Marinesco, 1929; Samorajski, 1960; Thomas and Pearse, 1961; Thomas, 1963; Kumamoto and Bourne, 1963). In a study of dehydrogenases, Gerebtzoff and Brotchi (1966) showed that succinic and glutamic dehydrogenase have similar but not parallel localization in the cat spinal ganglion cells. The activity of these enzymes is mainly localized in the neuroplasm of ganglion cells and the perinuclear cytoplasm of the capsular cells (satellite). Sotelo (1966) studied the distribution of phosphorylase, NADH-tetrazolium reductase, succinic dehydrogenase, and 6-phosphogluconate dehydrogenase in the rabbit dorsal root ganglia and observed that the phosphorylase activity was higher in the astrocytes than in the oligodendrocytes and neurons. This is essentially similar to our observations on the squirrel monkey (Shantha *et al.*, 1967). Robins (1960) also showed that the levels of nine enzymes, including hexokinase, glycolytic and citric acid enzymes, and glutamate enzymes, are only slightly different in the two kinds of cells. Our results differ with regard to the LDH activity, which, according to Robins, is approximately one-half as high in the anterior horn cells as in the dorsal root ganglion cells.

The majority of neurons show homogeneous distribution of enzyme activity in the cytoplasm, but in a few cells the activity may be more concentrated at one of the poles of the cell. Tewari and Bourne (1962a, 1962b) interpreted homogeneous or perinuclear distribution of enzyme reaction in the rat as representing cycles of metabolic activity and correlated it with similar stages of mitochondrial distribution. Sharma (1967), studying the superior cervical sympathetic ganglion of the rat, held a similar view. In our preparations, the size of the cell seems to be no criterion for moderate or strong enzyme reactions. Small cells with strong reactions and large cells with moderate activity and vice versa have been observed in all the preparations and probably reflect the functional state of a cell at a particular time. All the neurons showed the enzyme reaction, whether it was mild or strong. Tewari and Bourne (1962a, 1962c), however, reported the complete absence of cytochrome oxidase and SDH in a number of spinal ganglion cells of the rat, but no such observations have been made in squirrel or rhesus monkey ganglion. Our studies have also indicated that there is no correlation between the area of enzyme concentration in the cytoplasm and the position of the nucleolus in the nucleus, as reported by Tewari and Bourne (1962a, 1962b). The position of the nucleolus in a particular cell is probably determined by the angle of the knife with respect to the cell at the time of sectioning. There is also a possibility that perinuclear concentration and rings of enzyme activity may be artificially produced due to slow freezing. This results in the formation of large ice crystals which displace the enzymatically active organelle on either side. In some of the ganglion cells, most of the Nissl material along with the endoplasmic reticulum is found in the peripheral part of the cytoplasm, thus physically displacing the mitochondria into the perinuclear position. These aggregations, rich in mitochondrial enzymes, may give an erroneous impression of perinuclear concentration of enzyme activity having some functional importance such as the exchange of material between the nucleus, nucleolus, and cytoplasm. For further discussion, see Chapter IX on the cerebellum.

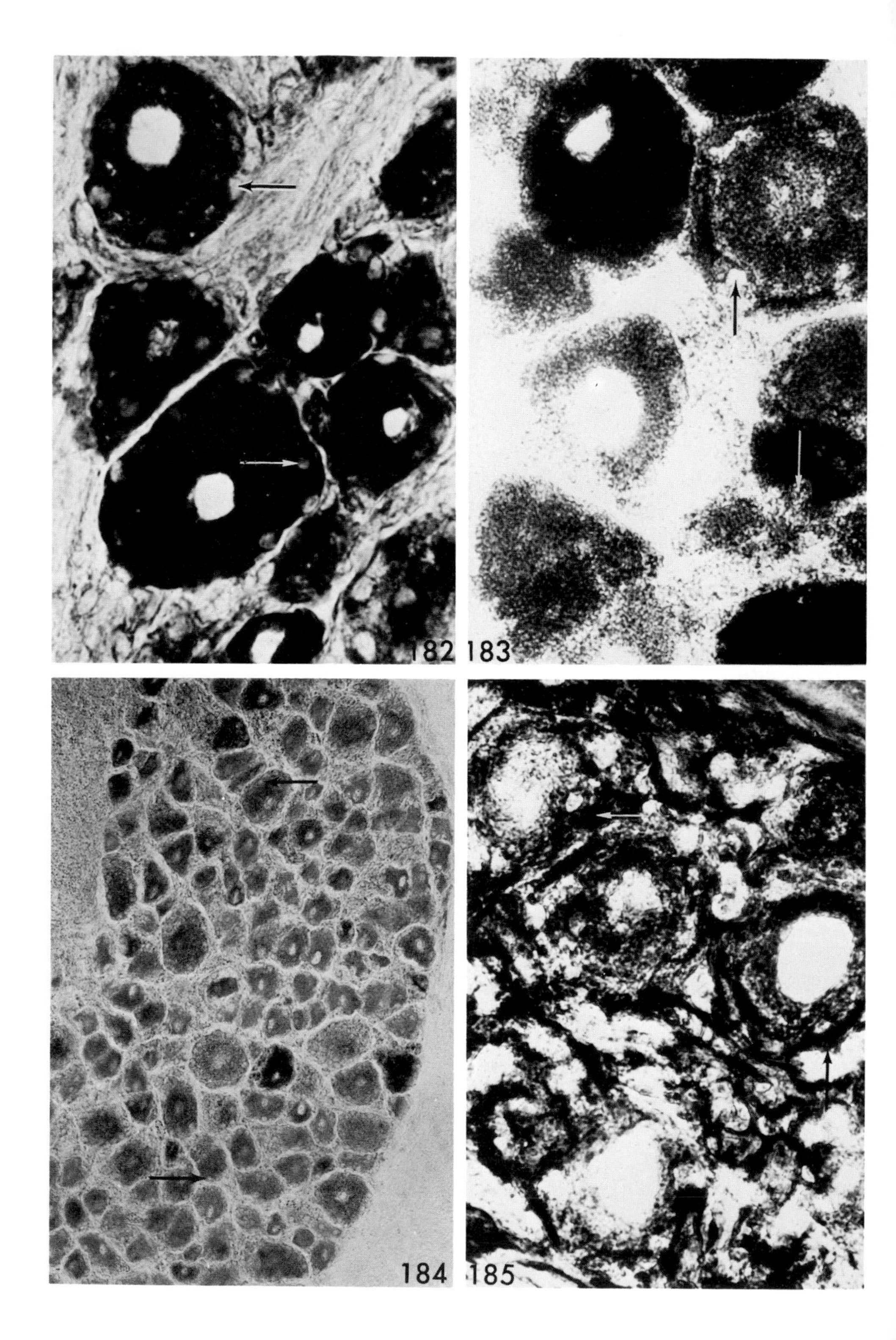
182
183
184
185

FIGURES 182–185. Dorsal root ganglion cells from the cervical region showing LDH, SE, SDH, and ATPase activity in these neurons. Note that the intensity of enzyme reaction varies from cell to cell with no relation to cell size. The satellite cells (arrows) surrounding the ganglion cells also show positive enzyme reactions in all the preparations. ×592, ×592, ×148, ×592.

The axons in the ganglion are strong in LDH, moderate in SDH, and mild in MAO. The myelin sheath is moderate in MAO and negligible in LDH and SDH preparations; no radiating bands of enzyme activity have been observed as reported by Tewari and Bourne (1962c). The Schwann cells are moderate in LDH and mild in SDH and MAO.

The activity of MAO in the ganglion cells ranges from mild to moderate and varies in individual neurons, irrespective of their size and shape. Similar observations have been made in the squirrel monkey, in rodents (Shantha *et al.*, 1967), and also in sympathetic and intestinal parasympathetic ganglia (Shimizu *et al.*, 1959; Smith, 1963). Glenner *et al.* (1957), however, showed that the nerve cells are unstained except for a faint reaction in the posterior root ganglion. A number of valuable studies have been made, not only on monoamine oxidase, but also on various other amines and cholinesterase in order to ascertain the adrenergic or cholinergic nature of the ganglion cells. The dorsal root ganglion is equipped with enzymes involved in destruction of catecholamines. Extensive studies regarding the adrenergic nature of the ganglion have been made on the sympathetic ganglion. For example, Norberg *et al.* (1966) reported the small sympathetic ganglion cells which store catecholamines probably belonging to a primitive catecholamine-storing cell system. The presence of significant amounts of noradrenaline in the sympathetic ganglion has also been reported by using a fluorescent method (Costa *et al.*, 1961; Kirpekar *et al*, 1962). Jacobowitz and Woodward (1968), studying the superior cervical ganglion of the cat, showed the presence of extensive adrenergic nerve terminals close to cell bodies. All these studies and that of Holmstedt *et al.* (1963) indicate that the sympathetic system is composed of predominantly adrenergic terminals with a few cholinergic ones. Decrease in MAO content in these MAO-containing neurons usually results after nerve transection.

ESTERASES

The distribution and pattern of intensity of cholinesterase are quite variable in different ganglion cells. The AChE activity is strong in the cytoplasm of some neurons, moderate in others, and mild or negligible in a few cells (Fig. 186). This compares with a strong and very strong enzyme activity in the SE preparations (Fig. 183), in which the enzyme reaction is uniformly distributed all over the cytoplasm and is diffuse as well as granular in character. The AChE activity, however, tends to accumulate more in the peripheral part of the cytoplasm, although in a number of neurons, mild to moderate reaction is observed in the main body of the cell as well. The neuropil in the SE and AChE is positive, but shows stronger reaction in the latter. The nucleus and nucleolus are negative in both enzymes. The satellite cells are strongly positive in SE, whereas they are moderate in AChE. A similar AChE reaction is given by the nerve fibers with a negative reaction in myelin sheath and Schwann cells. In SE, however, the axons are moderately reactive, and the Schwann cells and myelin sheath show less activity.

The cholinesterases have been extensively studied in the stellate, mesenteric, coeliac, sympathetic, dorsal root, and trigeminal ganglia in a variety of animals such as the rat, cat, mouse, guinea pig, hedgehog, chicken, frog, squirrel monkey, man and others. Klingman *et al.* (1968) recently made a detailed study of the cholinesterase content of the sympathetic ganglia of rats and observed that, of the total enzyme activity, 55–63% was due to AChE and 31–39% due to pseudocholinesterase. The superior cervical, stellate, and superior mesenteric ganglia contained greater amounts of specific and nonspecific cholinesterase than did the thoracic chain, coeliac, and cardiac (abdominal) ganglia.

As observed in this study on the spinal ganglion cells, most previous workers have described variable ganglionic AChE content at a particular time (Koelle, 1951, 1954, 1955, 1963; Koelle and Koelle, 1959; Gerebtzoff, 1959; Chacko and Cerf, 1960; Cauna *et al.*, 1961; Tewari and Bourne, 1963a; Shantha *et al.*, 1967). There is some difference of opinion on the degree of variability in the AChE content in the different cells and what it represents. Giacobini (1956, 1959) showed two different groups of ganglion cells, one having relatively higher AChE activity compared to others, irrespective of cell diameter, cell surface, and cell volume. Koelle (1955) also demonstrated moderate and variable intensity in the ganglion cells of rodents and rhesus monkey. We believe that this variable enzyme activity represents different functional states of the neurons compared to each other. Similarly, Cauna and Naik (1963), after studying the sensory nerve cells, suggested that variation in AChE content may signify the existence of several types of sensory ganglia, and postulated that this enzyme activity is related more to cell composition than secretory activity. Sjöqvist (1963) linked sympathetic ganglion cells with strong AChE activity to peripheral cholinergic function, i.e., sweat secretion. Eränkö (1967) pointed out that the AChE of the cytoplasm of these ganglion cells reflects their choline optive rather than cholinergic nature.

Tewari and Bourne (1963a) in AChE studies of the trigeminal ganglion classified the neurons in this ganglion into (a) cells with positive cytoplasm, which synthesize AChE in the cytoplasm, (b) cells with peripheral enzyme localization, which may be due to AChE activity in the capsular region or cell wall, and (c) cells totally devoid of AChE activity. It is possible that the cells belonging to the last category may have such low enzyme activity that the presently available techniques are not sensitive enough to detect it. The first category includes those which synthesize AChE in the cytoplasm and is associated with the endoplasmic reticulum, as originally suggested by Koelle and Steiner (1956) and Koelle (1959). The second category includes those cells which show AChE activity on the cell wall or the capsular region, which may be the areas of synaptic contacts receiving cholinergic fibers. Gerebtzoff (1959) remarked that most sympathetic ganglia are surrounded by synaptic rings rich in specific cholinesterase while cells presenting these structures are a minority of the spinal ganglion cells. In the latter he observed a few cells showing the AChE activity just inside the "pericellular capsule" in the dog.

The electron microscopic studies of Novikoff *et al.* (1966) and Novikoff (1967) in the dorsal root ganglia have shown AChE localization mainly in the endoplasmic reticulum (ER), with highest levels of this enzyme in smaller neurons. This indicates that the ER is probably involved in the synthesis of AChE, as previously suggested by Koelle and Steiner (1956) and Koelle (1959). This AChE from the ER is transferred through the canalicular system through the perikaryon and along the length of the axon and dendritic termination. Thus "the reserve cholinesterase" becomes the functional cholinesterase (Fukuda and Koelle, 1959; Koelle, 1962). Snell (1957) also showed that the nerve fibrils of the proximal stump in regenerating nerves contain a large amount of AChE. Cauna and Naik (1963) believed that the cytoplasmic AChE not only provides a reserve enzyme for nerve endings but also may participate in propagation of impulses in which AChE is involved. It is, however, well known that the AChE functions to remove the released ACh and to restore the nervous mechanism for repeated activity (Eccles, 1957).

Hamberger *et al.* (1963), using a fluorescent method, showed intense catecholamine fluorescence in the majority of the cat sympathetic ganglion cells, irrespective of weak, moderate, or strong AChE activity, indicating both the adrenergic and cholinergic nature of these cells. Burn and Rand (1959, 1965) believed that the release of norepinephrine from

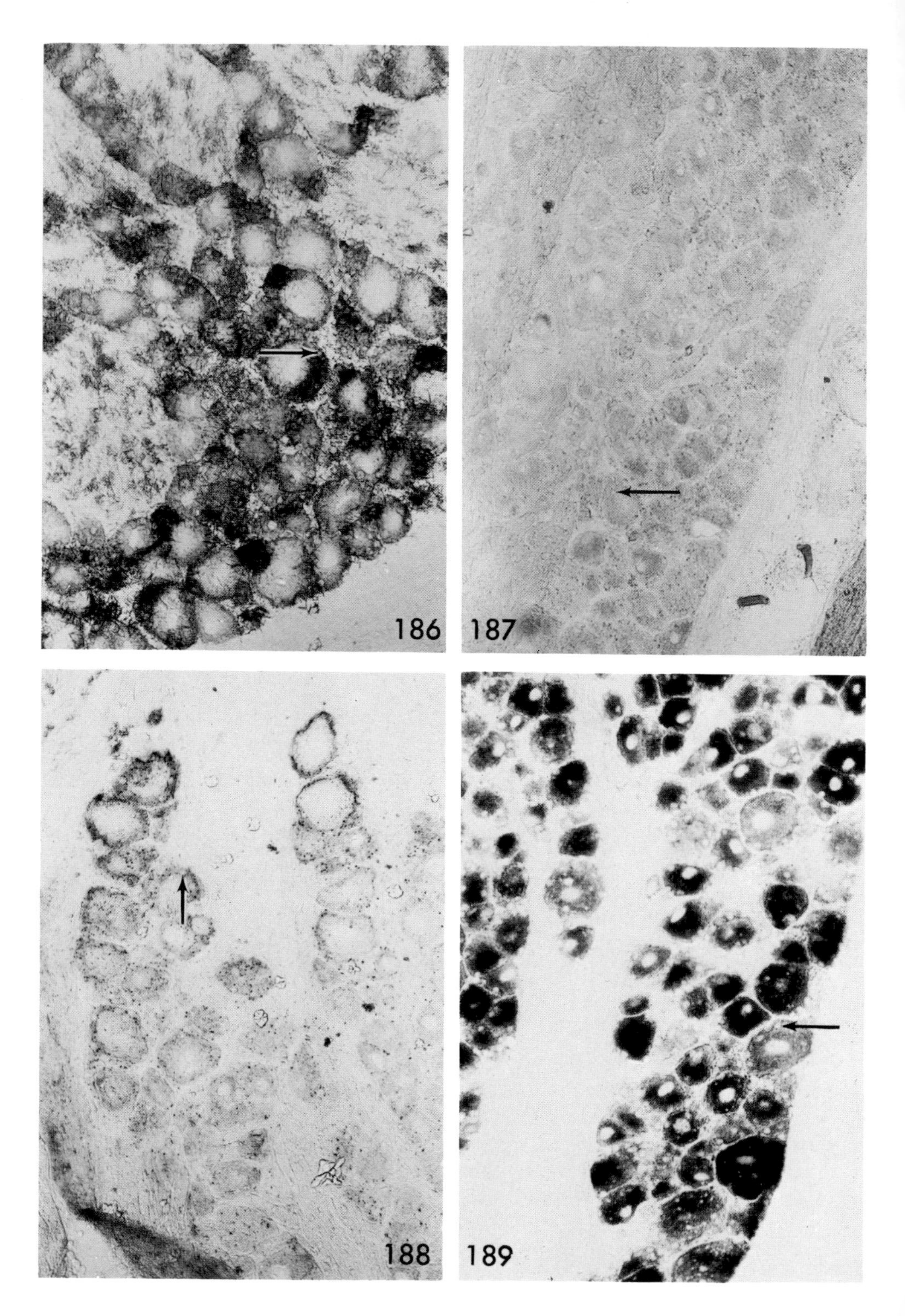
186
187
188
189

FIGURES 186–189. Dorsal root ganglion cells from the cervical region showing AChE, MAO, AK, and AC activity. The variation in these enzymes from cell to cell and enzyme to enzyme is evident. In the AChE preparation the enzyme activity is found at the periphery of the ganglion cells as well as in the satellite cells (arrows), whereas in AK, the activity appears to be mainly localized in the satellite cells with negligible to mild activity in the ganglion cells. The satellite cells are negative in the MAO and mild to moderate in the AC preparations. ×148, ×148, ×148, ×148.

adrenergic fibers is mediated by acetylcholine. Furthermore, BChE has been shown by Härkönen (1964) to be inversely proportional to the catecholamine fluorescence in the rat superior cervical ganglion. The final decision as to whether a neuron or nerve terminal is cholinergic or adrenergic in nature is hard to say. As Eränkö (1967) remarked, "Final solution of the problems can be expected only when acetylcholine and catecholamines can be demonstrated at ultrastructural level in a single synapse." In view of the rapid development of electron histochemical techniques, the solution may be found in the not too distant future.

The AChE content of the developing chick embryo has shown that the AChE activity appears at an early stage in the ganglion, increasing in amount up to hatching (with the maturation of cells); thereafter it alters, mostly falling off or disappearing, and with full development only a few cells show intense staining (Gerebtzoff, 1959; Hamburger, 1962; Strumia and Baima–Ballone, 1962). These workers believed that the AChE activity is closely related to the neurofibrillary differentiation of these cells.

PHOSPHATASES

The azo dye methods for the demonstration of acid phosphatase were preferred to the lead nitrate method because in the latter some unspecific black precipitate is constantly found when the tests are repeated. The nucleoli give some reaction in the Gomori method not given by the azo dye techniques. Some earlier reports have also cautioned about the use of lead because of the affinity of nucleoli for lead ions which may result in a false reaction product. Deane (1963) believed that even if an enzyme were present in the nuclei, an extensive proof of the validity of the reaction should be undertaken if the metallic ions used were lead or the product of a metal salt. For alkaline phosphatase, observations have been taken from both the calcium cobalt and the azo dye methods.

The alkaline phosphatase reaction is prominent in the neuropil, cell membrane, peripheral part of the cytoplasm, and the satellite cells (Fig. 188). The reaction is granular in nature. The blood vessels react positively. The nuclei and nucleoli of the cells are negative, as are the axons and the myelin. In the acid phosphatase preparations, the nucleus is negative, whereas the nucleolus, axons, and the Schwann cells are mildly positive (Fig. 189). The satellite cells give a moderate reaction. The cytoplasm shows a granular type of reaction and varies from moderate in some cells to strong in others, and very strong in the rest. The AC activity in the cytoplasm is particularly prominent in those areas which show concentration of lipofuscin pigments. In the spinal ganglion cells of the rat (Colmant, 1959; Tewari and Bourne, 1964) and bat (Thakar and Tewari, 1967), two types of cell populations with light or dark AC-positive activity have been reported. Recently, Kalina and Bubis (1968) observed in a light and electron microscopic study of the distribution of acid phosphatase that the enzyme activity in the small neurons of trigeminal and spinal ganglia appeared as granules, whereas in the larger ones the enzyme activity was in the form of a network of filaments in scattered granules (Golgi apparatus in association with the rough endoplasmic reticulum). These authors suggested the presence of at least two acid phosphatases based on their differences in sensitivity in the small and large cells toward fixatives and inhibitors. It is obvious, therefore, that such differences in the distribution of AC activity in the different neurons are of great functional significance. Thakar and Tewari (1967) believed that the cells showing high acid phosphatase activity represent those cells in which active protein synthesis is going on at a high rate. On the contrary, we think that high AC activity, especially in neurons, reflects high lysosomal and lipofuscin pigment content in the cytoplasm.

In a histochemical study of the distribution of molybdate-activated acid phosphatase, Franks *et al.* (1967) showed its presence in the satellite cells of some nerve ganglia. Norberg *et al.* (1966) observed that the satellite cells show stronger α-glycerophosphate activity than the neurons.

The ATPase activity in the dorsal root ganglion cells is stronger in the neuropil than in the neurons (Fig. 185). In the latter, the cytoplasm gives a moderate and moderately strong reaction, and in some neurons the enzyme activity is concentrated at the periphery of the nucleus. The nuclei of these cells are negative, whereas the nucleoli are moderately positive. The position of the nucleolus may vary in the nucleus but does not seem to show any relationship to a particular type of enzyme activity in the cytoplasm, as discussed below. The satellite cells give a strong reaction in the cytoplasm, mild in the nucleus, and moderate in the nucleolus. The nuclear activity may be a diffusion artifact.

As discussed earlier, most of the cells in a ganglion show varying reactions at any given time, and this phenomenon, in our opinion, reflects the functional state of the cell at that time. The nuclear wall and cell membrane also show ATPase activity. Tewari and Bourne (1963b), however, believed that ATPase reaction in the cytoplasm depended on the position of the nucleolus in the nucleus and they noted that: "The ATPase metabolic cycles start near the nuclear membrane and pass as waves towards the periphery of the cytoplasm. There is even evidence that the enzyme passes across the cell membrane." The present studies on the dorsal root ganglion cells of the rhesus monkey fail to confirm the above observations. Kumamoto and Bourne (1963) in their studies of dorsal root ganglion cells of the rat did not report any enzymatic cycles as described by Tewari and Bourne. The ATPase activity is not revealed in areas where there is pigment accumulation. Such areas, however, show a large amount of β-glucuronidase activity (Tewari and Bourne, 1962b). Since ATPase is concerned with the transport of sodium ions across the cell membrane (Jarnefelt, 1961), the localization of ATPase at the nuclear wall may play a significant role in active transport of various materials between the nuclear and cytoplasmic areas.

The AK reaction is mainly localized near the cell surface in the neurons and in the satellite cells. This observation conforms to the view of a number of workers (Emmell, 1946; Tewari and Bourne, 1964; Thakar and Tewari, 1967) that AK near the cell surfaces is associated with the phenomenon of permeability and plays an important role in the exchange of ions, as well as Hokin and Hokin's (1960) suggestion that the phosphatidic cycle is primarily concerned with the exchange of sodium ions.

The distribution of acid phosphatase, unlike that of AK, is mainly localized in the cytoplasm and is believed to be associated with the lysosomes. In a study of histochemical alterations in autonomic ganglion cells associated with aging, Sulkin and Kuntz (1952) observed that the AC activity in the neuroglial elements in the ganglia of dogs greatly diminishes during senility. In dogs under 30 days of age the ascorbic acid content is very low. It increases during maturity and diminishes during senility. It may be interesting to add that the thiamine pyrophosphatase (TPPase)-positive Golgi apparatus is absent or is present in very small quantities in the areas of lipofuscin accumulation (Shantha *et al.*, 1967).

The amount and morphological form of Golgi apparatus varies in the different neurons belonging to the same ganglion as well as in the same section. Such a conspicuous variation in the form of Golgi apparatus in the ganglion cells of the rat and the squirrel monkey has been observed by Shanthaveerappa and Bourne (1965), Shantha and Bourne (1966a, 1966b) and Shantha *et al.* (1967). They have described the Golgi apparatus of the spinal, nodosal, trigeminal, and sympathetic ganglion cells under four different categories based on their morphological shape:

1. Cells containing a TPPase-positive Golgi network formed by darkly stained vesicular enlargements interconnected by lightly stained thin strands; separate vesicles and granules can also be observed embedded in the network.

2. Cells containing medium and small-sized vesicles and granules.

3. Cells containing large and small, oval or rounded masses with small scattered granules and vesicles.

4. Cells containing a mixture of vesicles, granules, comma-shaped masses, cylindrical units, dumbbell-shaped units, ring-shaped units, saucer-shaped units, etc.

It is common to see a darkly stained cell lying by the side of a cell which contains lightly stained, morphologically different types of TPPase-positive material. The blood vessels and the nucleoli of neurons have been diffusely stained. The satellite cells show the Golgi material in the form of small granules located at one pole of the cytoplasm. Shantha *et al.* (1967) believed that those neurons which have greater amounts of Golgi apparatus are probably those which contain fewer mitochondria and thus have smaller amounts of oxidative enzymes; however, those neurons which have less Golgi apparatus may have more mitochondria and may be those showing a stronger reaction of oxidative enzymes. Since the Golgi apparatus and lysosomes are intimately related to each other, the degree of positive activity for various lysosomal enzymes, including acid phosphatase, is directly proportional to the amount of Golgi material present.

REFERENCES

Burn, J. H., and Rand, M. J. Sympathetic post ganglionic mechanism. *Nature* **184**:163–175 (1959).

Burn, J. H., and Rand, M. J. Acetylcholine in adrenergic transmission. *Ann. Rev. Pharmacol.* **5**:163–182 (1965).

Cauna, N., and Naik, N. T. The distribution of cholinesterases in the sensory ganglia of man and some mammals. *J. Histochem. Cytochem.* **3**:129–138 (1963).

Cauna, N., Naik, N. T., Leaming, D. B., and Alberti, P. The distribution of cholinesterases in the autonomic ganglia of man and some mammals. *Bibl. Anat.* (*Basel*) **2**:90–96 (1961).

Chacko, L. W., and Cerf, J. A. Histochemical localization of cholinesterase in the amphibian spinal cord and alterations following ventral root section. *J. Anat.* (*London*) **94**:74–81 (1960).

Colmant, H. J. Activitätsschwankungen der sauren Phosphatase im Rückenmark und den spinal Ganglien der Ratte nach Durchschneidung des Nervus ischiadicus. *Arch. Psychiat. Nervenkr.* **199**:60–71 (1959).

Costa, E., Revzin, A. M., Kuntzman, R., Spector, S., and Brodie, B. B. Role for ganglionic norepinephrine in sympathetic synaptic transmission. *Science* **133**:1822–1823 (1961).

Deane, H. W. Nuclear location of phosphatase activity: fact or artifact? *J. Histochem. Cytochem.* **11**:443–444 (1963).

Eccles, J. C. *The Physiology of Nerve Cells.* Oxford Press, London, 1957.

Emmell, V. M. The intracellular distribution of alkaline phosphatase activity following various methods of histologic fixation. *Anat. Rec.* **95**:159–175 (1946).

Eränkö, O. Histochemistry of nervous tissues: catecholamines and cholinesterases. *Am. Rev. Pharmacol.* **7**:203–222 (1967).

Franks, L. M., Maggi, V., and Carbonell, A. W. A molybdate-activated phosphatase in arterio venous anastomoses and in nerve ganglia satellite cells *J. Anat. (London)* **101**:777–782 (1967).

Fukuda, T., and Koelle, G. B. The cytological localization of intracellular neuronal acetylcholinesterase. *J. Biophys. Biochem. Cytol.* **5**:433–440 (1959).

Gerebtzoff, M. A. *Cholinesterases.* International Series of Monographs on Pure and Applied Biology (Division: Modern Trends in Physiological Sciences), (P. Alexander and Z. M. Bacq, eds.). Pergamon Press, New York, 1959.

Gerebtzoff, M. A., and Brotchi, J. Localisations et activités d'enzymes oxydoréducteurs dans le nerf et le ganglion rachidien. *Ann. Histochim.* **11**:63–69, (1966).

Giacobini, E. Histochemical demonstration of AChE activity in isolated nerve cells. *Acta Physiol. Scand.* **36**:276–290 (1956).

Giacobini, E. Quantitative determination of cholinesterase in individual spinal ganglion cells. *Acta Physiol. Scand.* **45**:238–254 (1959).

Glenner, G. G., Burtner, H. J., and Brown, G. W., Jr. The histochemical demonstration of monoamine oxidase activity by tetrazolium salts. *J. Histochem. Cytochem.* **5**:591–600 (1957).

Hamberger, B., Norberg, K. A., and Sjöqvist, F. Cellular localization of monoamines in sympathetic ganglia of the cat: A preliminary report. *Life Sci.* **2**:659–661 (1963).

Hamburger, V. Experimental analysis of the development of the trigeminus ganglion in the chick embryo. *J. Cell Comp. Physiol.* **60**: Suppl. I, 81–92 (1962).

Härkönen, M. Carboxylic esterases, oxidative enzymes and catecholamines in the superior cervical ganglion of the rat and the effect of pre- and post ganglionic nerve division. *Acta Physiol. Scand.* **63**: Suppl. 237 (1964).

Hokin, L. E., and Hokin, M. R. The role of phosphotidic acid and phosporinositide in transmembrane transport elicited by acetylcholine and other humoral agents. *Intern. Rev. Neurobiol.* **2**:99–136 (1960).

Holmstedt, B., Lundgren, G., and Sjöqvist, F. Determination of acetylcholinesterase activity in normal and denervated sympathetic ganglia of the cat. A biochemical comparison. *Acta Physiol. Scand.* **57**:235–247 (1963).

Jacobowitz, D., and Woodward, J. K. Adrenergic neurons in the cat superior cervical ganglion and cervical sympathetic nerve trunk: A histochemical study. *J. Pharm. Exptl. Therap.* **162**:213–227 (1968).

Jarnefelt, J. Mechanism of sodium transport in cellular membranes. *Nature* **190**:694–697 (1961).

Kalina, M., and Bubis, J. J. Histochemical studies on the distribution of acid phosphatases in neurones of sensory ganglia; light and electron microscopy. *Histochemie* **14**:103–113 (1968).

Kirpekar, S. M., Cervoni, P., and Furchgott, R. F. Catecholamine content of the cat nictitating membrane following procedures sensitizing it to norepinephrine. *J. Pharmacol. Exptl. Therap.* **136**:180–190 (1962).

Klingman, G., Klingman, J., and Foliszczuk, A. Acetyl- and pseudocholinesterase activities in sympathetic ganglia of rats. *J. Neurochem.* **15**:1121–1131 (1968).

Koelle, G. B. The elimination of enzymatic diffusion artifacts in the histochemical localization of cholinesterases and a survey of their cellular distributions. *J. Pharmacol. Exptl. Therap.* **103**:153–171 (1951).

Koelle, G. B. The histochemical localization of cholinesterases in the central nervous system of the rat. *J. Comp. Neurol.* **100**:211–228 (1954).

Koelle, G. B. The histochemical identification of acetylcholinesterases in cholinergic, adrenergic and sensory neurons. *J. Pharmacol. Exptl. Therap.* **114**:167–184 (1955).

Koelle, G. B. The histochemical localization of the cholinesterase in the central nervous system of the rat. *J. Comp. Neurol.* **100**:211–255 (1959).

Koelle, G. B. A new general concept of the neurohumoral functions of acetylcholine and acetylcholinesterase. *J. Pharm. Pharmacol.* **14**:65–90 (1962).

Koelle, G. B. (ed.) Cytological distribution and physiological functions of cholinesterases. In: *Handbuch der Experimente Pharmakologie,* pp. 187–298. Springer-Verlag, Berlin, 1963.

Koelle, W. A., and Koelle, G. B. The localization of external or functional acetylcholinesterase at the synapses of autonomic ganglia. *Exptl. Ther.* **126**:1–8 (1959).

Koelle, G. B., and Steiner, E. C. The cerebral distributions of a tertiary and a quaternary anticholinesterase agent following intravenous and intraventricular injection. *J. Pharmacol. Exptl. Therap.* **118**:420–434 (1956).

Kumamoto, T., and Bourne, G. H. Experimental studies on the oxidative enzymes in spinal neurons. I. A histochemical and biochemical investigation of the spinal ganglion and the spinal cord following sciatic neurotomy in the guinea pig. *Acta Anat.* **55**:255–277 (1963).

Marinesco, C. L. Recherches histo-chimiques sur le rôle des ferments oxydants dans les phénomènes de la vie à l'état normal et pathologique. *Ann. Anat. Path. Medico.* **1**:121–162 (1929).

Norberg, K. A., Ritzen, M., and Ungerstedt, U. Histochemical studies on a special catecholamine-containing cell type in sympathetic ganglia. *Acta Physiol. Scand.* **67**:260–270 (1966).

Novikoff, A. B. Enzyme localization and ultrastructure of neurons. In: *The Neuron* (H. Hydén, ed.), pp. 255–318. Elsevier, Amsterdam, 1967.

Novikoff, A. B., Quintana, N., Villaverde, H., and Forschirm, R. Nucleoside phosphatase and cholinesterase activities in dorsal root ganglia and peripheral nerve. *J. Cell. Biol.* **29**:525–545 (1966).

Robins, E. The chemical composition of central tracts and of nerve cell bodies. *J. Histochem. Cytochem.* **8**:431–436 (1960).

Samorajski, T. The application of diphosphopyridine nucleotide diaphorase methods in a study of dorsal ganglia and spinal cord. *J. Neurochem.* **5**:349–353 (1960).

Shantha, T. R., and Bourne, G. H. Thiamine pyrophosphatase technique as an indicator of morphology of the Golgi apparatus in the neurons. IV. Studies on spinal cord, hippocampus and trigeminal ganglion. *Ann. Histochem.* **11**:337–351 (1966a).

Shantha, T. R., and Bourne, G. H. Thiamine pyrophosphatase technique as an indicator of morphology of the Golgi apparatus in the neurons. V. Studies on the sympathetic ganglion cells. *Cytologia* **31**:132–143 (1966b).

Shantha, T. R., Manocha, S. L., and Bourne, G. H. Enzyme histochemistry of the mesenteric and dorsal root ganglion cells of cat and squirrel monkey. *Histochemie* **10**:234–245 (1967).

Shanthaveerappa, T. R., and Bourne, G. H. The thiamine pyrophosphate technique as an indicator of the morphology of the Golgi apparatus in the neurons. *Acta Histochem.* **22**:155–178 (1965).

Sharma, N. N. Studies on the histochemical distribution of oxidative enzymes in sympathetic ganglion cells of rat. *Acta Anat.* **68**:416–432 (1967).

Shimizu, N., Morikawa, N., and Okada, M. Histochemical studies of monoamine oxidase of the brain of rodents. *Z. Zellforsch.* **49**:389–400 (1959).

Sjöqvist, F. The correlation between the occurrence and localization of acetylcholinesterase rich cell bodies in the stellate ganglion and the outflow of cholinergic sweat secretory fibers to the fore paw of the cat. *Acta Physiol. Scand.* **57**:339–351 (1963).

Smith, B. Monoamine oxidase in the pineal gland, neurohypophysis, and brain of the albino rat. *J. Anat.* (*London*) **97**:81–86 (1963).

Snell, R. S. Histochemical appearance of cholinesterase in the normal sciatic nerve and the changes which occur after nerve section. *Nature* **180**:378–379 (1957).

Sotelo, C. Histo-enzymological studies of the metabolism of glucides in the neuroglia. II. Phosphorylase and oxido-reduction enzymes. *Arch. Anat. Micr. Moph. Exp.* **55**:571–602 (1966).

Strumia, E., and Baima-Ballone, P. L. Accrescimimento ed attivata acetilcolinesterasica nei neuroni dei gangli spinali, nel pollo. *Monit. Zool. It.* **70**: Suppl. 94 (1962).

Sulkin, N. M., and Kuntz, A. Histochemical alterations in autonomic ganglion cells associated with aging. *J. Gerontol.* **7**:533–543 (1952).

Tewari, H. B., and Bourne, G. H. The morphological and chemical identity of the intracellular organelles and inclusions in the spinal ganglion cells of the rat. *La Cellule* **63**:25–50 (1962a).

Tewari, H. B., and Bourne, G. H. The histochemistry of the nucleus and nucleolus with reference to nucleo-cytoplasmic relations in the spinal ganglion neuron of the rat. *Acta Histochem.* **13**:323–350 (1962b).

Tewari, H. B., and Bourne, G. H. Histochemical evidence of metabolic cycles in spinal ganglion cells of rat. *J. Histochem. Cytochem.* **10**:42–64 (1962c).

Tewari, H. B., and Bourne, G. H. Histochemical studies on the distribution of simple esterase, specific and non-specific cholinesterase in trigeminal ganglion cells of rat. *Acta Anat.* **53**:319–332 (1963a).

Tewari, H. B., and Bourne, G. H. Histochemical studies on the distribution of adenosine triphosphatase in the trigeminal ganglion cells of the rat. *J. Histochem. Cytochem.* **11**:511–519 (1963b).

Tewari, H. B., and Bourne, G. H. Histochemical studies on the distribution of alkaline and acid phosphatases and 5-nucleotidase in the trigeminal ganglion cells of rat. *Acta Histochem.* **17**:197–207 (1964).

Thakar, D. S., and Tewari, H. B. Histochemical studies on the distribution of alkaline and acid phosphatases amongst the neurons of the cerebellum, spinal cord, and trigeminal ganglia of bat. *Acta Histochem.* **28**:359–367 (1967).

Thomas, E. Dehydrogenasen und Esterasen in unveranderten und geschadigten spinal Ganglienzellen vom Menschen. *Acta Neuropathol.* **2**:231–245 (1963).

Thomas, E., and Pearse, A. G. E. The fine localization of dehydrogenases in the nervous system. *Histochemie* **2**:266–282 (1961).

IX

CEREBELLUM

The structure and subdivisions of the rhesus monkey cerebellum are similar to those of other primates, including the chimpanzee and man (Shantha and Manocha, 1969a, 1969b). The cerebellar cortex is made up of white matter (medullary layer) containing numerous myelinated nerve fibers and a thin layer of gray matter covering the medullary layer. This gray layer is made up of three cytoarchitectonically different layers, namely (from top to bottom) the molecular layer, Purkinje layer, and granule cell layer.

The molecular layer is the outermost layer of the cerebellar cortex and consists of only a few small round or oval neurons (Figs. 190–193). The cells in this superficial layer are smaller in size than cells in the deeper layers and are called stellate and basket cells. The rest of the molecular layer is made up of nerve fibers and glial cells.

The Purkinje cells are large flask-shaped neurons with thick dendrites at the apex and an axon originating from the base. The base of the Purkinje cell is covered by basket fibers. As described by Fox and Barnard (1957), as many as 270,000 axons from other neurons may pass through the dendrite tree of one Purkinje cell. Similarly, one large motor neuron may have around 10,000 synaptic knobs on its surface (Hydén, 1967). In between the Purkinje cells are numerous glial cells which have large vesicular nuclei and belong to the category of Cajal–Bargmann glial cells.

The granule layer is formed mainly of compactly arranged granule cells, which have large nuclei, very small nucleoli, and thin rims of cytoplasm. Interposed between the masses of granule cells are the irregularly shaped cerebellar glomeruli. These glomeruli are the synaptic areas between the mossy fibers and granule cell processes. In Nissl stained preparations, these areas look blank and unstained. In addition to these elements, this layer also contains neuroglial cells, Golgi type II neurons and occasional displaced Purkinje cells.

Because of the difference in cytoarchitectonics, four layers of the cerebellum stand out very well in all the histochemical preparations (Figs. 190–193). The neuropil in the molecular layer exhibits moderate SE, SDH, LDH, and ATPase; negligible AChE; mild AC and MAO; and negative AK activity. The stellate cells and basket cells give moderately strong SE; moderate LDH; mild SDH, ATPase, and MAO; negligible AC; and negative AK and AChE reactions. The basket cells show much more enzyme activity than the stellate cells.

The Purkinje cells show very strong SE and LDH, strong AC and SDH, moderate ATPase and MAO, and negative AChE and AK activity. The degree of positive activity varies from cell to cell in the various enzyme preparations. For example, in SDH preparations some cells are more positive than the others, and there is no uniform positive activity in any one preparation. This clearly demonstrates that the enzyme equipment of the individual Purkinje cells varies greatly. The basket formation around the Purkinje cells shows strong SDH and LDH, moderate ATPase, mild SE and MAO, negligible AC, and negative AK and AChE activity. The positive activity in most of these preparations extends to the dendritic processes for some distance but not into the axons. The Bargmann's glial cells show strong LDH; moderate SDH, AC, SE, and ATPase; mild MAO; and negative AK and AChE activity.

The granule cells show moderately strong LDH and SE, moderate SDH and MAO, mild AC and ATPase, and negative AChE and AK activity. It is interesting to note that the positive activity is in the form of fine granules restricted to the periphery of the cell cytoplasm. The cerebellar glomeruli exhibit very strong LDH; strong SDH; moderate AC, SE, ATPase, AChE, and MAO; and negative AK activity. The glomeruli also show diffuse as well as granular activity, especially for oxidative enzymes.

The nerve fiber layer shows moderately strong LDH and SE, moderate ATPase and MAO, mild AChE, negligible AC and SDH, and negative AK activity. The glial cells in this layer as well as in the molecular and granule cell layers show strong LDH; moderate ATPase; mild SDH, AC and MAO; mild to moderate SE; and negative AK and AChE activity. The blood vessels in the granule cell layers (but not in other layers) show AChE activity. On the other hand, AK-positive blood vessels are found in all the layers, with largest numbers in the granule cell layers. The nerve fiber layer has the least number of blood vessels.

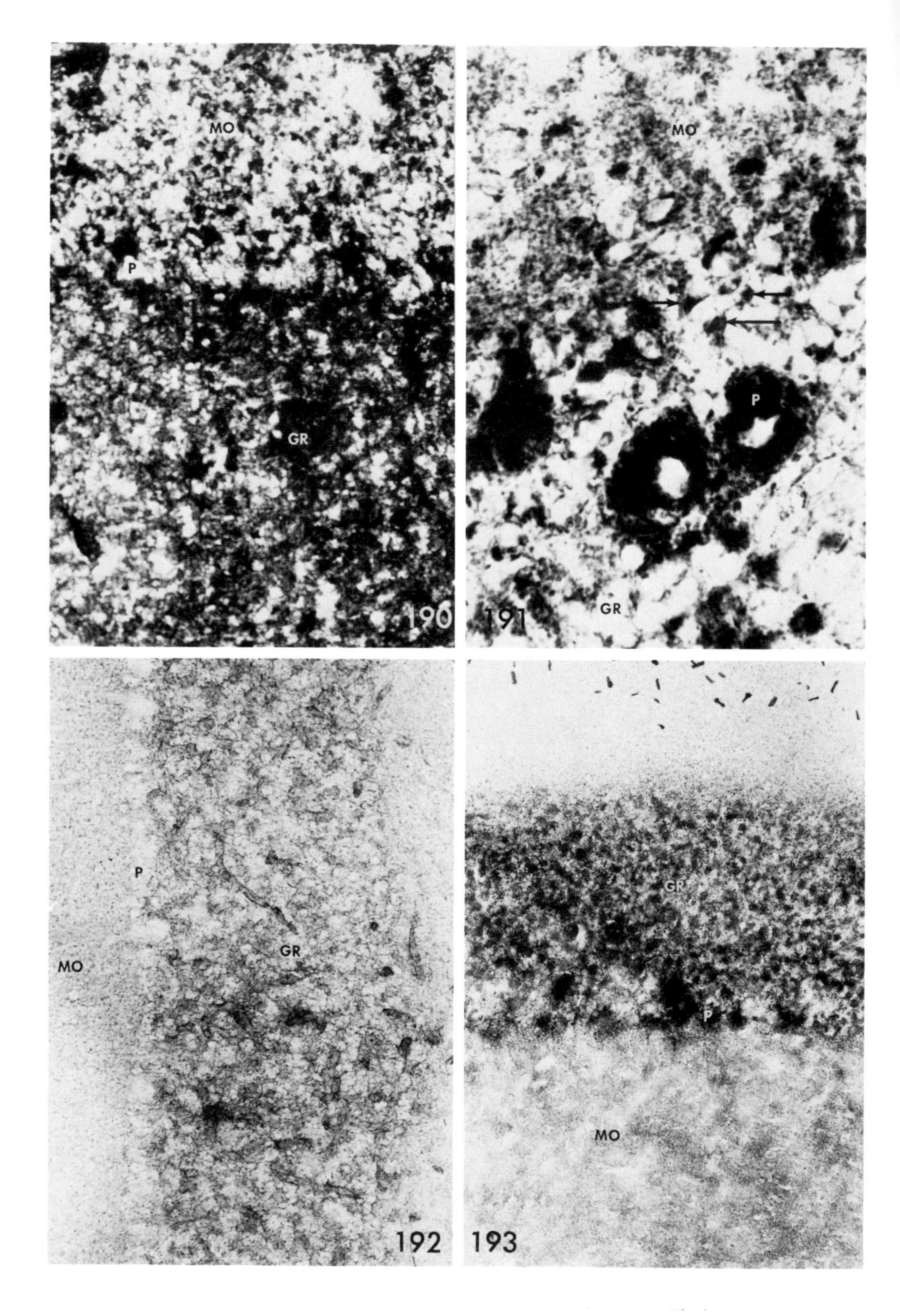
MO
P
GR
190
MO
P
GR
191
P
MO
GR
192
GR
P
MO
193

FIGURES 190–193. The cerebellar cortex showing ATPase, LDH, AChE, and AC activity, respectively, in the various layers. The stellate and basket cells of the molecular layer show negligible ATPase, moderate to moderately strong LDH, mild AC, and negative AChE activity. The Purkinje cells are mostly mild with some moderately strong activity in ATPase, strong to very strong activity in LDH and AC, and negative activity in AChE preparations. The Bargmann glial cells (arrows) show maximum activity in the LDH preparations. The glomeruli are strongly positive in LDH and moderately strong in AC, AChE, and ATPase preparations. The granule cells show comparatively stronger activity in the LDH preparations. ×148, ×592, ×148, ×148.

GENERAL DISCUSSION

A great quantity of published material is available on the biochemistry and histochemistry of the cerebellum in different species of animals. Arvy (1966) and Shantha *et al.* (1967a) have reviewed the pertinent histochemical literature on this subject. There appears to be great variation in the distribution of AChE in various components of the cerebellum of different animals. Shen *et al.* (1955) reported the presence of AChE mainly in the molecular layer and in some fiber components of the granule cell layer of the frog cerebellum. Kasa and Silver (1969) found differences in distribution of choline acetyltransferase and AChE in rat and guinea pig archi- and paleocerebellum. Koelle (1954) has noted its presence in the granule cell layer, nerve fiber layer and deep cerebellar nuclei of the rat, whereas Tewari and Bourne (1962b) reported its occurrence exclusively in the granule cell nuclei of the rat cerebellum. Gerebtzoff (1959) localized AChE in the molecular layer, glomeruli, and dendrites of granule cells of the rat, rabbit, and guinea pig. The results obtained in the present study on the rhesus as well as in the squirrel monkey (Shantha *et al.*, 1967a) are similar to the findings of Koelle and Gerebtzoff. It is interesting that Joo *et al.* (1965), by using a modified Koelle and Friedenwald method, showed the presence of AChE in the Purkinje cells of the kitten for the first time; this may very well be an artifact due to lead salt in the substrate, as it is well known that lead salts have an affinity for nervous tissue which results in coloration.

Csillik *et al.* (1963) and Friede and Fleming (1963) commented on the differences in the AChE activity among the different folia of the cerebellum. They found AChE reaction in the granular layer of the nodule, lower uvula, and flocculi compared to negative activity in the hemispheres. This is in contrast to the findings of Robins and Smith (1952), who believed that there was no significant difference between the vermis and the hemispheres. The species differences in the AChE content of the cerebellar cortex have been summarized by Friede and

Fleming as follows: no enzyme reaction in both layers (monkey, hamster), strong reaction in both layers (squirrel, guinea pig), strong reaction in molecular layer only (bird, man), and strong reaction in granular layer only (cat). According to Bennett *et al.* (1966), AChE activity in the cerebellum is very high in the dog, and very low in the rat. It is evident that the species difference in AChE content is more apparent in the cerebellum than in any other area of the central nervous system and needs further investigation.

AChE reaction is also prominent in the glomeruli of the granular layer containing mossy Golgi endings and in parallel axons in the molecular layer as well as in the basket synapses around and below the Purkinje cells (Gerebtzoff, 1959; Friede and Fleming, 1964; Austin and Phillis, 1965; Phillis, 1965; Kasa *et al.*, 1965). As explained by these authors, the presence of AChE in presynaptic position and in the excitatory and inhibitory synapses (Austin and Phillis, 1965) suggests that acetylcholine plays the role of pretransmitter or modulator substance, promoting the activity of the excitatory and inhibitory chemical transmitters. On the other hand, Curtis and Crawford (1965) and Crawford *et al.* (1966), using iontophoretic techniques, found no evidence for cholinergic synapses in the cat cerebellum, and they believed that AChE is not an indicator of cholinergic transmission.

The cerebellum, therefore, is a very controversial tissue with regard to cholinergic or noncholinergic transmission. Crawford *et al.* (1966) suggested that granule cells are cholinergic, whereas the Purkinje cells are cholinoceptive. In the latter the AChE may appear or disappear in the developing brain (Csillik *et al.*, 1964; Silver, 1967), and Crawford *et al.* (1966) suggested that the AChE activity in the Purkinje cells is not connected with the hydrolysis of acetylcholine. Kasa *et al.* (1966) believed that AChE is involved during development in the protein synthesis. Possibly the Purkinje cells are neither cholinergic nor cholinoceptive, but the fact that the granular layer stains even after deafferentation may mean that the Golgi cells are cholinergic.

The electron microscopic studies of Torack and Barrnett (1962) showed that AChE is present in some synapses and absent in others, and De Robertis *et al.* (1961) believed that there are more noncholinergic endings than cholinergic ones. The AChE in the cerebellum may have a nonsynaptic role, as has been indicated by a number of workers (Austin and Phillis, 1965; Curtis and Crawford, 1965; Crawford *et al.*, 1966). Such a role might be concerned with permeability and/or ion transport (Koelle, 1963). The recent studies of McCance and Phillis (1968) show that cholinergic synaptic transmission is evident in the cerebellar granule cell layer. Phillis (1968) described that the mossy afferent fibers and deep nuclear cells, and possibly many cells in the granule cell layer, parallel fibers, and association pathways as well, are cholinergic. Lewis *et al.* (1967) believed that experimental evidence has not disproved the association of heavy staining for AChE with cholinergic mechanisms. The interpretation of Friede and Fleming (1963) seems sound to us, that in the cerebellum there is a possible choice between the cholinergic and noncholinergic types of transmission and either type can be used by a given species to operate the same circuit. Phillis (1965) observed the persistence of AChE-staining cells, fibers, and synaptic glomerular areas in isolated areas of the cerebellar cortex of the cat and suggested that there may be cholinergic interneuronal circuits within this region.

Gerebtzoff (1959) proposed that in the cerebellum, transmission might be affected by very small quantities of acetylcholine, and Kasa *et al.* (1965) speculated that there might be present an amplifier mechanism for acetylcholine. Silver (1967) believed that "if the cerebellar receptors can be fired by minute quantities of ACh, either because they are especially sensitive or because the morphological arrangement of pre- and postsynaptic elements increases the effectiveness of ACh, then a high AChE activity might be a necessary safeguard against anomalous interactions between the tightly packed cells. It would be particularly important to prevent the spread of unhydrolyzed transmitter if, as suggested by Kasa and Csillik (1965), ACh could excite both excitatory and inhibitory cells. Although there is no experimental evidence to suggest that ACh is exceptionally effective in the cerebellum, this may itself be an indication of the protection afforded by AChE against iontophoretically applied ACh." For details the reader is referred to this review.

The present study on the distribution of MAO in the rhesus monkey and those done on rat and squirrel monkeys (Tewari and Bourne, 1963c; Shantha *et al.*, 1967a; Manocha and Bourne, 1968) indicate that this enzyme is present in all components of the cerebellum. Shimizu *et al.* (1959) showed the absence of MAO in Purkinje cells, faint activity in granule cells, and mild to moderate activity in the molecular layer of mouse cerebellum (Smith, 1963). The presence of both MAO and AChE in the cerebellar glomeruli and deep cerebellar nuclei may indicate that they are both cholinergic and adrenergic in function. Also, there is an inverse ratio between the distribution of AChE and MAO in various components of the cerebellum as observed in the olfactory bulb, spinal cord, and other parts of the nervous system (Shanthaveerappa and Bourne, 1965a, 1965b; Manocha *et al.*, 1967). Aprison *et al.* (1964) showed that in the pigeon cerebellum, the catabolic enzymes (AChE and MAO) are present in great excess over the synthetic enzymes (cholineacetylase and 5-hydroxytryptophan decarboxylase). The ratio of AChE/cholineacetylase in the cerebellum is particularly high. This may suggest the importance of the regulation of acetylcholine in the CNS and that the cerebellum may have a special role of protecting both the brain (Crossland, 1960) and itself against any free acetylcholine (Aprison *et al.*, 1964).

The simple esterases have been found in the cell components of squirrel monkey cerebellum (Shantha *et al.*, 1967a) as is the case in the rhesus monkey, including the Cajal glial cells. The significance of this enzyme complex is not well understood, but there is every possibility that it may be concerned with the hydrolysis of some transmitter substance other than acetylcholine.

AK is localized mainly in the blood vessels, while the glomeruli are mildly active. The nucleoli in Purkinje cells, granule cells, and Golgi type II cells are stained when Gomori's method is used for AK localization, but they are negative when Burstone's method is used. On the basis of AK studies by Rogers (1961, 1963) on chick embryo of different ages, newly hatched chicken, and on 7-day-old Agelains cerebellum, Arvy (1966) observed that the increase of AK activity in the embryo seems to preceed the morphological and functional development of the tissue with an increase in RNA, proteins, and phosphatides.

Purkinje cells in the rhesus monkey show somewhat less ATPase than those in the rat and squirrel monkey (Tewari and Bourne, 1963a, 1963b; Shantha *et al.*, 1967a). The nerve fiber layers are moderately ATPase-positive in the rhesus and squirrel monkeys in contrast to the negative activity reported by Tewari and Bourne (1963b) in the rat. The presence of positive activity for AK, AC, AMPase, and ATPase in cerebellar as well as in olfactory glomeruli indicates that these enzymes in some way facilitate the synaptic transmission of impulses in the glomeruli.

Hess and Pope (1959) showed higher amounts of calcium-activated ATPase in areas rich in dendritic processes compared to the neurons and believed that magnesium-activated ATPase is more closely related to cell bodies. This is in contrast to the observations of Naidoo and Pratt (1951, 1956) who observed minimum activity of magnesium-activated ATPase in the cell bodies, which were well stained with the calcium-activated ATPase.

Another important enzyme that has been studied in detail in the cerebellum by Scott (1963, 1967) is 5′-nucleotidase. Its role is not precisely defined and suggestions vary, from its involvement in nucleic acid catabolism or breakdown of NADP (Dixon and Webb, 1964), to its role in the regulation of glycolysis (Reis, 1951). Its distribution in the mouse differs in the anterior and posterior portions of the molecular layer of the cerebellum (Scott, 1963), the latter showing more activity than the former. The transition between the two zones occurs at the apex of the folium immediately posterior to the fissura prima. Scott showed that 5′-nucleotidase activity is present in fair amounts in the Golgi cells and Purkinje cells as well as in the baskets and their dendrites; all these cells are inhibitory in function (Eccles *et al.*, 1966). The excitatory neurons (granule cells) do not show as much activity of 5′-nucleotidase. Elaborate studies are needed to shed more light on the function of 5′-nucleotidase in the cerebellum.

Shantha *et al.* (1966) developed a histochemical method for the localization of cyclic 3′5′-nucleotide phosphodiesterase; they described large amounts of this enzyme in the molecular layer, nerve fiber layers, and glial cells. The Purkinje cells, granule cells, stellate and basket cells, glomeruli, and Golgi type II cells also show some activity mainly localized in their cell membranes. A wide variety of biochemical functions are believed to be mediated by cyclic 3′5′-adenosine monophosphate (cyclic 3′5′-AMP), and the enzyme which forms this, adenyl cyclase, is said to be associated with cell membranes. The rate of breakdown of the cyclic 3′5′-AMP depends upon the presence of the enzyme 3′5′ AMPase. It appears from the above study that the glial cells and molecular layer of the cerebellum play a major role in the breakdown of cyclic 3′5′-AMP and that the neuronal components have a very small quantity of this enzyme.

The acid phosphatase level in the molecular layer is higher than that of the granular layer, and the latter shows more AC activity than the white matter. The presence of AC activity of the Golgi and basket cells of the molecular layer has been reported in a 2000 gm human fetus. At birth, most of the cellular elements, including the dentate nucleus, showed the presence of this enzyme (Arvy, 1966). Its localization in various components of the cerebellum has been studied by various workers in the adult human (Olsen and Petri, 1963; Iijima, 1963a), rat (Tewari and Bourne, 1963a), and squirrel monkey (Shantha *et al.*, 1967a). Olsen and Petri (1963) showed that the bulk of AC activity is located in the cytoplasm of Bargmann cells. Some activity is also observed in the Purkinje cells and Golgi cells. The granule cells give a fine granular AC reaction in the perinuclear position. In the adult human brain, positive activity has been described by a number of workers in Purkinje cells, Bargmann glial cells, granule cells, Golgi type II cells, astrocytes, basket fibers, and molecular layers.

More significant differences in the localization of AC have been reported among rat, squirrel, and rhesus monkey cerebellum. In the rat, except for the Purkinje cells, all other elements (i.e., stellate cells, basket cells, Cajal [Bargmann] epithelial cells, granule cells, Golgi Lugaro cells, cerebellar glomeruli) are said to be AC-negative (Tewari and Bourne, 1963a). On the other hand, in the squirrel monkey (Shantha *et al.*, 1967a) and rhesus monkey all these elements including Purkinje cells are moderately to strongly AC-positive. The thiamine pyrophosphatase (TPPase) positive Golgi apparatus and AC activity are at the same site of the cell cytoplasm as seen in our earlier preparations (Shanthaveerappa and Bourne, 1965c).

Detailed studies on the distribution of amylophosphorylase (AP) in various species of animals have been made by Iijima (1964) and Gentscher (1967). Iijima described the absence of AP activity in rabbit and cat Purkinje cells and the presence of AP activity in rat Purkinje cells. He has also described the presence of this enzyme in glial cells of various layers of the cerebellum in all these animals. On the other hand, Tewari and Bourne (1962a) reported the absence of glucosan phosphorylase in rat Purkinje cells and did not report the presence of this enzyme in various glial cells. Studies on squirrel monkey cerebellum (Shantha *et al.*, 1967a) showed its presence in various components of cerebellum. Cajal's epithelial cells, glial cells in various layers of cerebellum, and nerve fiber layers were moderately to strongly AP-positive. The Purkinje cells, molecular layer, deep cerebellar neurons, and glomeruli showed negligible activity, whereas the neuropil of the deep cerebellar nuclei, synapses around these neurons, and the nucleoli in deep cerebellar nuclei neurons were strongly AP-positive. It has also been observed (Shantha *et al.*, 1967a) that the fibers of the nerve fiber layer show moderate AP activity toward the cortex which increases progressively toward the deep cerebellar nuclei. A difference in the degree of enzyme activity in various components of some nerve tracts has also been reported by Friede (1966). The positive activity in these nerve fibers was mainly observed in the axons and in the cytoplasm of myelin-forming glial cells which surround the axons. The myelin sheath showed little or no activity. High AP activity in the white matter of the central nervous system has been recorded by other workers (Okada, 1957; Shimizu and Okada, 1957; Amakawa, 1959). The presence of AP in the nucleoli indicates that these organelles can utilize glycogen for their energy source. The absence of glycogen-synthesizing and breakdown capacity (due to lack of enzymes concerned with this process) and the absence of glycogen (Shimizu and Kubo, 1957; Shanklin *et al.*, 1957; Iijima, 1963b) in basket cells and basket formation in squirrel monkey suggest that these components cannot act as energy donors to Purkinje cells, as proposed by Tewari and Bourne (1962b).

Gentscher (1967) studied phosphorylase activity in the cerebellar cortex using thick, fresh, unfrozen slices as well as squash preparations and found two types (small and large) of newly synthesized glycogen granules. The smaller glycogen granules showed morphological resemblance to the mitochondria, whereas the larger ones, which showed characteristic distribution on the surface of the Purkinje cells and their dendrites, indicated their association with presynaptic terminals. This work needs further investigation.

The basket cells have lower levels of the same enzyme complement as the Purkinje cells, and some basket formations show high oxidative enzymes compared to the Purkinje cells themselves. In addition, the basket formation shows mild AChE activity. It appears from these findings that their function must be somehow concerned with the nerve impulse, in its modification and/or transmission. Contrary to these findings, Tewari and Bourne (1962b) described the virtual absence of oxidative enzymes in the basket cells.

As one would expect, the Purkinje cells are metabolically very well equipped. The G6PD activity is an indication of an active hexosemonophosphate shunt mechanism which by-passes the main glycolytic pathway and provides reduced TPN and ATP, also indicating RNA synthesis. Thus, the cell is equipped for oxidation of carbohydrates and synthesis of ATP and protein. The absence of phosphorylases and the low glycogen level indicate that these cells are not engaged in the production or storage of this fuel. The presence of G6P is possibly related to the movement of glucose into the cell.

The glial cells in the cerebellum are positive for AP, G6PD, AD, LDH, SDH, and CYO. The degree of positive activity varies from one enzyme to another, the highest activity being observed for AP and LDH. The presence of AP, SDH, CYO, and MAO in the glial cells of several mammals such as the mouse, rat, guinea pig, rabbit, dog, cat, and squirrel monkey has also been reported by various workers (Shimizu and Morikawa, 1957; Wolfgram and Rose, 1959; Potanos *et al.*, 1959; Iijima 1964; Iijima and Nakajima, 1964; Manocha and Bourne, 1966, 1968; Manocha *et al.*, 1967; Iijima *et al.*, 1967a, 1967b; Shantha *et al.*, 1967a). Shimizu and Kumamoto (1952) described the presence of glycogen in the glial cells. All these studies suggest that these cells are equipped for glycolytic pathways and the TCA cycle. In addition to the presence of SDH, a fair amount of AC, ATPase, DPNH, and TPNH activity has also been observed in the neuroglial cells (Becker *et al.*, 1960). There is a relationship between the thickness of an axon and enzyme content of the enveloping glial cell. For example, the thick axons show marked enzymatic activity compared to the surrounding glial cells, whereas the thin axons show little activity of oxidative enzymes as compared to a fair amount in the surrounding glial cells (Friede, 1961).

Arvy (1966) reported that the white matter of the cerebellum in the rat was stained selectively for β-glucuronidase when the Seligman *et al.* (1954) techniques were used. We agree with Arvy that in view of the discrepancy existing in these methods, the localization of this enzyme should be reconsidered. The presence of this enzyme in Purkinje cells has also been described by Pearse (1961) and Waltimo and Talanti (1965). Since acid phosphatase and β-glucuronidase are both said to be lysosomal and since the acid phosphatase of the Purkinje cell is contained in lysosomes, it appears likely that the β-glucuronidase is also present in these structures (Arvy, 1966). β-glucuronidase is normally said to be present in tissues involved in secretory activity, such as neurosecretory neurons and the choroid plexus, but the function this enzyme performs in Purkinje cells is not known. Further investigations are necessary to evaluate the functional significance of this enzyme in these cells.

In the granule cells, the presence of small granules of enzyme activity around the nucleus in the thin rim of cytoplasm has been reported for most of the oxidative enzymes tested. Small, enzymatically active granules represent the enzyme activity in the mitochondria. It is of interest that some workers describe the absence of oxidative enzymes in granule cells (Tewari and Bourne, 1962a; Iijima and Nakajima, 1964).

The distribution of various oxidative enzymes in the rhesus and squirrel monkeys indicates that the molecular layer is more active metabolically than the granular layer. To put it in another way, the cytoplasm of the dendritic branches in the molecular layer is more active; this makes the molecular layer stain strongly compared to the granule cell layer, which is conspicuous by the presence of numerous large nuclei and scanty cytoplasm in the granule cells (Manocha and Bourne, 1966; Shantha *et al.*, 1967a). The white matter is markedly different enzymatically from the gray matter, being completely negative in SDH (Sotelo and Wegmann, 1964). Robins and Smith (1952) observed the quantity of enzymes concerned with carbohydrate metabolism; aldolase (of the glycolytic cycle) was much lower than LDH in the granular layer, and fumarase (of the tricarboxylic acid cycle) was of disproportionately lower activity than malic dehydrogenase in the white matter. They suggested that the different enzymes might act as the rate-limiting step in the same metabolic pathways of different histologic locations. Lodin *et al.* (1968) showed that there are no qualitative differences in dehydrogenase activity between the excitatory and inhibitory neurons and also observed a high dehydrogenase level in the spaces of the granular layer—the glomeruli cerebellosi. These authors concluded that glutamate dehydrogenase, glucose-6-phosphate dehydrogenase, and β-hydroxybutyrate dehydrogenase activity is higher in glomerular structures, whereas lactic dehydrogenase is higher in the Purkinje cells.

Our observations in the Purkinje cells agree with those of Sotelo (1967) that, in general, the Purkinje cells are very active enzymatically but that all cells (perikarya and dendrites) are not active at the same time. As he observed, 90% of the cells do show high SDH activity, whereas about 10% showed varying degrees of enzyme reaction. This, as we have suggested, may be due to different functional states at a particular time. Purkinje cells cultured *in vitro* give similar results. Kim (1966) showed that the cultured neurons show strong activity of α-glycerophosphate, lactic, malic, isocitrate, succinic, and glucose-6-phosphate dehydrogenase. This indicates the active performance of glycolysis, oxidation through the citric acid cycle, and the hexosemonophosphate shunt. He concluded that the high activity of enzymes associated with carbohydrate metabolism points out that the glucose is the main fuel of the neurons grown in culture.

There are a number of reports on the localization of oxidative enzymes in glomeruli of the granule cell layer (Friede, 1959; Burstone, 1960; Thomas and Pearse, 1961; Lazarus *et al.*, 1962; Tewari and Bourne, 1962a; Iijima and Nakajima, 1964; Manocha and Bourne, 1966; Shantha *et al.*, 1967a). The glomeruli give very strong positive reactions for most of the oxidative enzymes tested, especially for SDH and LDH. The positive activity extends for some distance into the mossy fibers and dendritic processes of granule cells starting from the glomeruli. Knob-like, strongly positive areas on the surface of the glomeruli indicate the areas of synaptic contact between the terminal end of the mossy fibers and dendritic processes of granule cells (Shantha *et al.*, 1967a).

The Purkinje cells also give positive reactions for almost all the oxidative enzymes. The dendritic processes and the synapses on these processes are positive as well. Strong activity of SDH (Neumann and Koch, 1953; Shimizu and Morikawa, 1957; Tewari and Bourne, 1962a; Shantha *et al.*, 1967a) and LDH, MDH, and G6PD (Lazarus *et al.*, 1962; Schiffer and Vesco, 1963) has been described in Purkinje cells of the cerebellum in the mouse, guinea pig, rabbit, rat, dog, human, and primates. The basket cells and the basket formation around the Purkinje cells as well as stellate cells also show positive activity for various oxidative enzymes, though this was not observed in these cells in the rat cerebellum (Tewari and Bourne, 1962a). Negligible to negative activity for AP and UDPG and moderate to strong positive activity for AD, G6PD, LDH, SDH, and CYO in Purkinje cells, stellate cells, granule cells, Golgi type II cells, basket cells, and basket formation around the Purkinje cells have been found in squirrel monkey studies. LDH and SDH localization in these components of the rhesus monkey has indicated that these structures are equipped for the Embden-Meyerhof-Parnas pathway and the TCA cycle.

Tewari and Bourne (1962a, 1962b, 1962c; 1963a, 1963b, 1963c), in their studies on the distribution of AC, ATPase, MAO, CYO, SDH, and SE, have described the presence of cycles of enzyme activity in the distribution of mitochondria and nucleoli in the Purkinje cells of the rat cerebellum and dorsal root ganglion. They observed that wherever the nucleolus touched the nuclear membrane, there was always a perinuclear concentration of these enzymes, and when the nucleolus was in the center of the nucleus, all these enzymes were uniformly distributed in the cytoplasm. Kumamoto and Bourne (1963) and Thomas (1963), in the dorsal root ganglion cells of rat and man, did not find such cycles of enzyme activity. Iijima and Nakajima (1964) in the cerebellum, Manocha *et al.* (1967) in the spinal cord, and Shanthaveerappa and Bourne (1965a, 1965b) and Shantha *et al.*, (1967a, 1967b) in the olfactory bulb, dorsal root ganglion, mesentric ganglion, and cerebellum did not find any such cycles of enzyme activity.

Our observations in the squirrel and rhesus monkeys on the distribution of a number of enzymes (AC, ATPase, TPPase, SE, CYO, MAO, LDH, SDH, AD, G6PD) give the following results with regard to the nucleolus and its relation to cytoplasmic enzyme activity:

1. When the nucleolus is in the center of the nucleus, the cytoplasmic enzyme activity is found distributed uniformly throughout the cell in some neurons, whereas in others (very few) it is much more concentrated at one pole of the cell.

2. When the nucleolus is touching the nuclear membrane, the cytoplasmic enzyme activity is uniformly distributed in some cells; there are perinuclear concentrations at the site of the nucleolar contact in a very small number of cells. Still fewer are the cells which show a concentration of these enzymes at the opposite pole of the nucleus, where the nucleolus is not touching.

3. Irrespective of the nucleolar-cytoplasmic relationship, it is generally observed in the majority of neurons that some cells are mildly positive, others moderately positive, and still others strongly positive. This is also true of the cells of the dorsal root ganglion and other areas of the nervous system, except the small cells which show little variation in enzyme activity. The variations described above are observed in the same enzyme preparation. None of the neurons show rings of positive activity such as have been observed in the rat by Tewari and Bourne. The presence of a variability in the degree of positive activity in the same type of neurons for various enzyme reactions indicates that some cells are

metabolically much more active than others. This also suggests that individual neurons undergo a series of phasic activities in which the enzyme contents of a cell increase or decrease, presumably depending on the functional state of that cell. The existence of such apparently phasic activity has been described in the CNS in thiamine pyrophosphatase positive Golgi apparatus (Shanthaveerappa and Bourne, 1965c). Studies in this laboratory suggest the possibility that the formation of large ice crystals in slowly frozen tissue may result in displacement of intracytoplasmic organelles, in various directions. This may result in the formation of rings of enzyme activity or perinuclear concentration of enzyme activity in histochemical preparations in which such tissue was used.

4. It is also commonly observed that the concentration of enzyme activity, especially for SDH, LDH, SE, and ATPase, appears to be greater toward the dendritic than the axonic pole of the neurons, although the neurons of the spinal and mesenteric ganglion cells do not show a similar tendency. This suggests a high level of metabolic activity at the dendritic pole of the cell. Less oxidative enzymes are found at the areas of lipofuscin pigment accumulation in neurons.

Synthesis of RNA by the nucleoli and its subsequent passage to the cytoplasm have been suggested by a number of workers (Marshak and Dalvet, 1949; Barnum and Huseby, 1950; Jeener and Szafarz, 1950; Bonner, 1959; Woods and Taylor, 1959; Bogoroch and Siegel, 1961). Our studies indicate that the nucleolus may make at least two functionally different types of movement toward the nuclear membrane. The first movement might have to do with the synthesized material to be emptied into the cytoplasm. The second might lead to an exchange of enzymes or compounds which have been synthesized (e.g., ATP by the mitochondria) across the nuclear membrane. Generally, this last movement may occur when the mitochondria congregate around the nucleus and the nucleolus is touching the nuclear membrane.

Under the electron microscope it is not uncommon to see invaginations of the nuclear membrane into the nucleoplasm and the presence of mitochondria in these invaginations. In some cases the nucleolus may be close to these folds, which would provide an ideal situation for the transfer of substances. In tissue culture cells, using time-lapse microphotography, Frederic (1954) and Lettre and Siebs (1954) observed that occasionally there was a decrease in volume and contrast of the nucleolus following its journey toward the nuclear membrane, which provides at least circumstantial evidence of the movement of material into the cytoplasm. The passage of enzymes or synthetic products of metabolism is also largely circumstantial. It appears that in primates,

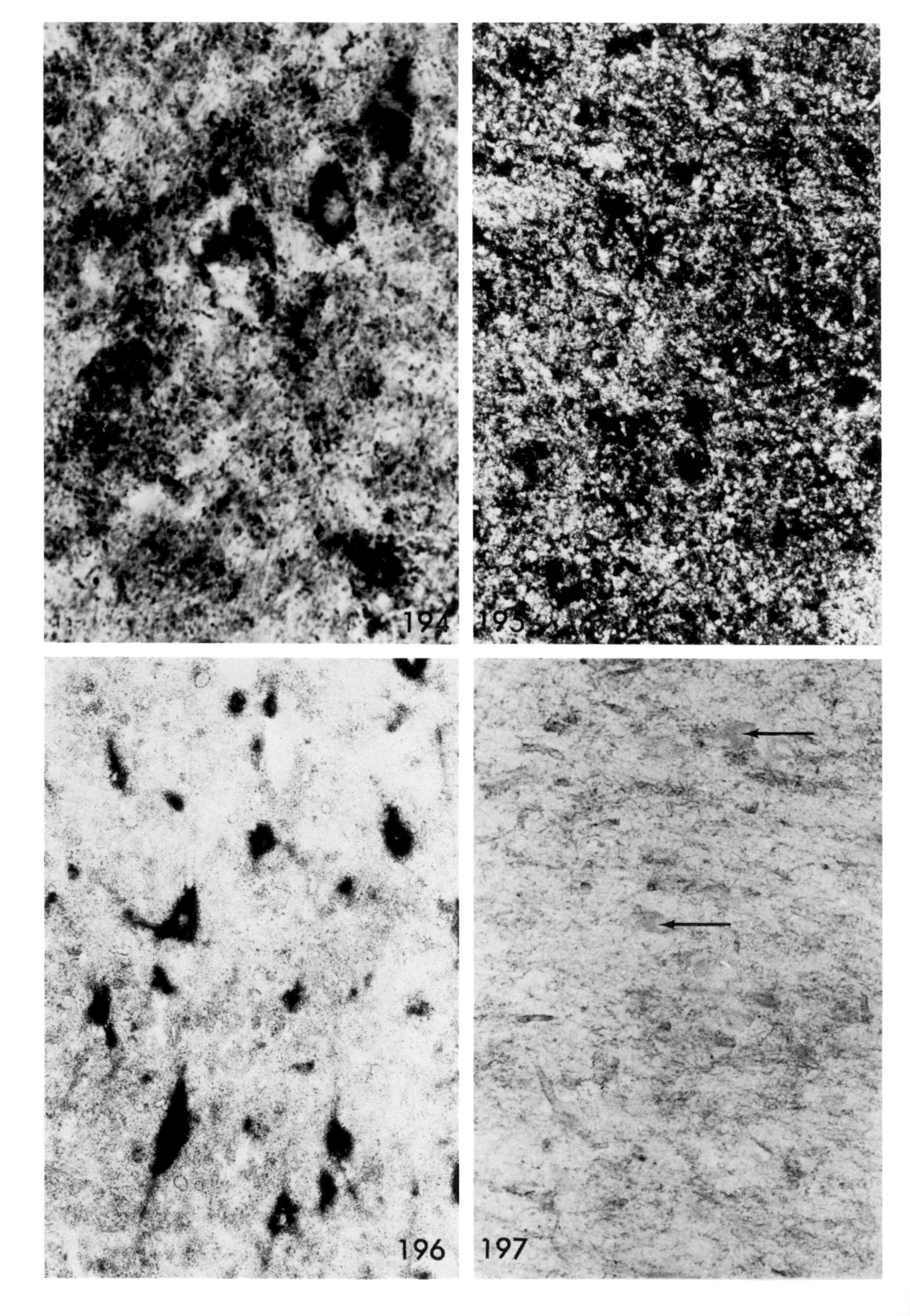
194
195
196
197

FIGURES 194–197. Deep cerebellar nuclei. Dentate nucleus (Figs. 194, 195, and 197), nucleus fastigii (Fig. 196). Note the variations in the enzyme preparations for SDH, LDH, SE, and AChE, respectively. The neurons give a mild reaction for AChE (arrows) and strong reactions for other enzymes. ×592, ×148, ×148, ×148.

including man, this type of movement is less frequent than in the rat. It has been suggested by Shantha *et al.* (1967a) that in higher species of animals some of the nucleolar activities might be located in the cytoplasmic organelles, implying that exchange of material between the nucleolus and cytoplasm, and therefore the movement of the nucleolus back and forth, is considerably reduced.

We hope that the development of electron histochemistry will explain some of the controversies pertaining to the type of nucleolar-cytoplasmic relationships under discussion.

DEEP CEREBELLAR NUCLEI

The deep cerebellar nuclei are situated in the white matter of the cerebellum, dorsal and dorsolateral to the IVth ventricle, and are grouped into nuclei (1) Fastigii, (2) globosus, (3) emboliformis, and (4) dentatus (Figs. 139, 141–143, 194–197). Embryonically, all these nuclei are derived from a single gray mass which later develops into medial and lateral parts. In higher animals such as the primates, these two masses further divide and give rise to two more nuclei. The medial mass divides into the medial nucleus Fastigii and lateral nucleus globosus while the lateral mass develops into the lateral nucleus dentatus and medial nucleus emboliformis. In reptiles and lower mammals, these two masses remain undivided and are referred to as the medial and lateral cerebellar nuclei. In recent years, the dentate nucleus has been called the nucleus lateralis cerebelli, the nuclei emboliformis and globosus have been referred to as the nucleus intermedius (interpositus) cerebelli, and the nucleus fastigii has been called nucleus medialis cerebelli. We have used this terminology in our description. In the rhesus monkey, one can observe the presence of both globosus and emboliformis as separate nuclei. Snider and Lee (1961) included both of them under nucleus interpositus. Recent studies on the chimpanzee and man show that the dentate nucleus is laminated and separates from the nucleus emboliformis farther laterally than in the macaque (Holloway, 1968; Shantha and Manocha, 1969b). Though the laminations are present in the rhesus dentate nucleus, they are not as complex as in apes and man.

The nuclei lateralis, intermedius, and medialis are composed of mostly medium-sized and large neurons with a few small neurons having oval, round, triangular, or fusiform shapes. The nucleus lateralis has few large neurons compared to the other two nuclear groups and is more cellular. The order of decreasing cellularity is lateral, medial, and intermediate. Histochemically, these three components of the deep cerebellar nuclei show similar types of histochemical activity, unless otherwise mentioned.

The neurons exhibit strong LDH and SE, moderately strong AC and SDH, moderate ATPase and MAO, mild AChE, and negative AK activity. The glial cells show strong LDH; moderate SE and ATPase; mild AC, SDH, and MAO; and negative AChE and AK activity. The nucleus lateralis cerebelli contains many more neuroglial cells than the other two complexes. The blood vessels are all ATPase-, AK-, and AChE-positive, and the nucleus lateralis cerebelli contains the largest number of them. AChE activity is found in the cytoplasm of the neurons and is diffuse in nature. The synapses around the perikarya show LDH, SDH, ATPase, SE, and MAO activity. These synapses are easily recognizable in SDH and LDH preparations.

Despite the numerous histochemical investigations of cerebellum, very little attention has been paid to the histochemical nature of the deep cerebellar nuerons. The present study on the rhesus monkey and the work of Shantha *et al.* (1967a) have shown the presence of most of the oxidative and phosphorylating enzymes and esterases except for BChE and UDPG. This indicates that these neurons are metabolically very active. The presence of AP, AD, G6PD, SDH, LDH, and CYO in these neurons and the synaptic boutons found on them suggests that all these structures are well equipped with enzymes of the glycolytic pathway and the TCA cycle which are needed to supply energy for the vigorous activities of these neurons and their synapses.

REFERENCES

Amakawa, F. Studies on the polysaccharide synthesis of the central nervous tissue with special reference to the phosphorylase reaction and several enzymes (in Japanese). *Kumamoto Igakkai Zasshi* **33**: Suppl. 8, 1–37 (1959).

Aprison, M. H., Takahashi, R., and Folkerth, T. L. Biochemistry of the avian central nervous system. I. The 5-hydroxytryptophan decarboxylase-monoamine oxidase and cholineacetylase-acetylcholinesterase systems in several discrete areas of the pigeon brain. *J. Neurochem.* **11**:341–350 (1964).

Arvy, L. Cerebellar enzymology. *Intern. Rev. Cytol.* **20**:277–359 (1966).

Austin, L., and Phillis, J. W. The distribution of cerebellar cholinesterases in several species. *J. Neurochem.* **12**:709–727 (1965).

Barnum, C. P., and Huseby, R. A. The intracellular heterogeneity of pentose nucleic acid as evidenced by the incorporation of radiophosphorus. *Arch. Biochem. Biophys.* **29**:7–26 (1950).

Becker, N. H., Goldfischer, S., Shin, Woo-Yung, and Novikoff, A. B. The localization of enzyme activities in the rat brain. *J. Biophys. Biochem. Cytol.* **8**:649–663 (1960).

Bennett, E. L., Diamond, M. C., Morimoto, H., and Herbert, M. Acetylcholinesterase activity and weight measures in fifteen brain areas from six lines of rats. *J. Neurochem.* **13**:563–572 (1966).

Bogoroch, R., and Siegel, B. V. Some metabolic properties of the nucleolus as demonstrated by recent radioisotope experiments. *Acta Anat.* **45**:265–287 (1961).

Bonner, J. Protein synthesis and control of plant processes. *Am. J. Bot.* **46**:58–62 (1959).

Burstone, M. S. *Calcification in Biological Systems* (R. F. Sogmaess, ed.). Amer. Assoc. Adv. Sci., Washington, D.C., 1960.

Crawford, J. M., Curtis, D. R., Voorhowe, P. E., and Wilson, V. J. Acetylcholine sensitivity of cerebellar neurones in the cat. *J. Physiol.* (*London*) **186**:139–165 (1966).

Crossland, J. Chemical transmission in the central nervous system. *J. Pharm. Pharmacol.* **12**:1–36 (1960).

Csillik, B., Joo, F., and Kasa, P. Cholinesterase activity of archicerebellar mossy fiber apparatuses. *J. Histochem. Cytochem.* **11**:113–114 (1963).

Csillik, B., Joo, F., Kasa, P., Tomity, A., and Kalaman, G. Development of acetylcholinesterase-active structures in the rat archibellar cortex. *Acta Biol. Acad. Sci. Hung.* **15**:11–17 (1964).

Curtis, D. R., and Crawford, J. M. Acetylcholine sensitivity of cerebellar neurones. *Nature* **206**:516–517 (1965).

De Robertis, E., Pellegrino de Iraldi, A., Rodriquez, de L. A., and Salganicoff, F. Cholinergic and non-cholinergic nerve endings in the rat brain. *J. Neurochem.* **9**:23–35 (1961).

Dixon, M., and Webb, E. C. *Enzymes.* Academic Press, New York and London, 1964.

Eccles, J. C., Llinas, R., and Sasaki, K. The inhibitory interneurones within the cerebellar cortex. *Exptl. Brain Res.* **1**:1–16 (1966).

Fox, C. A., and Barnard, J. W. A quantitative study of the Purkinje cell dendritic branchlets and their relationship to afferent fibers. *J. Anat.* (*London*) **91**:299–313 (1957).

Frederic, J. Action of various substances on the mitochondria of living cells, cultivated *in vitro*. *Ann. N.Y. Acad. Sci.* **58**:1246–1263 (1954).

Friede, R. L. Histochemical investigations succinic dehydrogenase in the central nervous system. III. Atlas of the midbrain of the guinea pig including pons and cerebellum. *J. Neurochem.* **4**:290–303 (1959).

Friede, R. L. A histochemical study of DPN-diaphorase in human white matter; with some notes on myelination. *J. Neurochem.* **8**:17–30 (1961).

Friede, R. L. A quantitative mapping of alkaline phosphatase in the brain of the rhesus monkey. *J. Neurochem.* **13**:197–203 (1966).

Friede, R. L., and Fleming, L. M. A mapping of the distribution of lactic dehydrogenase in the brain of the rhesus monkey. *Am. J. Anat.* **113**:215–236 (1963).

Friede, R. L., and Fleming, L. M. A comparison of cholinesterase distribution in the cerebellum of several species. *J. Neurochem.* **11**:1–7 (1964).

Gentscher, T. Histochemical determination of phosphorylase in the cerebellum. *Brain Res.* **5**:350–365 (1967).

Gerebtzoff, M. A. *Cholinesterases.* International Series of Monographs on Pure and Applied Biology (Division: Modern Trends in Physiological Sciences), (P. Alexander and Z. M. Bacq, eds.). Pergamon Press, New York, 1959.

Hess, H., and Pope, A. Intralaminar distribution of adenosine triphosphatase activity in rat cerebral cortex. *J. Neurochem.* 3:287–299 (1959).

Holloway, R. L. The evolution of the primate brain: some aspects of quantitative relations. *Brain Res.* **7**:121–172 (1968).

Hydén, H. (ed.) Dynamic aspects of the neuron-glia relationship. A study with microchemical methods. In: *The Neuron.* Elsevier, Amsterdam, 1967.

Iijima, K. Histochemical studies on the glomeruli cerebellosi (eosin bodies) of human and mammalian cerebellar cortex. *Bull. Tokyo Med. Dent. Univ.* **10**:121–144 (1963a).

Iijima, K. Histochemical studies on the phosphatases and cholinesterases of the human cerebellum. *Bull Tokyo Med. Dent. Univ.* **10**:145–180 (1963b).

Iijima, K. On the distribution of the amylophosphorylase of the mammalian cerebelli. *Bull. Tokyo Med. Dent. Univ.* **11**:77–104 (1964).

Iijima, K., and Nakajima, Y. Histochemical studies on the oxidative enzymes of the human cerebellum. *Bull. Tokyo Med. Dent. Univ.* **11**:103–135 (1964).

Iijima, K., Shantha, T. R., and Bourne, G. H. Enzyme-histochemical studies on the hypothalamus with special reference to the supraoptic and paraventricular nuclei of the squirrel monkey (*Saimiri sciureus*). *Z. Zellforsch.* **79**:76–91 (1967a).

Iijima, K., Bourne, G. H., and Shantha, T. R. Histochemical studies on the distribution of some enzymes of the glycolytic pathways in the olfactory bulb of the squirrel monkey (*Saimiri sciureus*). *Acta Histochem.* **27**:1–9 (1967b).

Jeener, K., and Szafarz, D. Relation between the rate of renewal and the intracellular localization of ribonucleic acid. *Arch. Biochem. Biophys.* **26**:54–67 (1950).

Joo, F., Savay, G., and Csillik, B. A new modification of the Koelle-Friedenwald method for the histochemical demonstration of cholinesterase activity. *Acta Histochem.* **22**:40–45 (1965).

Kasa, P., and Csillik, B. Cholinergic excitation and inhibition in the cerebellar cortex. *Nature* **208**:695–696 (1965).

Kasa, P., and Silver, A. The correlation between choline acetyltransferase and acetylcholinesterase activity in different areas of the cerebellum of rat and guinea pig. *J. Neurochem.* **16**:389–396 (1969).

Kasa, P., Csillik, B., and Joo, F. Histochemical localization of acetylcholinesterase in the cat cerebellar cortex. *J. Neurochem.* **12**:31–35 (1965).

Kasa, P., Csillik, B., Joo, F., and Knyichas, E. Histochemical and ultrastructural alterations in the isolated archicerebellum of the rat. *J. Neurochem.* **13**:173–178 (1966).

Kim, S. U. Histochemical demonstration of oxidative enzymes associated with carbohydrate metabolism in cerebellar neurons cultured *in vitro. Arch. Histol. Jap.* **27**:465–471 (1966).

Koelle, G. B. The histochemical localization of cholinesterases in the central nervous system of the rat. *J. Neurol.* **100**:211–228 (1954).

Koelle, G. B. (ed.) Cytological distribution and physiological functions of cholinesterases. In: *Handbuch der Experimente Pharmakologie*, pp. 187–298. Springer-Verlag, Berlin, 1963.

Kumamoto, T., and Bourne, G. H. Experimental studies on the oxidative enzymes and hydrolytic enzymes in spinal neurons. I. A histochemical and biochemical investigation of the spinal ganglion and the spinal cord following a sciatic neurotomy in the guinea pig. *Acta Anat.* **55**:255–277 (1963).

Lazarus, S. S., Wallace, B. J., Edgar, G. W. F., and Volk, B. W. Enzyme localization in rabbit cerebellum and effect of post mortem autolysis. *J. Neurochem.* **9**:227–232 (1962).

Lettre, R., and Siebs, W. Beobachtungen am Nucleolus *in vitro* gezühteter Zellen. *Z. Krebsforsch.* **60**:19–30 (1954).

Lewis, P. R., Shute, C. C. D., and Silver, A. Confirmation from choline acetylase analyses of a massive cholinergic innervation to the rat hippocampus. *J. Physiol. (London)* **191**:215–224 (1967).

Lodin, Z., Muller, J., and Fattin, J. Distribution of some dehydrogenase in the cerebellar cortex. *Nature* **217**:655–657 (1968).

Manocha, S. L., and Bourne, G. H. Histochemical mapping of succinic dehydrogenase and cytochrome oxidase in the spinal cord, medulla oblongata and cerebellum of squirrel monkey (*Saimiri sciureus*). *Exptl. Brain Res.* **2**:216–229 (1966).

Manocha, S. L., and Bourne, G. H. Histochemical mapping of lactate dehydrogenase and monoamine oxidase in the medulla oblongata and cerebellum of squirrel monkey (*Saimiri sciureus*). *J. Neurochem.* **15**:1033–1040 (1968).

Manocha, S. L., Shantha, T. R., and Bourne, G. H. Histochemical studies on the spinal cord of the squirrel monkey (*Saimiri sciureus*). *Exptl. Brain Res.* **3**:25–39 (1967).

Marshak, A., and Dalvet, F. Specific activity of P32 in cell constituents of rabbit liver. *J. Cellular Comp. Physiol.* **34**:451–455 (1949).

McCance, I., and Phillis, J. W. Cholinergic mechanisms in the cerebellar cortex. *Intern. J. Neuropharmacol.* **7**:447–462 (1968).

Naidoo, D., and Pratt, O. E. The localization of some acid phosphatases in brain tissue. *J. Neurol. Neurosurg. Psychiatry* **14**:287–294 (1951).

Naidoo, D., and Pratt, O. E. The effect of magnesium and calcium ions on adenosine triphosphatase in the nervous and vascular tissues of the brain. *Biochem. J.* **62**:475–479 (1956).

Neumann, K. H., and Koch, G. Übersicht über die feinere Verteilung der Succinodehydrogenase in Organen und Gewben verschiedener Säugetiere, besonders des Hundes. *Hoppe-Seylers Z. Physiol. Chem.* **295**:35–61 (1953).

Okada, M. Histochemical studies of brain phosphorylase. I. Histochemical distribution of phosphorylase in the brain of normal adult rodents. *Arch. Histol. Jap.* **12**:493–508 (1957).

Olsen, S., and Petri, C. Histochemical localization of acid phosphatase in the human cerebellar cortex. *Acta Neurol. Scand.* **39**:112–122 (1963).

Pearse, A. G. E. *Histochemistry, Theoretical and Applied.* Little, Brown, Boston, 1961.

Phillis, J. W. Cholinesterase in the cat cerebellar cortex, deep nuclei, and peduncles. *Experientia* **21**:266–268 (1965).

Phillis, J. W. Acetylcholinesterase in the feline cerebellum. *J. Neurochem.* **15**:691–698 (1968).

Potanos, J. N., Wolf, A., and Cowen, D. Cytochemical localization of oxidative enzymes in the human nerve cells and neuroglia. *J. Neuropathol. Exptl. Neurol.* **18**:627–636 (1959).

Reis, J. The specificity of phosphomonoesterase in human tissues. *Biochem. J.* **48**:548–551 (1951).

Robins, E., and Smith, D. E. A quantitative histochemical study of eight enzymes of the cerebellar cortex and subjacent white matter in the monkey. *Res. Publ., Ass. Res. Nerv. Ment. Dis.* **32**:305–327 (1952).

Rogers, K. T. Studies on chick brain of biochemical differentiation related to morphological differentiation and onset of function. III. Histochemical localization of alkaline phosphatase. *J. Exptl. Zool.* **145**:49–59 (1961).

Rogers, K. T. Studies on chick brain differentiation. IV. Comparative biochemical alkaline phosphatase studies on an altricial brain and chick kidney and intestine. V. Comparative histochemical alkaline phosphatase studies on the chick retina and blackbird, mouse, rabbit, cat, and human brains. *J. Exptl. Zool.* **153**:15–35 (1963).

Schiffer, D., and Vesco, C. Histochemical observations about the pattern of tetrazolium reduction, with different substrates, in glia cells of normal and pathological human nervous system. *J. Histochem. Cytochem.* **11**:335–341 (1963).

Scott, T. G. A unique pattern of localization within the cerebellum. *Nature* **200**:793 (1963).

Scott, T. G. The distribution of 5′-nucleotidase in the brain of the mouse. *J. Comp. Neurol.* **129**:94–113 (1967).

Seligman, A. M., Tsou, K. C., Rutenberg, S. H., and Cohen, R. B. Histochemical demonstration of *β-d*-glucuronidase with a synthetic substrate. *J. Histochem. Cytochem.* **2**:209–229 (1954).

Shanklin, W. M., Issidores, M., and Nassar, T. K. Neurosecretion in the human cerebellum. *J. Comp. Neurol.* **107**:315–337 (1957).

Shantha, T. R., Woods, W. D., Waitzman, M. B., and Bourne, G. H. Histochemical method for localization of cyclic 3′5′-nucleotide phosphodiesterase. *Histochemie* **7**:177–190 (1966).

Shantha, T. R., Iijima, K., and Bourne, G. H. Histochemical studies on the cerebellum of squirrel monkey (*Saimiri sciureus*). *Acta Histochem.* **27**:129–162 (1967a).

Shantha, T. R., Manocha, S. L., and Bourne, G. H. Enzyme histochemistry of the mesenteric and dorsal root ganglion cells of cat and squirrel monkey. *Histochemie* **10**:234–245 (1967b).

Shantha, T. R., and Manocha, S. L. The brain of chimpanzee (*Pan troglodytes*). I. External morphology. In: *Handbook of Chimpanzee* (G. H. Bourne, ed.), Vol. I, pp. 200–249. S. Karger, Basel and New York, 1969a.

Shantha, T. R., and Manocha, S. L. The brain of chimpanzee (*Pan troglodytes*). III: Midbrain, pons, medulla oblongata and cerebellum. In: Handbook of Chimpanzee (G. H. Bourne, ed.), Vol. I, pp. 318–380. S. Karger, Basel and New York, 1969b.

Shanthaveerappa, T. R., and Bourne, G. H. Histochemical studies on distribution of dephosphorylating and oxidative enzymes and esterases in olfactory bulb of the squirrel monkey. *J. Nat. Cancer Inst.* **35**:153–165 (1965a).

Shanthaveerappa. T. R., and Bourne, G. H. Histochemical studies on the olfactory glomeruli of squirrel monkey. *Histochemie* **5**:125–129 (1965b).

Shanthaveerappa, T. R., and Bourne, G. H. The thiamine pyrophosphatase technique as an indicator of the morphology of the Golgi apparatus in the neurons. IV. Studies on the cerebellum of rat and squirrel monkey. *Z. Zellforsch.* **68**:699–710 (1965c).

Shen, S. C. Greenfield, P., and Boell, J. The distribution of cholinesterase in the frog brain. *J. Comp. Neurol.* **102**:717–742 (1955).

Shimizu, N., and Kubo, Z. Histochemical studies on brain glycogen of the guinea pig and its alteration following electric shock. *J. Neuropathol. Exptl. Neurol.* **16**:40–47 (1957).

Shimizu, N., and Kumamoto, T. Histochemical studies on the glycogen of the mammalian brain. *Anat. Rec.* **114**:479–498 (1952).

Shimizu, N., and Morikawa, N. Histochemical studies of succinic dehydrogenase of the brain of mice, rats, guinea pigs, and rabbits. *J. Histochem. Cytochem.* **5**:334–345 (1957).

Shimizu, N., and Okada, M. Histochemical distribution of phosphorylase in the rodent brain from newborn to adults. *J. Histochem. Cytochem.* **5**:459–471 (1957).

Shimizu, N., Morikawa, N., and Okada, M. Histochemical studies of monoamine oxidase of the brain of rodents. *Z. Zellforsch.* **49**:389–400 (1959).

Silver, A. Cholinesterases of the central nervous system with special reference to the cerebellum. *Intern. Rev. Neurobiol.* **10**:57–109 (1967).

Smith, B. Monoamine oxidase in the pineal gland, neurohypophysis and brain of the albino rat. *J. Anat.* (*London*) **97**:81–86 (1963).

Snider, R. S., and Lee, J. C. *A Stereotaxic Atlas of the Monkey Brain* (*Macaca mulatta*). The Univ. of Chicago Press, Chicago, 1961.

Sotelo, C. Cerebellar neuroglia: morphological and histochemical aspects. *Progr. Brain Res.* **25**:226–250 (1967).

Sotelo, C., and Wegmann, R. Differences du metabolism des glucides de la substance blanche et de la substance grise du cervelot. *Acta Histochem.* **18**:125–316 (1964).

Tewari, H. B., and Bourne, G. H. Histochemical studies on the distribution of oxidative enzymes in the cerebellum of the rat. *J. Histochem. Cytochem.* **10**:619–627 (1962a).

Tewari, H. B., and Bourne, G. H. Histochemical studies on the distribution of specific and non-specific cholinesterases and simple esterase in the cerebellum of the rat. *Acta Anat.* **51**:349–368 (1962b).

Tewari, H. B., and Bourne, G. H. Histochemical evidence of metabolic cycles in spinal ganglion cells of rat. *J. Histochem. Cytochem.* **10**:42–64 (1962c).

Tewari, H. B., and Bourne, G. H. Histochemical studies on the distribution of acid phosphatases and 5′-nucleotidase in the cerebellum of the rat. *J. Anat.* (*London*) **97**:65–72 (1963a).

Tewari, H. B., and Bourne, G. H. Histochemical studies on the localization of adenosine triphosphatase in the cerebellum of the rat. *J. Histochem. Cytochem.* **11**:246–257 (1963b).

Tewari, H. B., and Bourne, G. H. Histochemical studies on the distribution of monoamine oxidase in the cerebellum of rat. *Acta Anat.* **52**:334–340 (1963c).

Thomas, E. Dehydrogenasen und Esterasen in unveränderten und geschädigten Spinal-ganglienzellen von Menschen. *Acta Neuropath.* **2**:231–245 (1963).

Thomas, E., and Pearse, A. G. E. The fine localization of dehydrogenase in the nervous system. *Histochemie* **2**:266–282 (1961).

Torack, R. M., and Barrnett, R. J. Fine structure and localization of cholinesterase activity in the rat brain stem. *Exptl. Neurol.* **6**:224–244 (1962).

Waltimo, O., and Talanti, S. Histochemical localization of β-glucuronidase in the rat brain. *Nature* **205**:499–500 (1965).

Wolfgram, F., and Rose, A. S. The histochemical demonstration of dehydrogenase in neuroglia. *Exptl. Cell Res.* **17**:526–530 (1959).

Woods, P. S., and Taylor, J. H. Studies of ribonucleic acid metabolism with tritium labeled as cytidine. *Lab. Invest.* **8**:309–318 (1959).

X

OLFACTORY BULB

The olfactory bulb in the rhesus monkey is considerably reduced in size compared to that of the rat, rabbit, and squirrel monkey. It is almost triangular, with its thin apex directed toward the cribriform plate of the ethmoid bone. From the posterior end the tractus olfactorius emerges to join the main part of the brain, then divides into lateral and medial olfactory stria near the olfactory tubercle at the base of the frontal cortex.

Histologically and histochemically, the following layers are easily identifiable (Figs. 198–205): (a) olfactory nerve fiber layer, (b) olfactory glomeruli formed mainly by synapses between the olfactory nerve fibers and dendrites of mitral cells, (c) external granular layer, (d) external plexiform (molecular) layer, containing tufted cells in addition to various nerve processes, (e) mitral cell layer, (f) internal plexiform (molecular) layer, and (g) internal granular layer.

The axons in the nerve fiber layer show a mild to moderate enzyme activity for AC, ATPase, SE, AChE, SDH and MAO. The reaction is somewhat stronger in the AK and LDH preparations. The myelin sheaths surrounding the axons show very little enzyme activity. On the other hand, the Schwann cells show positive reactions for all the above mentioned enzymes. It is interesting to note that the nerve fibers originating from the receptor cells of the rhesus monkey olfactory mucosa (which continue to form the external nerve fiber layer of the olfactory bulb) give similar enzyme activity (Shantha and Nakajima, 1970).

Histochemically, the AChE activity appears to be localized more in the peripheral part of the glomeruli than in the central part. The activity for SDH and LDH is quite granular in appearance, indicating that these enzymes are mainly localized in the mitochrondria of the synapses. The periglomerular glial cells show activity for all the enzymes except AChE and AK. Some glial cells show considerably more activity for SE than others. The activity for LDH is much stronger than for the other enzymes.

External and internal plexiform (molecular) layers show strong LDH and ATPase, moderately strong SDH, moderate SE and MAO, mild AC, negligible AChE, and negative AK activity. In overall comparison, the external plexiform layer shows more enzyme activity than the internal plexiform layer.

The mitral cells show strong LDH, SE, and AC; moderate ATPase, SDH, and MAO; negligible AChE; and negative AK activity. Most of the activity is localized in the dendritic processes of the cells rather than in the perikarya. In contrast, the AChE-positive material is observed in the peripheral part of the cell as well as in the dendritic processes, probably in the synaptic areas.

The tufted cells show activity similar to the mitral cells, except that the intensity of positive activity is considerably less than that of the mitral cells. Further concentration of the enzyme activity is not observed in the dendritic processes, as is the case in mitral cells.

The granule cells have a large nucleus surrounded by a thin rim of cytoplasm; therefore, they contain very few mitochondria and other intracellular organelles. These cells show strong LDH; moderate MAO, SDH, SE, and AC; mild ATPase; and negative AChE and AK activity. The AC activity is generally restricted to one part of the cell. The nucleus and the nucleolus show a diffuse type of positive activity. The blood vessels surrounding the glomeruli show AChE activity. Blood vessels in the rest of the olfactory bulb and those surrounding the glomeruli show strong ATPase and AK activity.

Histochemical studies on the olfactory bulb are not numerous. Barbera and Galletti (1963) studied AK in the dog and rabbit and found positive activity in most of the layers using Gomori's technique. On the other hand, the azo dye methods used in the present study stained only the olfactory glomeruli, nerve fiber layers, and blood vessels. Ochi (1966) and Sharma (1967a) described strong AK reactions in some mitral granule cells and in the glomeruli; this is in contrast to the observations of Nandy (1965) and Shanthaveerappa and Bourne (1965a, 1965b) who reported moderate AK activity in the glomeruli and negative reaction in the mitral and granular cells. Scott (1967) described strong 5′-nucleotidase activity in the inner and outer plexiform and the mitral and granular layers. Gerebtzoff (1959) also demonstrated strong positive activity in the olfactory glomeruli and moderate activity in the lamina

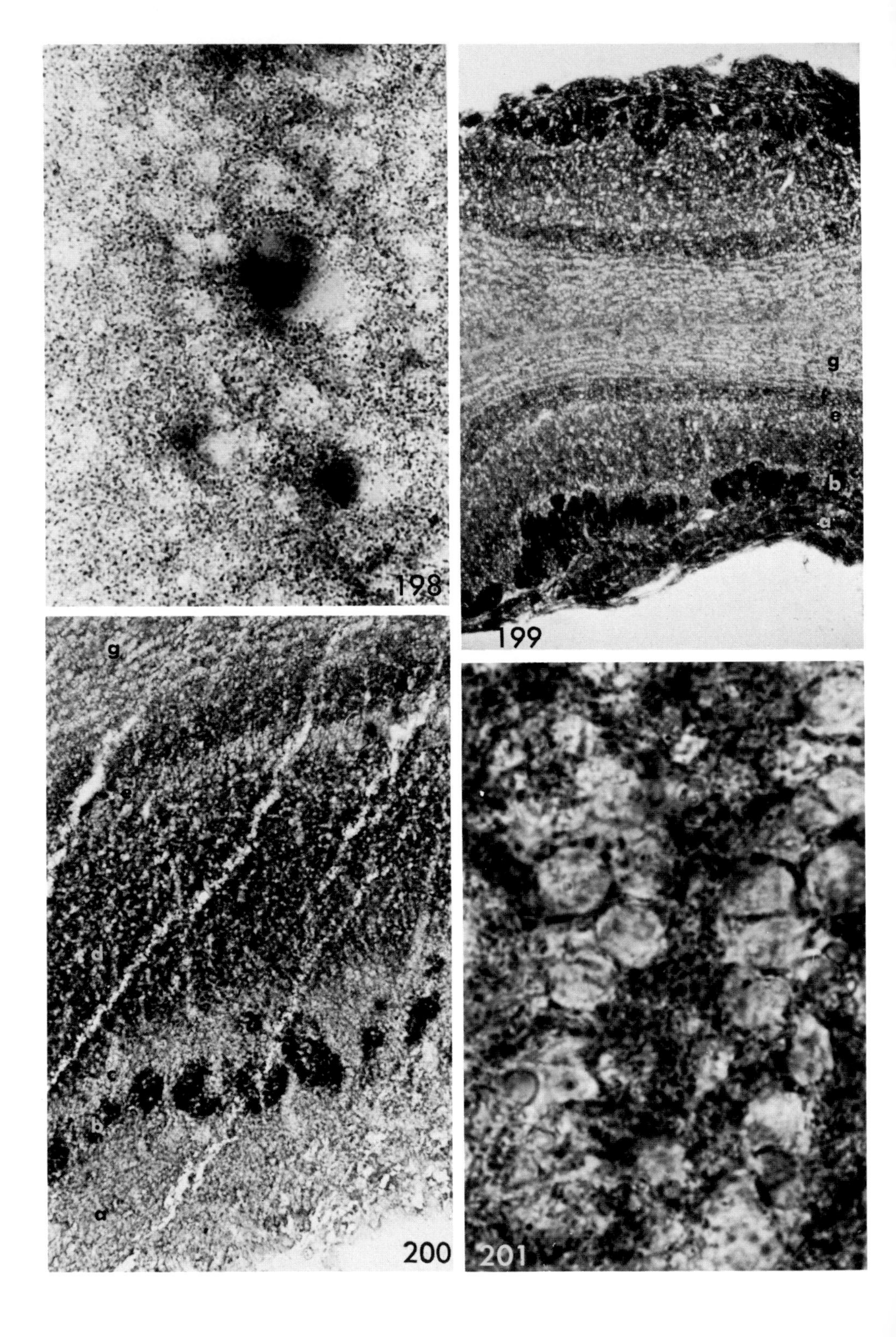
198
g
e
b
a
199
g
d
c
b
a
200
201

FIGURE 198. Olfactory bulb showing strong AC activity in mitral cells and mild activity in granule cells. ×592.

FIGURES 199, 200. LDH and SDH preparations, respectively, of the olfactory bulb showing varying enzyme activity in the different layers (a. olfactory nerve fiber layer, b. olfactory glomeruli, c. external granular layer, d. external plexiform layer, e. mitral cell layer, f. internal plexiform layer, g. internal granular layer). The glomerular and the external plexiform layers show stronger activity compared to other layers. The granule cell layer shows little activity. ×37, ×94.

FIGURE 201. LDH preparation of the granule cells in the olfactory bulb showing moderately strong enzyme activity in the thin rim of cytoplasm. ×1480.

glomerulosa in the rat. His observations on the distribution of AChE activity are similar to those of the present study. The activity is high in the synaptic junctions of the lamina glomerulosa and in the lamina granulosa interna, whereas the reaction is moderate in the lamina gelatinosa, and almost no activity can be seen in the lamina cellularum containing the mitral cells. Girgis (1967) studied the AChE activity in a primate (*Mycocaster coypus*) olfactory bulb and showed fairly good staining in the glomerular layer, perikarya of the mitral cells, outer and inner plexiform layers, and in the tufted mitral cells. Detailed histochemical studies on the olfactory bulb of the squirrel monkey have been made by Shanthaveerappa and Bourne (1965a, 1965b) and Iijima *et al.* (1967), which include the study of glucose-6-phosphatase, cytochrome oxidase, nonspecific cholinesterase, amylophosphorylase, uridine diphosphoglucose glycogen transferase, aldolase, and glucose-6-phosphate dehydrogenase, besides the eight enzymes studied in the rhesus monkey. It is clear from these investigations that the glomeruli are very active histochemically, suggesting, thereby, that they are the seat of active metabolic sequences. Nandy (1965) and Nandy and Bourne (1965, 1966) also studied the distribution of various phosphatases and oxidative enzymes in the rodent olfactory bulb. Sharma (1967a, 1967b) described similar results as earlier discussed by Shanthaveerappa and Bourne (1965a, 1965b) in the olfactory bulb.

The presence of a high complement of oxidative enzymes and many dephosphorylating enzymes in the glomeruli is not surprising, because the olfactory glomeruli are nothing but a mass of synapses with numerous mitochondria and synaptic vesicles (De Lorenzo, 1960). They are surrounded by numerous, strongly amylophosphorylase-positive neuroglial cells, which form a sort of capsule around them, and these glial cells seem to act as energy donors to the glomeruli (Iijima *et al.*, 1967). The glial cells also show considerable LDH and SE activity. There is a very close similarity between the olfactory glomeruli and glomeruli in the granule cell layer of the cerebellum, except that in the latter

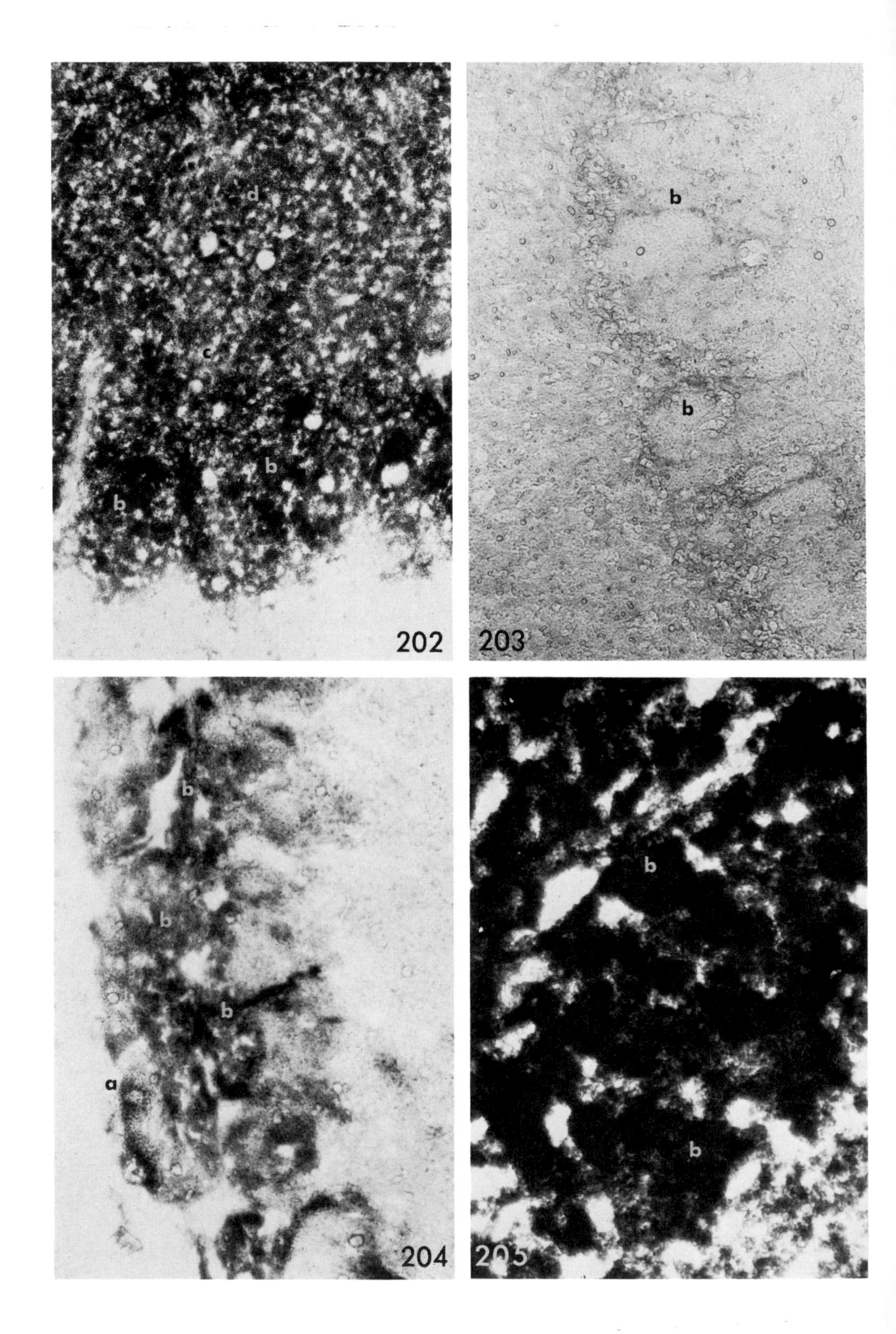
d
c
b
b
202
b
b
203
b
b
b
a
204
b
b
205

FIGURES 202–205. SE, AChE, AK, and ATPase preparations, respectively, of the olfactory bulb showing strong SE activity in glomeruli and external plexiform layers (Fig. 202), mild AChE activity in glomeruli, and negligible activity in other layers (Fig. 203). Also observed is moderately strong AK activity in the nerve fiber layer, moderate AK activity in the glomeruli (Fig. 204), and very strong. ATPase activity in the glomeruli (Fig. 205) [b = olfactory glomeruli]. ×148, ×148, ×148, ×592.

there are not as many surrounding glial cells as in the former. Presence of nonspecific cholinesterase has been described in the squirrel monkey blood vessels surrounding the glomeruli. Along with the peripherally concentrated specific cholinesterase, the nonspecific cholinesterase may hydrolyze any acetylcholine which tends to leave the glomeruli. It is interesting that AChE as well as MAO are present in the glomeruli, which may mean that they are both cholinergic and adrenergic. By the fluorescent histochemical methods, Dahlström *et al.* (1965) observed noradrenaline-containing fibers in the inner plexiform layer. They believed that such fibers arose from the brain stem because the fluorescence did not appear when the caudal portion of the olfactory bulb was sectioned. Vaccarezza and Saavedra (1968) also concluded that the "olfactory bulb seems to be one of the places in which more biochemical, histochemical, pharmacological, electrophysiological and ultrastructural evidences have been accumulated to support the existence of an adrenergic chemical mediator in the central nervous system."

Studies of the rat, squirrel monkey, and rhesus monkey indicate that the size and number of mitral cells are reduced with ascending species of animals and so is the complexity of the Golgi apparatus. This clearly suggests the lesser role played by the olfactory apparatus in microsmatic animals as opposed to the macrosmatic animals (Shanthaveerappa and Bourne, 1965b, 1965c). Also, there is great variation in the Golgi complex in the mitral cells of different species, but the granule cells show the same type of Golgi material, particularly in the rat and squirrel monkey. Shanthaveerappa and Bourne have also shown that the amount of AC present in a cell is directly proportional to the amount of Golgi apparatus. The sites of pigment accumulation, however, show a disproportionate amount of AC activity compared to the activity in Golgi complex.

Shanthaveerappa and Bourne (1965b) indicate that the strong positive activity for various enzymes in the external plexiform layer is due to the presence of abundant mitochondria in the dendritic processes of the mitral cells as well as the presence of a number of tufted cells. It is also observed that there are small streaks of AChE-positive areas, indicating enzymatic activity at the synaptic areas on the cell processes.

The presence of amylophosphorylase, glucose-6-phosphate dehydrogenase, cytochrome oxidase, and aldolase, in addition to LDH and SDH, indicates that the olfactory glomeruli are well equipped with enzymes of glycolytic pathways and the TCA cycle. The presence of AK and AC in the glomeruli is surprising. Although the glomeruli do not contain any thiamine pyrophosphatase (TPPase) positive Golgi apparatus, they show the presence of diffuse TPPase activity. All these results suggest that these three enzymes in the glomeruli are somehow concerned with diffusion permeability and facilitate the transmission of impulses across the synapses. Detailed analysis of TPPase-positive Golgi apparatus indicates that only the mitral cells, tufted cells, granule cells, and neuroglial cells contain the Golgi apparatus (Shanthaveerappa and Bourne, 1965c). Furthermore, the morphological appearance and quantity of Golgi apparatus varies from cell to cell, which suggests a functional difference between the different cells. The Golgi complex in the mitral and tufted cells is in the form of a complex network, whereas in the glial cells and granule cells, it is in the form of small vesicles and granules. In mitral cells the Golgi complex extends into the dendritic processes. The complexity and the amount of this Golgi material are directly proportional to the size of the mitral cell. Smaller neurons found in the mitral cell layer show little or no extension of Golgi material into dendritic processes. In most of the mitral cells, it is observed that the TPPase-positive Golgi material started from a darkly staining oval or round body called the "basal body" situated at the beginning of the dendrite (Shanthaveerappa and Bourne, 1965c).

Shanthaveerappa and Bourne (1966) observed a Pacinian corpuscle located at the inferior surface of the olfactory bulb of the squirrel monkey between the leptomeninges. It was oriented in a coronal plane to the bulb and had typical concentric lamellae of epithelial cells surrounding the central nerve core. It probably subserves pressure and possible vibration sensation. It appears that such a corpuscle may carry pressure sensation on the olfactory bulb exerted by the cerebrospinal fluid.

REFERENCES

Barbera, S., and Galletti, C. Histochemical contribution to the study of the olfactory bulb and the acoustic area in the rabbit and dog. (in Italian). *Otorinolaring. Ital.* **32**:163–176 (1963).

Dahlström, A., Fuxe, R., Olson, L., and Ungerstedt, N. On the distribution and possible function of monoamine nerve terminals in the olfactory bulb of the rabbit. *Life Sci.* **4**:2071–2074 (1965).

De Lorenzo, A. J. Electron microscopy of the olfactory and gustatory pathways. *Ann. Otol.* **69**:410–420 (1960).

Gerebtzoff, M. A. *Cholinesterases.* International Series of Monographs on Pure and Applied Biology (Division: Modern Trends in Physiological Sciences), (P. Alexander and Z. M. Bacq, eds.). Pergamon Press, New York, 1959.

Girgis, M. Distribution of cholinesterase in the basal rhinencephalic structures of the coypu (*Myocastor coypus*). *J. Comp. Neurol.* **129**:85–95 (1967).

Iijima, K., Bourne, G. H., and Shantha, T. R. Histochemical studies on the distribution of some enzymes of the glycolytic pathways in the olfactory bulb of the squirrel monkey (*Saimiri sciureus*). *Histochemie* **10**:224–229 (1967).

Nandy, K. Histochemical study of the olfactory bulb in rodents. *Ann. Histochem.* **10**:245–256 (1965).

Nandy, K., and Bourne, G. H. Histochemical study of the localization of adenosine triphosphatase and 5-nucleotidase in the olfactory bulb of rat. *Histochemie* **4**:488–493 (1965).

Nandy, K., and Bourne, G. H. Histochemical study on the oxidative enzymes in the olfactory bulb of rat and guinea pig. *Acta Histochem.* **23**:86–93 (1966).

Ochi, J. Über die Chemodifferenzierung des bulbus olfactorius von Ratte und Meerschweinchen. *Histochemie* **6**:50–84 (1966).

Scott, T. G. The distribution of 5-nucleotidase in the brain of the mouse. *J. Comp. Neurol.* **129**:97–113 (1967).

Shantha, T. R., and Nakajima, Y. Histological and histochemical studies on the rhesus monkey (*Macaca mulatta*) olfactory mucosa. *Z. Zellforsch.* **103**:291–319 (1970).

Shanthaveerappa, T. R., and Bourne, G. H. Histochemical studies on the olfactory glomeruli of squirrel monkey. *Histochemie* **5**:125–129 (1965a).

Shanthaveerappa, T. R., and Bourne, G. H. Histochemical studies on distribution of dephosphorylating and oxidative enzymes and esterases in olfactory bulb of the squirrel monkey. *J. Nat. Cancer Inst.* **35**:153–165 (1965b).

Shanthaveerappa, T. R., and Bourne, G. H. The thiamine pyrophosphatase technique as an indicator of morphology of the Golgi apparatus in the neurons. III. Studies on the olfactory bulb. *Exptl. Cell Res.* **40**:292–300 (1965c).

Shantaveerappa, T. R., and Bourne, G. H. Pacinian corpuscle of the olfactory bulb of the squirrel monkey. *Nature* **209**:1260 (1966).

Sharma, N. N. Studies on the histochemical distribution of 5-nucleotidase, acid and alkaline phosphatases in the olfactory bulb of rat. *Acta Histochem.* **27**:100–106 (1967a).

Sharma, N. N. Studies on the histochemical distribution of glucose-6-phosphatase, glucose-6-phosphate dehydrogenase, β-glucuronidase and glucosan phosphorylase in olfactory bulb of rat. *Acta Histochem.* **27**:165–171 (1967b).

Vaccarezza, O. L., and Saavedra, J. P. Granulated vesicles in mitral cells and synaptic terminals of the monkey olfactory bulb. *Z. Zellforsch.* **87**:118–130 (1968).

XI

EYE

In the present histochemical study the retina, optic nerve, iridocorneal angle, ciliary process, lens epithelium, cornea, conjunctiva, sclera, choroid, and intra- and extraocular muscles of the eye shall be described.

RETINA

The retina is the nervous coat of the eyeball. As in other primates, including man, the histochemical reactions in the rhesus monkey can be described by examining the individual anatomical layers (Figs. 206–213), which are (a) layer of pigmented epithelium, (b) layer of neuroreceptors (having outer and inner segments), (c) outer limiting membrane, (d) outer nuclear layer, (e) outer plexiform layer, (f) inner nuclear layer, (g) inner plexiform layer, (h) ganglion cell layer, (i) nerve fiber layer, and (j) inner limiting membrane. The intensity of enzyme activity varies from layer to layer and from enzyme to enzyme.

The pigmented epithelial layer shows the strongest positive activity for SE and AC, moderately strong positive activity for LDH and ATPase, moderately positive for SDH, negligible activity for MAO and AK, and negative for AChE.

The outer segments of the neuroreceptor layer are mild to moderately positive for MAO and AK and negative for the rest of the enzymes tested. On the other hand, the inner segments of this layer, which contain mitochondria and the Golgi complex, show strong activity for SDH, LDH, SE, and ATPase; moderately positive for AC and MAO; negligible for AK; and negative for AChE.

The external limiting membrane is difficult to identify in most histochemical preparations except for LDH. It is moderately strongly positive for LDH and mildly positive for SDH. In SE preparations, the positive activity is so strong in the adjoining layers that it is virtually impossible to recognize the limiting membrane.

The outer nuclear layer shows less enzyme activity compared to the inner nuclear layer. The cells here have only a thin rim of cytoplasm surrounding the large nucleus. These layers are much more strongly positive for SE than other layers, but are negative for AK and AChE. The outer nuclear layer shows mild activity for AC, MAO, and ATPase, while the inner layer is moderately positive for these enzymes. Both of these layers are moderately positive for SDH; the outer nuclear layer is moderately positive for LDH, while the inner layer is strongly positive for this enzyme.

The outer and inner plexiform layers show strong LDH, ATPase, and SE activity; moderately strong SDH, AC, and MAO; and mild AK and AChE activity. The enzyme activity is greater in the inner plexiform layer than the outer, except AK, which shows more activity in the latter.

The ganglion cell layer contains cells which vary greatly in size and show strong LDH, SE, and AC; moderately strong SDH; moderate ATPase; mild MAO; and negative AK and AChE activity.

The nerve fiber layer shows the strongest activity for LDH and SE. It is moderate for ATPase; mild for SDH, AC, AChE, and MAO; and shows negligible AK activity.

The internal limiting membrane appears to show negative activity for AK, MAO, AChE, SE, and AC. It is moderately positive for LDH, mildly positive for ATPase, and negligible for SDH.

Histochemical studies on the eye as a whole, and the retina in particular, have been made by a number of workers. Attention has been paid to the distribution of glycogen (Shimizu and Maeda, 1953; Kuwabara and Cogan, 1961; Eichner and Themann, 1962), lipids (Lillie, 1952), monoamines (Malmfors, 1963), oxidative enzymes (Francis, 1953; Wislocki and Sidman, 1954; Cogan and Kuwabara, 1959; Kuwabara and Cogan, 1959, 1960; Niemi and Merenmies, 1961a; Berkow and Patz, 1961a, 1961b; Pearse, 1961; Shanthaveerappa and Bourne, 1964a), monoamine oxidase (Glenner *et al.*, 1957; Shanthaveerappa and Bourne, 1964b), cholinesterases (Koelle *et al.*, 1952; Wislocki and Sidman, 1954; Eränkö *et al.*, 1961; Esila, 1963), β-glucuronidase (Shanthaveerappa and Bourne, 1963a, 1964c), phosphatases (Süllmann and Payot, 1949; Reis, 1954; Shanthaveerappa and Bourne, 1964a, 1964d, 1965c; Shantha *et al.*, 1966), and amylophosphorylase (Shanthaveerappa *et al.*, 1966).

The retina has proved a suitable material for histochemical investiga-

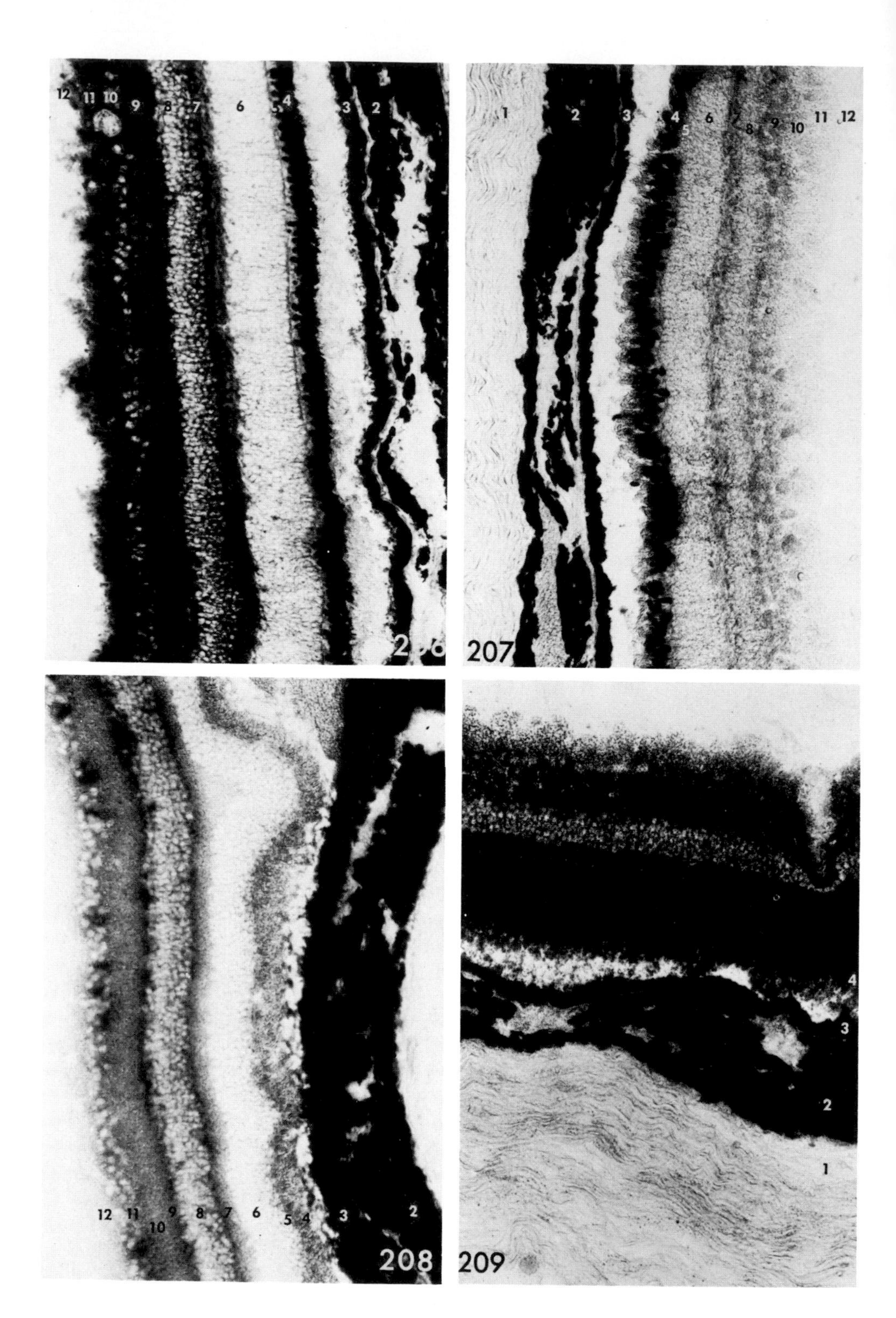
12 11 10 9 8 7 6 5 4 3 2
1 2 3 4 5 6 7 8 9 10 11 12
207
12 11 10 9 8 7 6 5 4 3 2
208
209
4
3
2
1

FIGURES 206–209. LDH, SDH, AC, and SE preparations, respectively, of the retina of the eye (1. sclera, 2. choroid, 3. pigment epithelium, 4. layer of neuroreceptors, 5. outer limiting membrane, 6. outer nuclear layer, 7. outer plexiform layer, 8. inner nuclear layer, 9. inner plexiform layer, 10. ganglion cell layer, 11. nerve fiber layer, 12. inner limiting membrane). The pigment epithelial cells are moderate in LDH, mild in SDH, and strong in AC and SE preparations. The outer half of the layer of neuroreceptors gives a negligible reaction in all these enzymes, whereas the inner half shows strong LDH, SDH, and SE and moderate AC activity. The outer nuclear layer shows a thin rim of positive activity in all the preparations. The inner nuclear layer is enzymatically more active than the outer one. The outer and inner plexiform layers give strong LDH, moderate SDH and AC, and strong SE activity. The ganglion cells are strong in LDH, AC, and SE and moderate in SDH activity. The nerve fiber layer is stronger in LDH, mild in SDH, negligible to mild in AC, and moderate in SE activity. ×148, ×148, ×148, ×148.

tions, particularly for the location of neuronal synapses. Nichols and Koelle (1968) demonstrated the presence of AChE in the inner plexiform layers and amacrine cells in pigeon, ground squirrel, rabbit, rat, and cat, whereas nonspecific cholinesterase was present in the horizontal cells of these animals. They concluded that the amacrine cells are the major source of acetylcholinesterase for the inner plexiform layers, and at this site the enzyme is concerned with synaptic transmission. They showed that in cone retinas (pigeon, ground squirrel) the cells corresponding morphologically to amacrine cells gave an AChE reaction, whereas in the rod retinas (rabbit, rat, cat) diffuse AChE staining was observed in the inner plexiform layer. The AChE-positive bands in the inner plexiform layer represent anatomically the stratified terminal arborizations of the amacrine cells. These, therefore, show somewhat better organization than rod retinas (Nichols and Koelle, 1968). Nichols and Koelle discussed the presence of AChE in amacrine cells in the following manner: ". . . the amacrine cells present an interesting analogy to two groups of inhibitory, AChE-containing neurons, the olivo-cochlear bundle of the inner ear (Churchill *et al.*, 1956; Vinnikov and Titova, 1958; Hilding and Wersall, 1962; Rossi and Cottesina, 1965) and the Renshaw elements of the spinal cord (Erulkar *et al.*, 1968), where acetylcholine does not appear to be the actual neurohumoral transmitter. This conclusion is based on pharmacological evidence that standard cholinergic blocking agents (e.g., atropine, dehydro-β-erythroidine) fail to block transmission at the terminals of the olivo-cochlear bundle (Desmedt and Monaco, 1960; Tanaka and Katsuki, 1966) and the Renshaw cells (Eccles *et al.*, 1954), whereas transmission at both sites is blocked by low concentrations of strychnine."

The outer plexiform layer shows a weak AChE reaction compared to a moderate reaction in the inner plexiform layer. Gerebtzoff (1959)

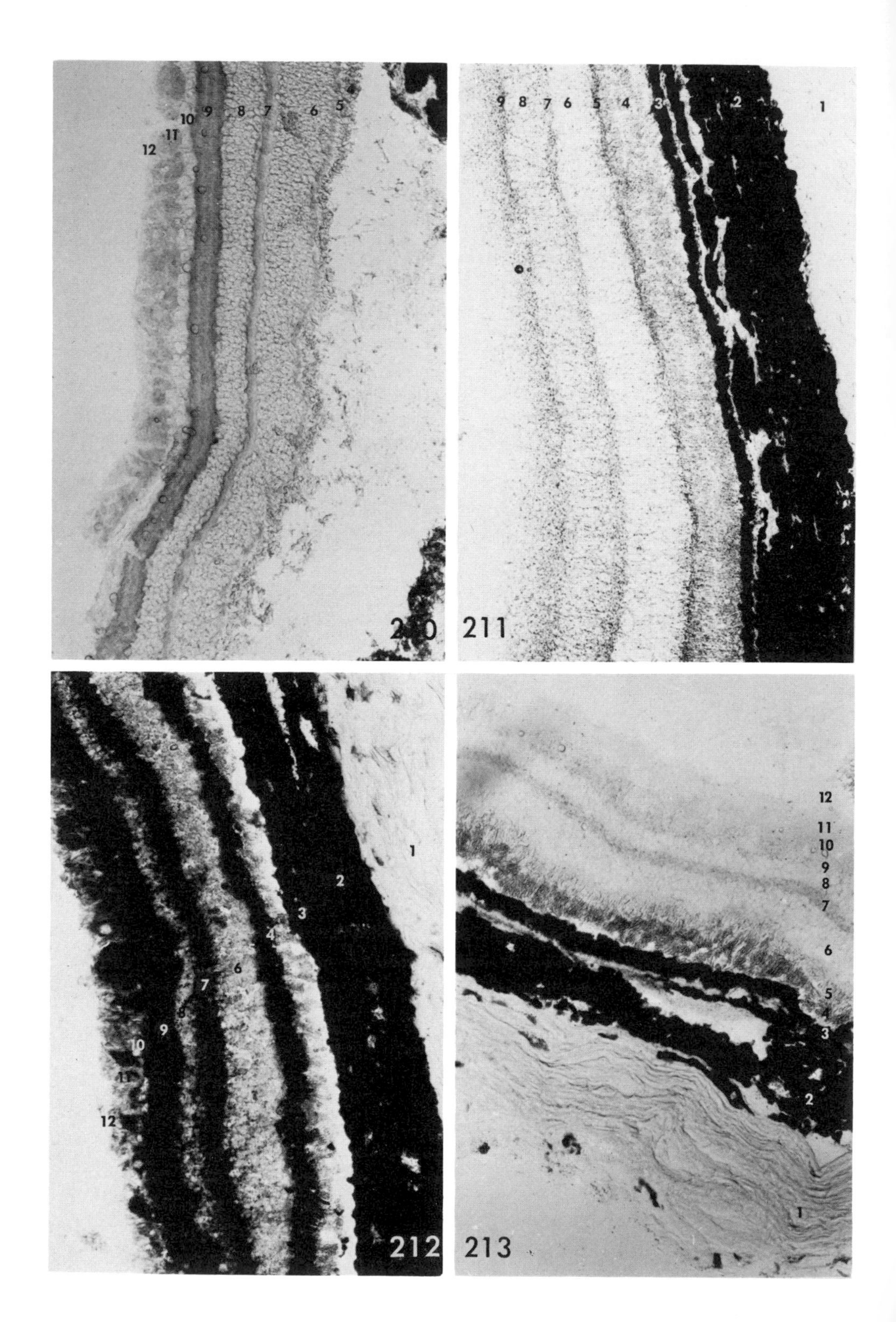

12
11
10
9
8
7
6
5
4
210
9 8 7 6 5 4 3 2 1
211
1
2
3
4
6
7
8
9
10
11
12
212
12
11
10
9
8
7
6
5
4
3
2
1
213

FIGURES 210–213. AChE, MAO, ATPase, and AK preparations, respectively, of the retina of the eye. The pigment epithelial cells are negative in AChE and AK, negligible in MAO, and strong in ATPase preparations. The outer half of the layer of neuroreceptors is negative in AChE and ATPase, mild in MAO and AK preparations. The inner half of the neuroreceptor layer is negligible in AChE and AK, moderate in MAO, and strong in ATPase preparations. The outer and inner nuclear layers show negative to negligible activity in AChE, MAO, and AK, and mild to moderate in ATPase preparations. The outer and inner plexiform layers show mild to moderate AChE and MAO, mild AK, and strong ATPase activity. The ganglion cells are negative in AChE and AK, mild in MAO, and moderate in ATPase preparations. The nerve fiber layer is mild to moderate in MAO and AChE, moderately strong in ATPase, and negative in AK preparations. ×148, ×148, ×148, ×148.

noted that in the inner synaptic layer, a horizontal striation is seen, which corresponds to the super-position of synaptic articulations between bipolar and ganglion cells in the neuropil. It may be that transmission between visual and bipolar cells is not cholinergic, at least in birds and mammals.

MAO activity in all the layers of the retina of the frog, rat, mouse, guinea pig, and rabbit was reported by Eränkö *et al.* (1961). In the present study fair amounts of MAO activity have been observed in the plexiform layer and the outer segments of the visual cells. Similar observations have been recorded earlier in our laboratory (Shanthaveerappa and Bourne, 1964b) and by Mustakallio (1967). The latter author observed the strongest reaction in the outer segments of the visual cells of the retina of cows, pigs, horses, and rats. The reaction in the outer plexiform layer is moderately strong, compared to moderate in the inner plexiform layer, and mild in the ganglion cell layer. The presence of MAO in the plexiform layers (the sites of synaptic contacts between neurons) is understandable. It is, however, interesting that MAO activity in the visual cells is located in the outer rather than the inner segments. No satisfactory answer is available at this time. Leplat and Gerebtzoff (1956) noted norepinephrine and epinephrine in the visual cell layer. The detailed study of Ehinger (1966a) showed that adrenergic nerve fibers in the retina are located at three different sites: between the inner nuclear and inner plexiform layer, in the middle of the inner plexiform layer, and between the inner plexiform and ganglion cell layer. The adrenergic neurons are located in the inner plexiform and the ganglion cell layers (Malmfors, 1963; Haggendal and Malmfors, 1963). This corresponds to the sites of MAO activity in our study. It is suggested that the adrenergic neurons contain noradrenaline or dopamine and, according to Haggendal and Malmfors (1963), dopamine is the dominating catecholamine in the retina of rabbits.

Müller fibers and the glial cells of the retina seem to contain abundant amounts of amylophosphorylase and lactic dehydrogenase. The additional presence of glycogen in them indicates that they act as energy donors to neuronal elements of the retina in times of need. Cogan and Kuwabara (1959) reported that lactate tetrazolium reductase increases in the retina undergoing gliosis. Kuwabara and Cogan (1963) indicated that the amount of glycogen is inversely proportional to the vascularity of the retina. For example, they noted that the rabbit retina has more glycogen and fewer basal blood vessels, while this situation is reversed in the rat and mouse. Lack of most of the enzymes in the internal limiting membrane indicates that this membrane is not very active metabolically.

Succinic dehydrogenase is localized in the inner segments of the photoreceptors; on the other hand, cytochrome oxidase has been reported to be localized in the outer segments (Niemi and Merenmies, 1961b). It is possible that the presence of cytochrome oxidase in the outer segments is due to nonspecific activity given by these lipid-rich segments. Niemi (1965) reported that this nonspecific localization can be excluded by the cyanide inhibition and lipid extraction tests.

Finding alcohol dehydrogenase in pigment epithelium is important because of the involvement of this enzyme in conversion of Vitamin A into retinene. The quantity of this enzyme is said to increase in dark-adapted eyes (Pearse, 1961).

Large concentrations of oxidative enzymes in the plexiform layers indicate a large number of mitochondria in these synaptic areas. Similarly, paucity of mitochondria is probably responsible for the low levels of enzyme activity in the outer and inner nuclear layers. Ganglion cells show large concentrations of LDH and SDH. The extent of SDH activity is slightly variable in these cells, with some of them showing more enzyme activity than others. Precise observations are always difficult to make in the LDH preparations of the ganglion cells due to the presence of high enzyme activity in these cells.

The histochemical observations of Kuwabara and Cogan (1959) and Shanthaveerappa and Bourne (1964a) have shown that two types of dehydrogenases predominate in the retina, one represented by the succinic dehydrogenase system and the other by the DPN- and TPN-linked lactic dehydrogenase. Not much difference of opinion exists as to their distribution. The activity of cytochrome oxidase is similar to that of SDH, and conforms to the distribution of mitochondria as shown by electron microscopic studies (Sjöstrand, 1953; De Robertis, 1956; Ladman and Young, 1958). Lowry *et al.* (1956) studied malic and lactic dehydrogenase microchemically and observed an inverse relationship in their activities, the highest malic dehydrogenase activity being found in the inner segments of rods and cones. Glucose-6-phosphate dehydrogenase (G6PD) and alcohol dehydrogenase (ADH) showed highest activity in the ganglion cell layer. The plexiform layers also showed ADH activity (Niemi and Merenmies, 1961a). The distribution of other dehydrogenases indicates that the ganglion cells are more active than the visual cells, and this conforms with the observation of Lowry *et al.* (1956) that the oxidative metabolism of the inner layers of the retina is characterized by a high rate of glycolysis.

Matschinsky (1968) measured the TPN^+ and TPNH in the different layers of the retinas of the monkey and rabbit and found that: (1) the profile for TPN is not related to the distribution of any of four major TPN-requiring dehydrogenases or their sum; (2) the ratio of TPN/TPNH in the mitochondrial layer is much higher than expected from studies with isolated mitochondria, and (3) the amount of total TPN is high in nonmitochondrial layers.

The distribution of acid phosphatase and thiamine pyrophosphatase (TPPase) reveals that the concentration of both these enzymes is complementary (Shanthaveerappa and Bourne, 1964d, 1965c). In neuroreceptors, for example, AC and the TPPase-positive Golgi apparatus are exclusively found at the junction of the inner and outer segments, whereas the rest of the neuroreceptors are free of any Golgi material or AC. The distribution of β-glucuronidase is also similar to that of AC. Although the plexiform layers do not have the Golgi apparatus, they show a diffuse type of AC activity, probably non-lysosomal and non-membrane-bound, indicating its involvement in permeability and facilitation of nerve impulse conduction in these layers. Mild and mild to moderate AK activity has been observed in the plexiform layers and neuroreceptor layer, respectively, in the present study. This agrees with the earlier observations of Yoshida (1957a), who also noted high alkaline phosphatase activity in the isolated nuclei from the ox retina compared to that of acid phosphatase, although no nuclei reacted positively for AK when stained by this histochemical method (Yoshida, 1957a, 1957b).

OPTIC NERVE

The optic nerve gives a moderately strong ATPase, moderate LDH, SE, and MAO, and mild to negligible SDH, AC, and AChE reaction. β-glucuronidase was found in the cytoplasm of glial cells enclosing the optic nerves (Shanthaveerappa and Bourne, 1963a, 1964c). Mustakallio (1967) observed high MAO activity in the optic nerve of the cat, cow, pig, horse, and rat. The reaction was localized in the myelin sheath.

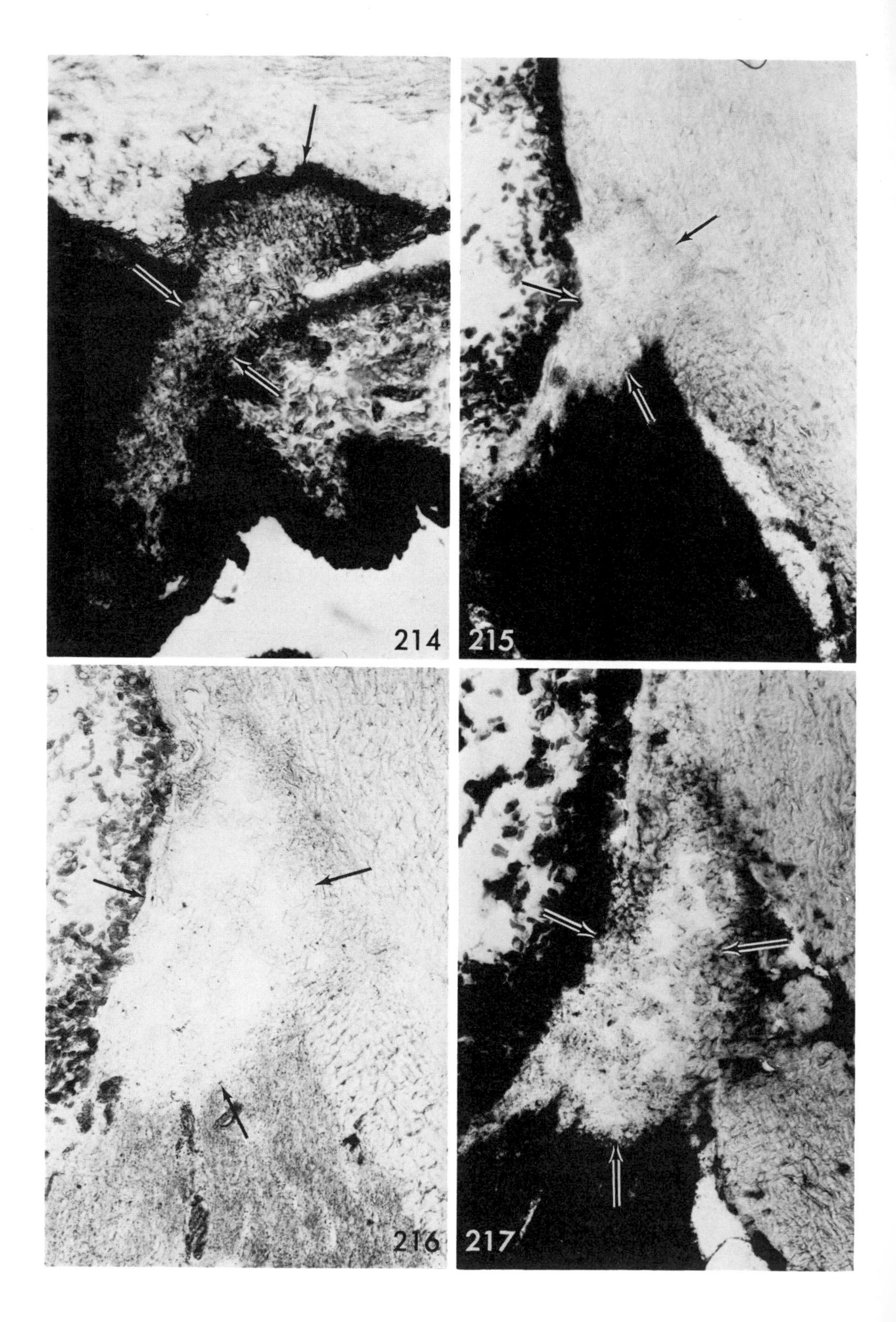
214
215
216
217

FIGURES 214–217. Irido-corneal angle (arrows) as shown in the LDH, SE, MAO, and ATPase preparations, respectively. The meshwork cells show moderately strong activity for LDH and ATPase, mild for SE, and negligible to mild for MAO. The ciliary muscle shows strong LDH, SE, and ATPase and moderate MAO activity. ×94, ×148, ×148, ×148.

The presence of radial bands has also been described in optic nerve myelin sheath (Shanthaveerappa and Bourne, 1962). Adrenergic nerve fibers in the optic nerve have not been reported (Ehinger, 1964, 1966e; Malmfors, 1965).

IRIDOCORNEAL ANGLE

The meshwork cells of this angle show moderate activity for LDH, ATPase, and AC; mild activity for SE, MAO, and SDH; and negative activity for AK and AChE (Figs. 214–219). The core of the meshwork, which is made up of collagen fibers, is negative for all the above enzymes. The degree of enzyme activity increases as we trace these cells toward the canal of Schlemm. In our sections of the eye, the meshwork appears to be made up of small square blocks of tissue. The histochemical reactions given by various components of iris for LDH, SE, MAO, ATPase, SDH, and AC have been shown in Figs. 214–219 and are almost similar to those observed in the rabbit (Shanthaveerappa and Bourne, 1964a, 1964b, 1964d).

The exact function of the endothelial cells of the meshwork is not known. Hitherto, it has been thought that they simply line the spaces in this meshwork. The presence of intercellular pores surrounded by a mass of cement substance in these cells has been demonstrated by Vrabec (1958a, 1958b) and Shanthaveerappa and Bourne (1963b). It is believed that the pores, which are present between the endothelial cells of the meshwork, may be concerned with passage of aqueous humor, and the enzymatic activity in these cells may be concerned with the formation of cement substance. It has been suggested by Vrabec (1958a, 1958b) and Shanthaveerappa and Bourne (1963b) that the substance formed by the meshwork cells may act as nutritive material for the corneal endothelium. The presence of MAO and β-glucuronidase activity in this angle indicates that MAO is concerned with detoxication of amines, and the β-glucuronidase may be concerned with steroid metabolism, hydrolysis of glucuronides, synthesis of mucoid material, and other processes. These meshwork cells also have high AC content and extensive Golgi apparatus (Shanthaveerappa and Bourne, 1965b, 1965c).

All this suggests that these meshwork cells are actively concerned with synthesis and secretory activity and play a major role in regulating the aqueous humor flow and nutrition of the cornea.

CILIARY PROCESS

The ciliary process is made up of two layers of cells, a heavily pigmented basal cell layer and a superficial apical layer which is comparatively less pigmented. As in the choroid, it is very difficult to assess the true enzyme activity in the basal cells except for AC, which shows strong positive activity (Figs. 220–222). Those apical cells which are said to be true secretory cells show strong SDH, LDH, and MAO; moderately strong SE, ATPase, and AC; mild AChE; and negative AK activity (Cogan and Kuwabara, 1959). Evan and Cole (1963) also described the presence of SDH activity in the ciliary process.

The role played by AChE and MAO in these cells is very difficult to determine. There is a strong possibility that these enzymes neutralize any monoamines and acetylcholine which enter the cells during the secretory process, so that very little of these substances will pass into the aqueous humor. MAO activity and the density of adrenergic nerve fibers are inversely proportional (Mustakallio, 1967); this is exemplified by the ciliary muscles, in which the MAO reaction is very strong but which are poor in adrenergic innervation (Ehinger, 1966a). The adrenergic network in the cat iris is abundant with concomitantly low levels of MAO (Ehinger, 1966c). In addition to the ciliary muscle, strong MAO activity is also observed in sphincter and dilator muscles, in the nerve bundles entering the ciliary body, in the nerve plexus within the ciliary muscle, in the walls of the ciliary vessels, and in the vessels of the iris (Lukas and Cech, 1965). Further, AChE may be concerned with the permeability of these cells. It is possible that this enzyme along with other enzymes may be involved in the secretion of aqueous humor. The presence of MAO, therefore, is of great physiological importance, particularly because there is increasing evidence that the autonomic nervous system and certain biogenic amines participate in the formation of the aqueous humor.

It was observed in the rabbit as well as in the rhesus monkey (Shanthaveerappa and Bourne, 1964a) that the intensity of oxidative enzyme activity varied from one group of cells to another, indicating that at any given time some cells are functionally more active than others. That is to say, some groups of cells are actively secreting the aqueous humor, whereas others are probably in a state of rest.

It is interesting that both AChE and MAO are present in the sphincter and dilator muscles of the iris. Niemi and Tarkkanen (1964) observed weak AChE activity in the dilator muscles compared to the MAO reaction, whereas in the sphincter muscle, the picture was reversed. Mustakallio (1967) also showed stronger MAO activity in the dilator muscles. But this does not alter the fact that cholinergic fibers are present in both the dilator and sphincter muscles (Eränkö and Räisänen, 1965; Ehinger, 1966b, 1967). It is also possible that some of the sympathetic fibers may be cholinergic and adrenergic at the same time, as believed by Koelle (1962) and Eränkö and Härkönen (1964).

Shantha *et al.* (1966) showed the presence of cyclic 3′5′ nucleotide phosphodiesterase (3′5′AMPase) in the ciliary processes as well as in various other components of the eye. There is evidence to suggest that norepinephrine and vasopressine stimulate the synthesis of 3′5′AMP which results in inhibition of ocular fluid secretion (Constant and Becker, 1956; Eakins, 1963). This clearly indicates the complex involvement and balance of mechanism of aqueous humor secretion through the ciliary process depending on MAO and 3′5′AMPase. If there is less or no MAO, then 3′5′AMP production is stimulated through the undestroyed adrenaline, resulting in inhibition of aqueous humor secretion. On the other hand, destruction of adrenaline by MAO results in inhibition of 3′5′AMP production, resulting in an increased aqueous humor secretion. The presence of large amounts of MAO in ciliary muscle appears to have control over aqueous humor production. There is every possibility that the amount of MAO in glaucomatous patients' eye ciliary muscle and processes is considerably high.

LENS EPITHELIUM

The epithelial cells of the lens show moderately strong LDH and ATPase; moderate AC, SE, SDH, and MAO; and negative AK and AChE activity (Figs. 224–227). The lens fibers show moderately strong activity for these enzymes. The positive activity rose particularly at the junction of lens epithelium and lens fibers. Very few investigations have been made on the lens epithelium. The present study on the rhesus monkey indicates that the lens epithelium is involved in both the Embden–Meyerhof pathway and TCA cycle. On the other hand, the presence of only LDH in lens fibers indicates that they derive their energy mostly through the Embden–Meyerhof pathway. Paucity of enzyme activity in lens fibers points out that this tissue is quite inactive metabolically.

In a detailed study of dehydrogenases, Cotlier (1964) showed high

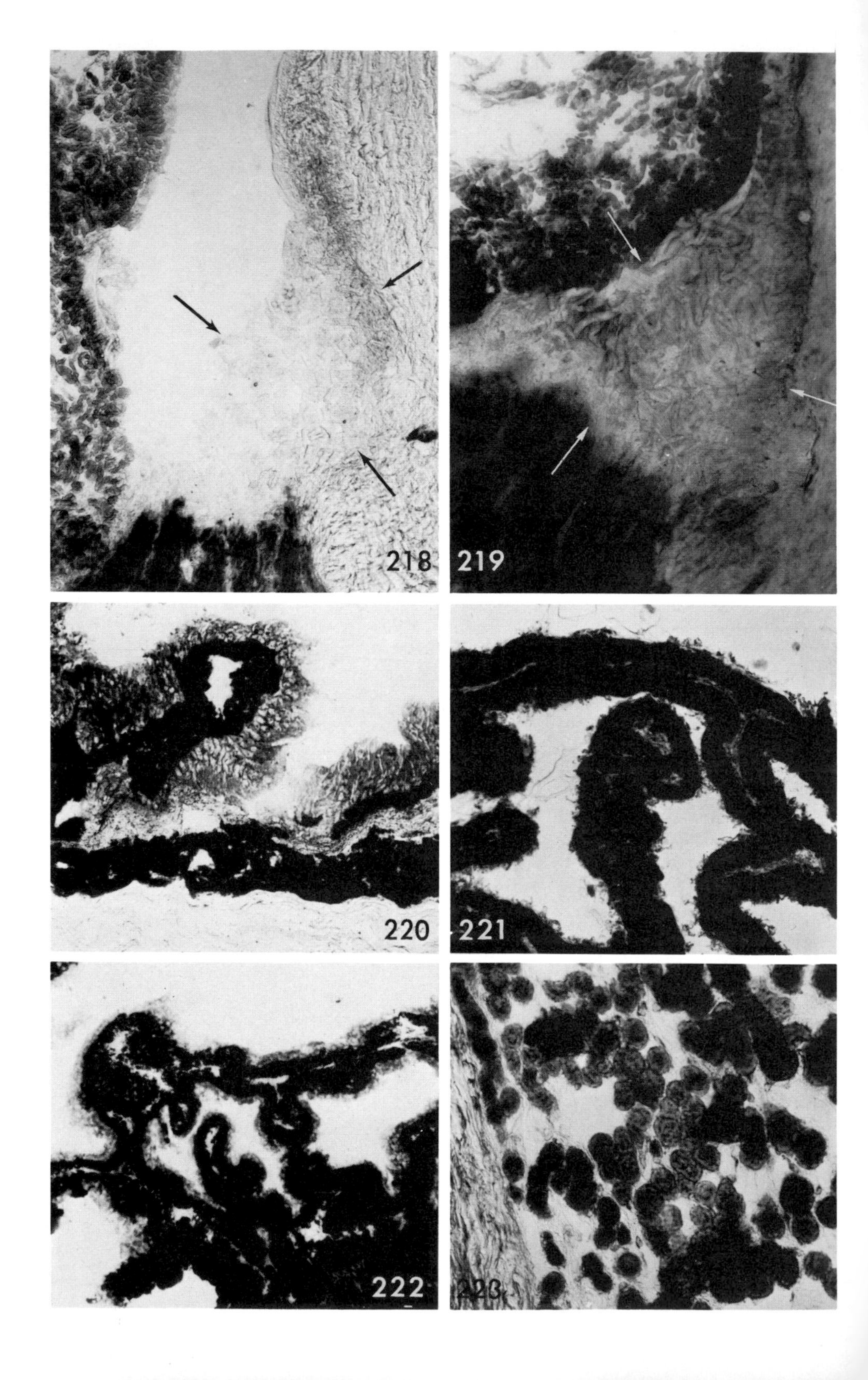
218
219
220
221
222
223

FIGURES 218, 219. Irido-corneal angle (arrows) as shown in the SDH and AC preparations, respectively. The meshwork cells show mild to moderate activity for these enzymes, whereas the ciliary muscle is strongly positive. ×148, ×148.

FIGURES 220–222. MAO, LDH, and SDH preparations, respectively, of the ciliary process showing moderately strong MAO and SDH and strong LDH activity in the apical cells. The basal cells are rich in pigment, and so it is difficult to interpret the true enzyme reaction. ×148, ×148, ×94.

FIGURE 223. External ocular muscles showing moderate, moderately strong, and strong LDH reaction in the various transversely cut fibers. ×148.

lactic and malic dehydrogenase and low uridine diphosphate glucose and glucose-6-phosphate dehydrogenase activities in the rabbit lens epithelium. Wortman and Becker (1956) also showed fair amounts of LDH, malic dehydrogenase, aldolase, fumarase, and glucose-6-phosphate isomerase and small quantities of glucose-6-phosphate dehydrogenase in the rabbit lens capsule epithelium. Small quantities of glucose-6-phosphate dehydrogenase indicate the presence of a direct oxidative system for glucose in the epithelium via the hexose-monophosphate shunt. Cotlier (1964) concluded that specific metabolic pathways operate in the rabbit lens epithelium as compared to the lens stroma.

CORNEA, CONJUNCTIVA, SCLERA, CHOROID

Corneal epithelium gives strong SE, AC, and ATPase; moderately strong LDH; mild to moderate SDH; mild MAO; and negative AK and AChE activity (Figs. 230–234). Basal cells show much more enzyme activity than the superficial cell layer. The corneal stroma as well as Bowman's and Descemet's membranes are negative; on the other hand, corneal stromal cells show positive activity for all the enzymes tested except AChE and AK. Corneal endothelium shows strong LDH and ATPase; mild to moderate SDH, AC, and SE; negligible MAO and AK; and negative AChE activity (Figs. 228, 229).

Mustakallio (1967) observed weak MAO activity in the corneal epithelium of a number of animals. According to him, it may be that the subepithelial network of adrenergic nerve fibers present in the stroma (Lukas and Cech, 1965, 1966; Ehinger, 1966d) is connected with the epithelium. Such a hypothesis would suppose that there is no MAO activity in the endothelium as shown by Mustakallio (1967) but this is not the case (Shanthaveerappa and Bourne, 1964b). A simpler explantation for a weak MAO reaction in the corneal epithelium is that the latter contains only a few scattered mitochondria (Jakus, 1961).

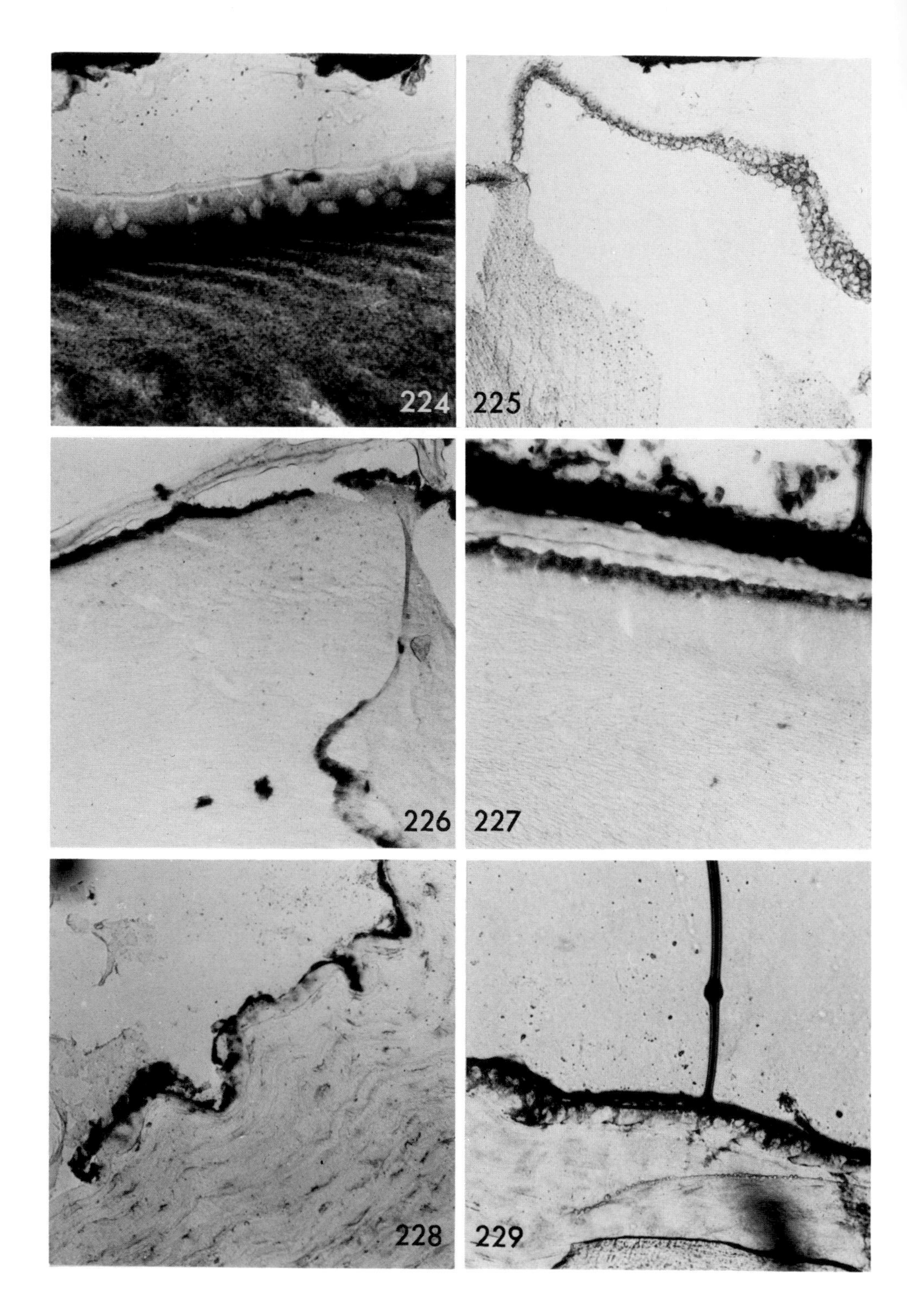
224
225
226
227
228
229

FIGURES 224–227. LDH, MAO, ATPase, and AC preparations, respectively, of the eye showing lens epithelium. Note the strong LDH and ATPase and moderate MAO and AC activity in the lining cells. ×370, ×148, ×148, ×148.

FIGURES 228, 229. ATPase and LDH preparations of the corneal endothelium showing strong enzyme activity in these cells. ×148, ×148.

The presence of large amounts of amylophosphorylase in the cornea of the rabbit and squirrel monkey has been described by Shanthaveerappa *et al.* (1966), indicating that corneal epithelial cells derive their energy through glycolysis.

The conjunctiva shows strong LDH, AC, SE, and ATPase; moderate SDH and MAO; and negative AK and AChE activity (Fig. 235). There are numerous cells rich in pigment (melanin) which are strongly AC-positive. The conjunctival stroma, like the scleral stroma, is virtually free from enzyme activity; however, stromal cells in the conjunctiva and sclera show the same type of positive activity as corneal stromal cells.

Due to the presence of many pigment-containing cells, it is very difficult to assess the true enzyme activity in the choroidal squamous cells and pigment cells (Figs. 208–213). In the AC preparations, the pigment cells seem to exhibit very strong positive activity. The presence of various oxidative and dephosphorylating enzymes has been demonstrated by Shanthaveerappa and Bourne (1965a) in the rabbit choroid. This study has also demonstrated, in a number of species of animals, that the leptomeninges covering the optic nerve continue on to the eye or choroid of the eye; they have homologized this structure with the perineural epithelium of peripheral nerves (Shanthaveerappa and Bourne, 1966). In a study on the optic nerve of man and the rhesus monkey, Shanthaveerappa and Bourne (1964f) found the arachnoid villi formation. They demonstrated the presence of intercellular pores in these villi. The cells of the villi in the optic nerve as well as in the sagittal sinus show high oxidative and dephosphorylating enzyme activity compared to choroid or arachnoid mater (Shantha, unpublished data).

Vincentis (1951) described the corneal AK and AC. Cogan and Kuwabara (1959) described the absence of tetrazolium reductase with succinate as substrate in the corneal stromal cells, corneal endothelial cells, choroid, and pigment epithelium and questionable activity in corneal, ciliary, and iris epithelium. The present studies as well as studies by Shanthaveerappa and Bourne (1964a) contradict these findings.

Histochemically, the most inactive layers of the eye are the stroma of the cornea and sclera (Figs. 207, 209, 213). Interestingly enough, the mesothelial cells covering the sclera show low enzyme activity, com-

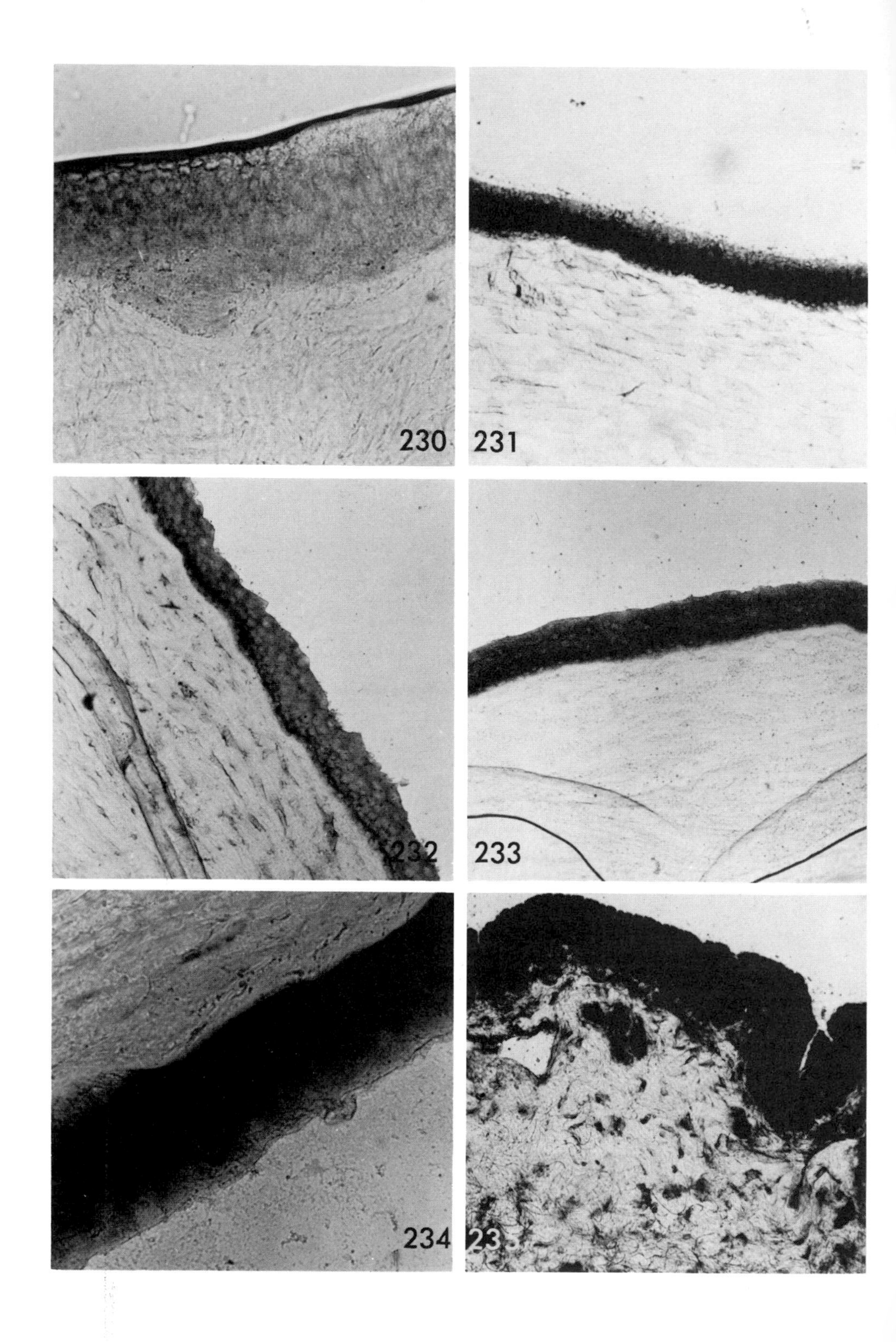

230
231
232
233
234
23

FIGURES 230–234. SDH, LDH, ATPase, SE, and AC preparations, respectively, of the corneal epithelium. Note moderate SDH (Fig. 230); strong LDH (Fig. 231), AC (Fig. 234), and SE (Fig. 233); and moderately strong ATPase (Fig. 232) activity in the epithelial cells. Varying degrees of enzyme activity in the stromal cells can also be seen in these preparations. ×370, ×148, ×148, ×148.

FIGURE 235. LDH preparation of the conjunctiva showing strong enzymatic activity in the lining cells. Note also the moderately strong reaction in the stromal cells. ×148.

pared to the similar peritoneal mesothelial cells which are enzymatically very active (Shanthaveerappa and Bourne, 1964e, 1965d). It appears that these layers act as insulators in separating the active internal and external layers of the eye. The presence of low oxidative enzyme activity in corneal epithelial cells in our preparations is in agreement with the work of Jakus (1961), who described the presence of a few small mitochondria in the corneal epithelial cells. Corneal epithelium shows abundant amounts of AK by Gomori's method, whereas it shows negligible activity by the azo dye method. It also shows the presence of glucose-6-phosphatase. There is every possibility that these enzymes may help in absorption of glucose from lacrimal fluid and from the aqueous humor which permeates the substantia propera. The presence of large amounts of amylophosphorylase indicates that the corneal epithelium draws its major source of energy from the breakdown and utilization of glycogen, especially at times when the endogenous supply of glycogen is low.

Corneal endothelium gives a much stronger positive reaction for the various oxidative and hydrolytic enzymes than does the epithelium. This indicates the former is metabolically much more active than the latter. This finding is further supported by the electron microscopic studies of Donn *et al.* (1961), who showed the presence of abundant mitochondria and numerous membrane-bound vesicles in the corneal endothelium.

INTRA- AND EXTRAOCULAR MUSCLES

The dilator and constrictor muscles of the iris are moderately positive for LDH, SDH, MAO, and AChE; strongly positive for AC, SE, and ATPase; and negative for AK. Similarly, the ciliary muscle exhibits the same type of activity except that the intensity of positive activity is much greater in the ciliary muscle (Figs. 214–219). In addition, it shows comparatively less AChE activity. The external ocular muscles are positive for all these enzymes except AChE and AK (Fig. 223). These muscle fibers exhibit at least three types of positive activity. Some are mild, some moderate, and some strongly positive. This is particularly observed in the LDH, SDH, and SE preparations. Such a difference in enzyme localization is not observed in the iris or ciliary muscles. The presence of large amounts of MAO and AChE in the iris and ciliary muscles is particularly interesting because these are smooth muscles influenced by the autonomic nervous system. High oxidative and dephosphorylating enzyme activity has also been demonstrated in previous studies (Shanthaveerappa and Bourne, 1964a) on these eye muscles, indicating considerable metabolic activity.

REFERENCES

Berkow, J. W., and Patz, A. Histochemistry of the retina. I. Introduction and methods. *Arch. Ophthal. (Chicago)* **65**:820 (1961a).

Berkow, J. W., and Patz, A. Histochemistry of the retina. II. Use of phenazine methosulfate to demonstrate the succino-oxidase system. *Arch. Ophthal. (Chicago)* **65**:828–831 (1961b).

Churchill, J. A., Schuknecht, H. F., and Doran, R. Acetylcholinesterase activity in the cochlea. *Laryngoscope* **66**:1–15 (1956).

Cogan, D. G., and Kuwabara, T. Tetrazolium studies on the retina. II. Substrate dependent patterns. *J. Histochem. Cytochem.* **7**:334–341 (1959).

Constant, M. A., and Becker, B. Experimental tonography. II. The effect of vasopressin chlorpromazine, and phentolamine methane sulfonate. *Arch. Ophthal. (Chicago)* **56**:19–25 (1956).

Cotlier, E. Histo- and cytochemical localization of dehydrogenases in rabbit lens epithelium. *J. Histochem. Cytochem.* **12**:419–423 (1964).

De Robertis, E. Electron microscope observations on the submicroscopic organization of the retinal rods. *J. Biophys. Biochem. Cytol.* **2**:319–330 (1956).

Desmedt, J. E., and Monaco, P. Suppression par la strychnine de l'effet inhibiteur centrifuge exercé par la faisceau olivocochléaire. *Arch. Int. Pharmacodyn.* **129**:244–248 (1960).

Donn, A. K., Gordon, I., Mallett, B. S., and Pappas, G. D. Pinocytosis in the rabbit corneal endothelium. *Arch. Ophthal. (Chicago)* **66**:93–104 (1961).

Eakins, K. The effect of intravitreous injections of norepinephrine, epinephrine, and isoproterenol on the intraocular pressure and aqueous humor dynamics of rabbit eyes. *J. Pharmacol. Exptl. Therap.* **140**:79–84 (1963).

Eccles, J. C., Fatt. P., and Korketsu, K. Cholinergic and inhibitory synapses in a pathway from motoraxon collaterals to motorneurones. *J. Physiol. (London)* **126**:524–562 (1954).

Ehinger, B. Distribution of adrenergic nerves to orbital structures. *Acta Physiol. Scand.* **62**:291–292 (1964).

Ehinger, B. Ocular and orbital vegetative nerves. *Acta. Physiol. Scand.* **67**:Suppl. 268, 1–35 (1966a).

Ehinger, B. Cholinesterases in ocular and orbital tissues of some mammals. *Acta Univ. Lund. Sect. II* No. 2 (1966b).

Ehinger, B. Distribution of adrenergic nerves in the eye and some related structures in the cat. *Acta Physiol. Scand.* **66**:123–128 (1966c).

Ehinger, B. Localization of ocular adrenergic nerves and barrier mechanism in some mammals. *Acta Ophthal.* (*Chicago*) **44**:814–822 (1966d).

Ehinger, B. Adrenergic nerves to the eye and to related structures in man and in the cynomolgus monkey (*Macaca irus*). *Invest. Ophthal.* **5**:42–52 (1966e).

Ehinger, B. Double innervation of the feline iris dilator. *Arch. Ophthal.* (*Chicago*) **77**:54–55 (1967).

Eichner, D., and Themann, H. Zurfrage des Netzhautglykogens beim Meerschweinchen. *Z. Zellforsch.* **56**:231–246 (1962).

Eränkö, O., and Härkönen, M. Noradrenaline and acetylcholinesterase in sympathetic ganglion cells of the rat. *Acta Physiol. Scand.* **61**:299–300 (1964).

Eränkö, O., and Räisänen, L. Fibers containing both noradrenalin and acetylcholinesterase in the nerve net of the rat iris. *Acta Physiol. Scand.* **63**:505–506 (1965).

Eränkö, O., Niemi, M., and Merenmies, E. Histochemical observations on esterases and oxidative enzymes of the retina. In: *The Structure of the Eye* (G. K. Smelser, ed.), pp. 159–171. Academic Press, New York and London, 1961.

Erulkar, S. D., Nichols, C. W., Popp, M. B., and Koelle, G. B. Renshaw elements: localization and acetylcholinase content. *J. Histochem. Cytochem.* **16**:128–135 (1968).

Esilä, R. Histochemical and electrophoretic properties of cholinesterases and nonspecific esterases in the retina of some mammals, including man. *Acta Ophthal.* (*Chicago*) **41**: Suppl. 77, 1–113 (1963).

Evan, C., and Cole, D. P. Succinic dehydrogenase in rabbit ciliary epithelium. *Exptl. Eye Res.* **2**:25–27 (1963).

Francis, C. M. Cholinesterase in the retina. *J. Physiol.* (*London*) **120**:435–439 (1953).

Gerebtzoff, M. A. *Cholinesterase.* International Series of Monographs on Pure and Applied Biology (Division: Modern Trends in Physiological Sciences), (P. Alexander and Z. M. Bacq, eds.). Pergamon Press, New York, 1959.

Glenner, G. G., Burtner, H. J., and Brown, G. W., Jr. The histochemical demonstration of monoamine oxidase activity by tetrazolium salts. *J. Histochem. Cytochem.* **5**:591–600 (1957).

Haggendal, J., and Malmfors, T. Evidence of dopamine-containing neurons in the retina of rabbits. *Acta Physiol. Scand.* **59**:295–296 (1963).

Hilding, D., and Wersall, J. Cholinesterase and its relation to the nerve endings in the inner ear. *Acta Otolaryng.* **55**:205–217 (1962).

Jakus, M. A. The fine structures of the human cornea. In: *The Structure of the Eye* (G. K. Smelser, ed.), pp. 343–366. Academic Press, New York and London, 1961.

Koelle, G. B. A new general concept of the neurohumoral functions of acetylcholine and acetylcholinesterase. *J. Pharm. Pharmacol.* **14**:65–90 (1962).

Koelle, G. B., Wolfand, L., Friedenwald, J. S., and Allen, R. A. Localization of specific cholinesterase in ocular tissues of the cat. *Am. J. Ophthal.* **35**:1580–1584 (1952).

Kuwabara, T., and Cogan, D. G. Tetrazolium studies on the retina. I. Introduction and technique. *J. Histochem. Cytochem.* **7**:329–333 (1959).

Kuwabara, T., and Cogan, D. G. Tetrazolium studies on the retina. III. Activity of metabolic intermediates and miscellaneous substrates. *J. Histochem. Cytochem.* **8**:214–224 (1960).

Kuwabara, T., and Cogan, D. G. Retinal glycogen. *Arch. Ophthal.* (*Chicago*) **66**:680–688 (1961).

Kuwabara, T., and Cogan, D. G. Glycogen in the retina. *Ann. Histochem.* **8**:223–228 (1963).
Ladman, A. J., and Young, W. C. An electron microscopic study of the ductuli efferentes and rete testis of the guinea pig. *J. Biophys. Biochem. Cytol.* **4**:219 (1958).
Leplat, G., and Gerebtzoff, M. A. Localisation de l'acétylcholinestérase et des médiateurs diphénoliques dans la rétine. *Ann. Oculist.* (*Paris*) **189**:121–128 (1956).
Lillie, R. D. Histochemical studies on the retina. *Anat. Rec.* **112**:477–495 (1952).
Lowry, O. H., Roberts, N. R., Chang, M. L. W. The analysis of single cells. *J. Biol. Chem.* **222**:97–107 (1956).
Lukas, Z., and Cech, S. Histochemical localization of monoamine oxidase in the anterior segment of the eye and the adrenergic innervation of its tissues. *Acta Histochem.* **21**:154–164 (1965).
Lukas, Z., and Cech, S. Adrenergic nerve fibers and their relation to monoamine oxidase distribution in ocular tissues. *Acta Histochem.* **25**:133–140 (1966).
Malmfors, F. Evidence of adrenergic neurons with synaptic terminals in the retina of rates demonstrated with fluorescence and electron microscopy. *Acta Physiol. Scand.* **58**:99–100 (1963).
Malmfors, F. Studies on adrenergic nerves. The use of rat and mouse iris for direct observations of their physiology and pharmacology at cellular and subcellular levels. *Acta Physiol. Scand.* **64**: Suppl. 248, 1–93 (1965).
Matschinsky, F. M. Quantitative histochemistry of nicotinamide adenine nucleotides in retina of monkey and rabbit. *J. Neurochem.* **15**:643–657 (1968).
Mustakallio, A. Monoamine oxidase activity in the various structures of the mammalian eye. *Acta Ophthal.* (*Chicago*) **45**: Suppl. 93, 1–62 (1967).
Nichols, C. W., and Koelle, G. B. Comparison of the localization of acetylcholinesterase and nonspecific cholinesterase activities in mammalian and avian retinas. *J. Comp. Neurol.* **133**:1–17 (1968).
Niemi, M. The retina and its diseases. In: *Neurohistochemistry* (C. W. M. Adams, ed.), pp. 599–621. Elsevier, Amsterdam, 1965.
Niemi, M., and Merenmies, E. Cytochemical localization of the oxidative enzyme systems in the retina. I. Diaphorases and dehydrogenases. *J. Neurochem.* **6**: 200–205 (1961a).
Niemi, M., and Merenmies, E. Cytochemical localization of the oxidative enzyme systems in the retina. II. Cytochrome oxidase. *J. Neurochem.* **6**:206–209 (1961b).
Niemi, M., and Tarkkanen, A. Cholinesterases, monoamine oxidase, and phosphorylase in the iris muscles. *Arch. Ophthal.* (*Chicago*) **72**:548–553 (1964).
Pearse, A. G. E. *Histochemistry, Theoretical and Applied.* Little, Brown, Boston, 1961.
Reis, J. L. Histochemical localization of alkaline phosphatase in the retina. *Brit. J. Ophthal.* **38**:35–38 (1954).
Rossi, G., and Cottesina, G. The efferent cochlear and vestibular system. *Acta Anat.* **60**:362–381 (1965).
Shantha, T. R., Woods, W. D., Waitzman, M. B., and Bourne, G. H. Histochemical method for localization of cyclic 3′,5′-nucleotide phosphodiesterase. *Histochemie* **7**:177–190 (1966).
Shanthaveerappa, T. R., and Bourne, G. H. Radial bands in the optic nerve myelin sheath. *Nature* **196**:1215–1217 (1962).

Shanthaveerappa, T. R., and Bourne, G. H. Further histochemical observations on the radial bands of the optic nerve myelin sheath. Studies on the distribution of β-glucuronidase. *Exptl. Cell Res.* **32**:196–199 (1963a).

Shanthaveerappa, T. R., and Bourne, G. H. Some observations on the corneal endothelium. *Acta Ophthal.* **41**:683–688 (1963b).

Shanthaveerappa, T. R., and Bourne, G. H. Histochemical studies on the distribution of oxidative and dephosphorylating enzymes of the rabbit eye. *Acta Anat.* **57**:192–219 (1964a).

Shanthaveerappa, T. R., and Bourne, G. H. Monoamine oxidase distribution in the rabbit eye. *J. Histochem. Cytochem.* **12**:281–287 (1964b).

Shanthaveerappa, T. R., and Bourne, G. H. Histochemical studies on the distribution of β-glucuronidase in the eye. *Histochemie* **3**:413–421 (1964c).

Shanthaverrappa, T. R., and Bourne, G. H. Histochemical studies on the distribution of acid phosphatase in the eye. *Acta Histochem.* **18**:317–327 (1964d).

Shanthaveerappa, T. R., and Bourne, G. H. Observations on the histological structure of the sclera of the eye. *Acta Anat.* **58**:296–305 (1964e).

Shanthaveerappa, T. R., and Bourne, G. H. Arachnoid villi in optic nerve of man and monkey. *Exptl. Eye Res.* **3**:31–35 (1964f).

Shanthaveerappa, T. R., and Bourne, G. H. Histological and histochemical studies of the choroid of the eye and its relation to the pia arachnoid mater of the central nervous system and the perineural epithelium of the peripheral nervous system. *Acta Anat.* **61**:379–398 (1965a).

Shanthaveerappa, T. R., and Bourne, G. H. Histochemical studies on the distribution of oxidative and dephosphorylating groups of enzymes in the meshwork cells of the anterior chamber angle of the eye. *Am. J. Ophthal.* **60**:49–55 (1965b).

Shanthaveerappa, T. R., and Bourne, G. H. Histochemical demonstration of thiamine pyrophosphatase and acid phosphatase in the Golgi region of the cells of the eye. *J. Anat. (London)* **99**:103–117 (1965c).

Shanthaveerappa, T. R., and Bourne, G. H. Histochemical studies on the localization of oxidative and dephosphorylating enzymes and esterases in the peritoneal mesothelial cells. *Histochemie* **5**:331–338 (1965d).

Shanthaveerappa, T. R., and Bourne, G. H. Perineural epithelium: a new concept of its role in the integrity of the peripheral nervous system. *Science* **154**:1464–1467 (1966).

Shanthaveerappa, T. R., Waitzman, M. B., and Bourne, G. H. Studies on the distribution of phosphorylase in the eyes of the rabbit and squirrel monkey. *Histochemie* **7**:80–95 (1966).

Shimizu, N., and Maeda, S. Histochemical studies of glycogen of the retina. *Anat. Rec.* **116**:427–438 (1953).

Sjöstrand, F. S. The ultrastructure of the outer segments of rods and cones of the eye as revealed by the electron microscope. *J. Cellular Comp. Physiol.* **42**:15–44 (1953).

Sullman, H., and Payot, P. Histochemische Untersuchungen über alkalische Phosphatase in der Cornea. *Ophthalmolgica* **118**:303–328 (1949).

Tanaka, Y., and Katsuki, Y. Pharmacological investigation of cochlear inhibition. *J. Neurophysiol.* **29**:94–108 (1966).

Vincentis, M. Ulteriore contributo istochimico allo studio della forfata nei Tessuti oculari. *Arch. Otall.* **55**:303–328 (1951).

Vinnikov, A., and Titova, L. K. Presence and distribution of AChE in the organ of

Corti in animals who are in a state of relative rest, and under conditions of acoustic effect. *Dokl. Akad. Nauk. SSSR.* **119**:164–168 (1958).

Vrabec, F. Studies on the corneal and trabecular endothelium. I. Cement substance of the corneal endothelium. *Brit. J. Ophthal.* **42**:529–534 (1958a).

Vrabec, F. Studies on the corneal and trabecular endothelium. II. Endothelium of the zone of transition. *Brit. J. Ophthal.* **42**:667–673 (1958b).

Wislocki, G. B., and Sidman, R. L. The chemical morphology of the retina. *J. Comp. Neurol.* **101**:53–91 (1954).

Wortman, B., and Becker, B. Enzymatic activities of the lens. *Am. J. Ophthal.* **42**: Part II, 342–345 (1956).

Yoshida, M. The alkaline phosphatase activities in the isolated nuclei of ox retina. *J. Physiol. Soc. Japan* **7**:190–194 (1957a).

Yoshida, M. The distribution of the alkaline phosphatase in the ox retina. *J. Physiol. Soc. Japan* **7**:195–198 (1957b).

XII

CHOROID PLEXUS, EPENDYMA, MENINGES, and PERINEURAL EPITHELIUM

CHOROID PLEXUS

The choroid plexus (CPL) of the lateral and IVth ventricle has been investigated in the present study. In the former, the CPL consists of oblong, lobulated masses, whereas it is more compact and extensive in the IVth ventricle. Each villus of the CPL consists of a layer of lining cells, enclosing the stroma on all sides except at the point of entry of the stromal blood vessels. The stroma is made up of connective tissue elements, blood vessels, etc. The epithelial lining of the plexus is generally single layered and quite active enzymatically. Kaluza *et al.* (1964) and Shantha and Manocha (1968) discussed the role of the choroid plexus after a detailed study of oxidative and hydrolytic enzymes and suggested a relationship between the enzyme content and the transport function. Similar conclusions have been drawn earlier by some morphological and experimental studies (Pappenheimer *et al.*, 1961; Welch, 1962). Kaluza *et al.* (1964) believed that there is a physicochemical process based on the movement of chemical particles across the epithelium of the choroid plexus.

Histochemically, the epithelial lining cells of the choroid plexus in the rhesus monkey show strong SDH activity in the cytoplasm, mild

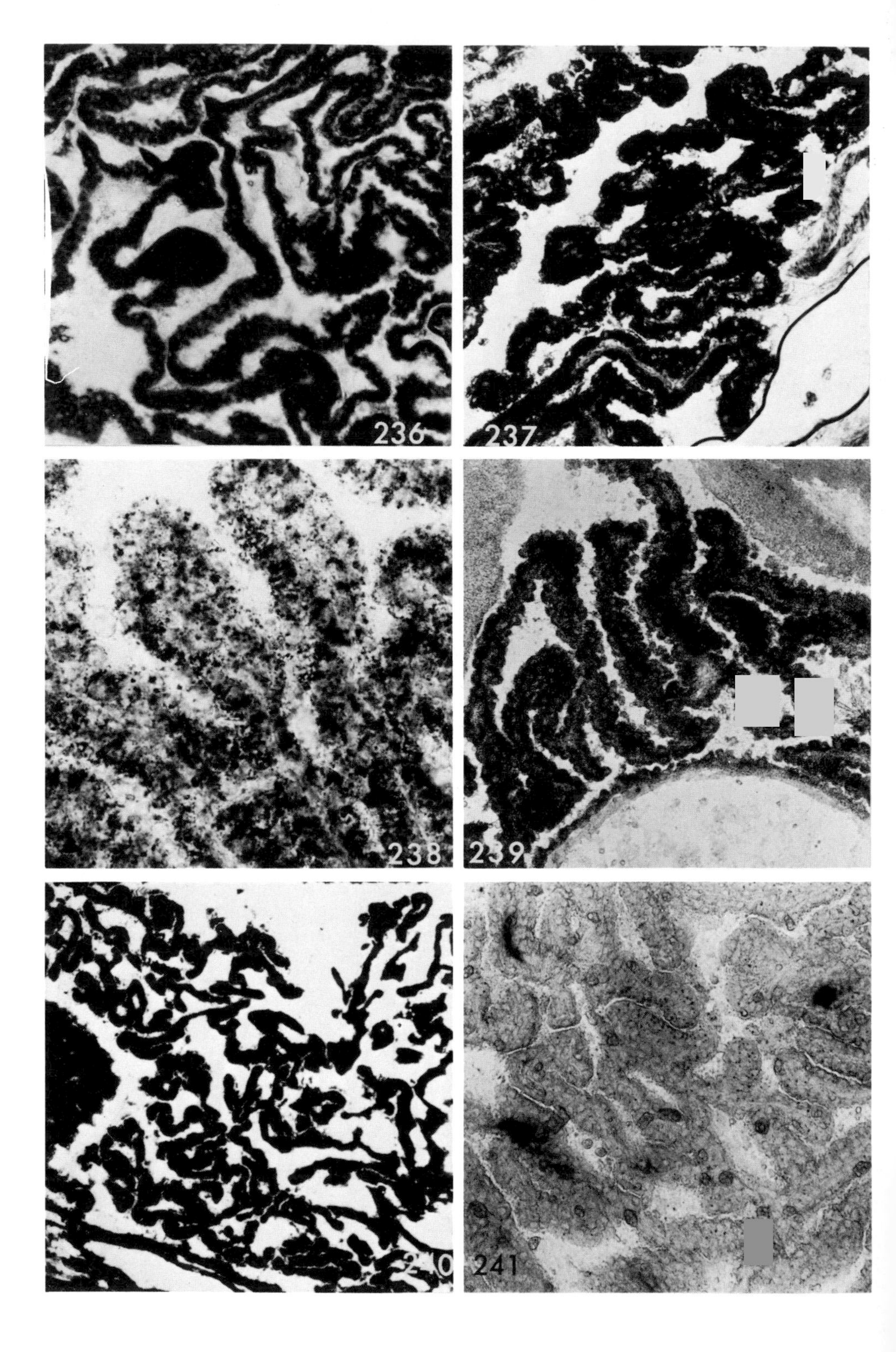
236
237
238
239
241

FIGURES 236–241. SDH, LDH, MAO, SE, ATPase, and AK preparations, respectively, of the choroid plexus. The lining epithelial cells show very strong activity for SDH, LDH, and ATPase; strong for SE; moderately strong for MAO; and moderate for AK preparations. The stroma is moderate in LDH, mild in SDH, moderately strong in ATPase, negligible in MAO, and mild to moderate in AK and SE preparations. The stromal blood vessels also give a strong reaction in AK and ATPase preparations. ×148, ×148, ×370, ×148, ×37, ×148.

in the nucleus, moderate in the nucleolus. Stromal activity in mild (Fig. 236). This compares with a very strong reaction in the lining cells and a moderate reaction in the stroma in the LDH preparations (Fig. 237). The positive activity is both diffuse and granular in character. The epithelial lining cells also show the presence of a large number of small, round and oval vesicles with negligible enzyme reaction. It is probable that these are the cerebrospinal fluid (CSF) vesicles, formed in the cytoplasm as a result of the secretory activity of the epithelial cells and later extruded into the ventricular cavities. The oxidative enzymes are not concentrated in the apical borders of the cells, as observed in the ependymal cells, although the choroid plexus cells originate from the ependyma. The monoamine oxidase preparations show a moderately strong reaction in the lining cells and mild activity in the stroma (Fig. 238). Similar observations have been made in the mouse (Arioka and Tanimukai, 1957), but somewhat stronger MAO activity in both the cytoplasm and nucleus of the lining cells was observed in the rat (Shantha and Manocha, 1968), which may be due to a species difference.

Moderately strong activity for simple esterases is seen in the lining cell cytoplasm, compared to a mild activity in the stroma (Fig. 239). Leduc and Wislocki (1952) described a similar reaction for nonspecific esterases in the choroid plexus lining cells of the rat. This compares to negative AChE activity. In the latter only some blood vessels give a positive reaction.

The alkaline phosphatase preparations show moderately strong activity in the stromal blood vessels, whereas the lining cells are mild (Fig. 241). There is slightly more activity at the free borders of the cells. The weak alkaline phosphatase activity reported in the present study was also reported earlier by Wislocki and Dempsey (1948), Fisher *et al.* (1959), and Rudolph and Sotelo (1962). AC activity is intensely strong in the lining cells, moderate in the stroma, and is spread throughout the cytoplasm (Fig. 248), but not at the luminal border as is the case in AK. The high activity of AC in the choroid plexus may imply a secretory function (Friede and Knoller, 1965).

ATPase activity is strong in the cells, moderately strong in the stroma, negligible in the nucleus, and moderate in the nucleolus (Fig. 240).

Scott (1967) showed that the choroid plexus exhibits the strongest activity of 5′-nucleotidase compared to any other part of the mouse brain.

The Golgi material in the rhesus monkey has been described by Wislocki and Dempsey (1948). The present authors showed in the squirrel monkey and the rat that the lining cells of the choroid plexus contain a well-developed Golgi apparatus distributed all over the cytoplasm, although it varies in size and shape (Shantha and Manocha, 1968). Some cells contain a darkly staining, highly fenestrated network compared to the lightly staining TPPase material in other cells. A few cells also show mixed vesicles, vacuoles, granules, and network types of Golgi apparatus. A similar network type of Golgi apparatus has been described earlier in the epithelial cells of the choroid plexus of human brain using the conventional Golgi techniques (Kopsch, 1926). The large quantity of reticular Golgi network of various types indicates a secretory activity of these cells. This function of secreting CSF has been suggested by a number of cytological studies (Schaltenbrand, 1955; Ariens–Kappers, 1958). It is probable that the ependymal cells, which are continuous with the lining cells of the choroid plexus, are also responsible for the production of CSF, but the experimental evidence suggests that the quantity produced by the choroid plexus is much greater than that produced by the ependyma. This explains why the latter has less Golgi material than the former. The regulation of the composition of the CSF was also attributed to the choroid plexus by Herlin (1956) while working on phosphate exchange and metabolic activity in barriers. This may indicate that the choroid plexus has within its structure some barrier mechanisms between the blood and the CSF.

Histochemically, the choroid plexus is a highly active tissue, as revealed by its high content of oxidative enzymes. The lining cells show much more activity than the stroma. Friede (1961a) found that the rate of respiration of these cells is twice that of the gray matter of the cortex and basal ganglia, and four times that of the white matter, although anaerobic glycolysis is less than that of the gray matter (Kemeny *et al.*, 1961). A comparative study of succinic dehydrogenase activity in these cells in various mammals showed that SDH is somewhat more active in species living in desert environments (Quay, 1960), although most other species also showed a high content of oxidative enzyme activity. Masai and Matano (1961) demonstrated high levels of SDH and cytochrome oxidase activity in amphibians and birds, and recently Paul (1968) showed distinct activity of SDH and LDH in the epithelium of the IIIrd and IVth ventricles of *Rana temporaria* uniformly distributed all over the cytoplasm. This is in contrast to apical accumulation of enzymatically active material in the ependymal cells. High levels of indophenol oxidase (Leduc and Wislocki, 1952; Shimizu *et al.*, 1957; Friede, 1961a, 1961b), SDH (Rutenburg *et al.*, 1953; Shimizu and Morikawa, 1957; Nachlas *et al.*, 1957; Becker *et al.*, 1960; Shantha and Manocha, 1968), NAD and NADP-linked dehydrogenases (Thomas and Pearse, 1961; Friede, 1961b; Chason and Pearse, 1961; Lazarus *et al.*, 1962), and glucose-6-phosphate dehydrogenase (Abe *et al.*, 1963; Shantha and Manocha, 1968) have been shown in the epithelial lining of the choroid plexus. It seems that the lining cells of the choroid plexus may show more of the anerobic pathway enzymes than the citric acid cycle enzymes, indicating some preference for the anaerobic glycolytic pathway enzymes. The citric acid cycle enzymes, however, show sudden increases in the postnatal developing brain (Friede, 1959; Quay, 1960; Kaluza *et al.*, 1964). For example, on the first day after birth the SDH activity of the epithelium is weak, but in the following days the enzyme activity increases so rapidly that adult levels are achieved within a week after birth (Friede, 1959; Quay, 1960).

In the rhesus monkey, Wislocki and Dempsey (1948) showed that the choroid plexus lining cells have a ciliated border. Cilia have also been shown by electron microscopic studies in embryonic as well as adult choroid plexus (Maxwell and Pearse, 1956; Miller and Rogers, 1956; Wislocki and Ladman, 1958; Tennyson and Pappas, 1961, 1964; Carpenter, 1966). However, our histochemical preparations show no cilia, probably because they lack the enzymes investigated in the present study.

EPENDYMA

The ependyma is made up of short columnar epithelial cells and lines the central canal of the spinal cord and ventricles of the brain. The LDH and SDH preparations show moderately strong to strong activity which is mostly restricted to the apical or luminal parts of the ependymal cells. The nucleus and nucleoli give a negative to negligible reaction (Figs. 242, 243). The cells are moderately MAO-positive, and the enzyme activity is distributed throughout the cytoplasm (Fig. 244). The AK and AC preparations show a negligible to mild and moderately strong activity, respectively, in the apical parts of the cells (Fig. 245), whereas the ATPase reaction is not only strong throughout the cytoplasm of the ependymal cells and in the nucleolus but also in the subependymal layer. The latter also shows moderately strong to strong MAO activity (Fig. 244). The SE activity is mild to moderate (Fig. 247) and the AChE is negative to negligible. It may be added that the AK preparation prepared by Gomori's method (Gomori, 1939) shows moderate

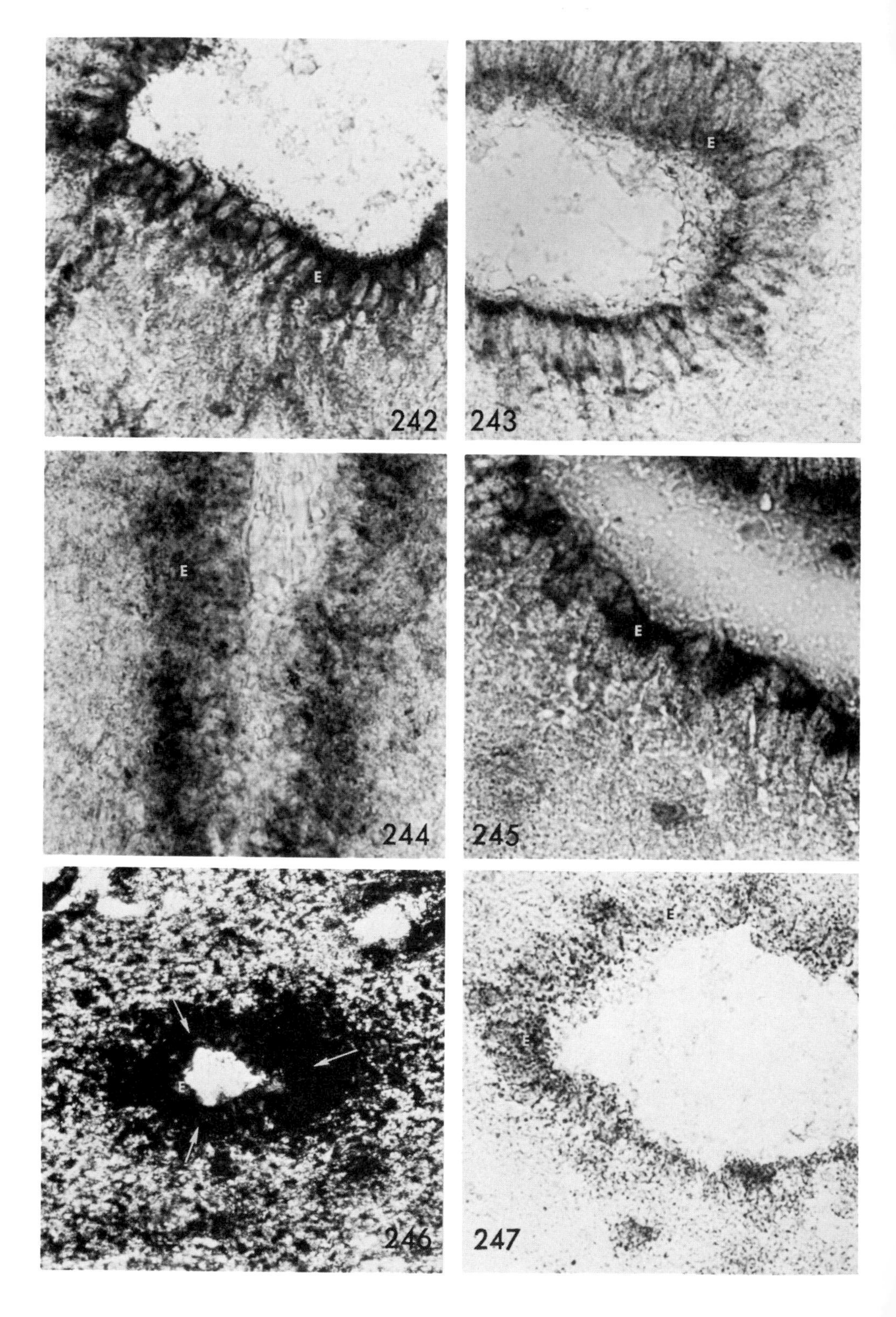
E
242
E
243
E
244
E
245
E
246
E
E
247

FIGURES 242–247. Photomicrographs of the ependymal cells lining the central canal and the third ventricle showing strong LDH (Fig. 242), moderately strong SDH (Fig. 243), moderate MAO (Fig. 244), strong AC and ATPase (Figs. 245, 246), and moderate SE (Fig. 247) reactions. It is interesting to note that most of the preparations show the enzyme activity mostly toward the luminal border. Also the subependymal layer in the ATPase preparation (Fig. 246) is very strongly positive (arrows). ×592, ×592, ×592, ×592, ×148, ×148.

AK activity in the apical parts of the ependymal cells, whereas in those sections stained by the azo dye technique (Burstone, 1958), the AK reaction is mild. Although negative to negligible activity of AChE is observed in the adult animals used in the present study, Krnjevic and Silver (1966) found AChE activity in the germinal layers of the striatal ependyma in the fetal kitten of 28 days gestation. These authors also observed that by 41 days, the activity of this region was greatly reduced, and by 50 days it had almost disappeared.

Histochemically, the ependymal cells have been studied by a number of workers (Pighini, 1912; Leduc and Wislocki, 1952; Rutenberg *et al.*, 1953; Shimizu *et al.*, 1957; Nachlas *et al.*, 1957; Shimizu and Morikawa, 1957; Becker *et al.*, 1960; Friede, 1961a, 1961b; Thomas and Pearse, 1961; Abe *et al.*, 1963; Nandy and Bourne, 1964, 1965; Albert *et al.*, 1966; Shuttleworth and Allen, 1966; Manocha *et al.*, 1967; Iijima *et al.*, 1967) in various parts of the central nervous system, including all the ventricles. Strong activity of SDH, cytochrome oxidase, TPN diaphorase, and DPN diaphorase in the ependymal lining of the ventricle of the cat, dog, guinea pig, rat, and rabbit and that of the spinal cord of the squirrel monkey has been observed by Friede (1961a), Thomas and Pearse (1961), and Manocha *et al.* (1967), thereby showing a good correlation between the enzyme activity and the distribution of mitochondria. The presence of oxidative enzymes in the apical parts of the cytoplasm indicates active metabolism in this part of the cell. Similar conclusions about the energy requirements of these cells can be drawn from their reactions for ATPase and acid phosphatase. The AC activity near the lumen of the ventricle is the same as that demonstrated by Becker *et al.* (1960), who showed a similar distribution of lysosomes. It is possible that this enzyme, along with AK, participates in the vital transport mechanisms of the cell walls such as pinocytosis, phagocytosis, and differential permeability of membranes, as described by Bourne (1958), DeDuve (1959), and Novikoff (1959a, 1959b). There is no doubt that these cells are metabolically very active and are not merely a mechanical lining or an anatomical limiting membrane; their function may be quite varied, such as secretion, absorption, and/or active transport of substances to and fro.

The strong ATPase reaction (Fig. 246) given by the nucleoli in the ependymal cells has also been described by Nandy and Bourne (1964). They suggest that the nucleoli and their precursors are important sites for the production of the enzyme, which then gradually passes into the cytoplasm. Our study on the ependyma of the rhesus monkey does not confirm these findings. The ATPase reaction at the cell membrane has been explained in terms of its role in ionic transport by Pope and Hess (1957).

The subcommissural organ (SCO) is a plate of modified ependyma attached to the ventral surface of the posterior commissure. The SCO forms the lining of the roof of the third ventricle near its junction with the cerebral aqueduct. It shows strong G6PD and TPNH diaphorase activity compared to a weak reaction in the adjoining ependymal cells, whereas DPNH diaphorase, lactic, succinic, and malic dehydrogenase activity is just the reverse (DeLong and Balogh, 1965). Barlow *et al.* (1967) showed that an increasing amount of gliosis and a concomitant decrease in endoplasmic secretory material accompanies the involution of the deep layer of the SCO in the adolescent and adult sheep.

It may be important that the Golgi apparatus, as revealed by the thiamine pyrophosphatase (TPPase) technique in the ependymal cells of the rat (Shantha and Manocha, 1968), shows vesicular, granular, and comma-shaped TPPase-positive Golgi material located only at the apical borders of the cell. Similar observations have been made by Brightman and Palay (1963) in their electron microscopical studies of rat ependyma. The presence of such a Golgi apparatus in the ependymal lining of the embryologically split double central canal has been described (Shanthaveerappa and Bourne, 1966b). Barlow *et al.* (1967) observed in the SCO that the Golgi apparatus in the superficial layer consists of vesicular configurations connected by fine threads arranged parallel to the long axis of the cell, whereas in the deep layer, the Golgi apparatus is closely applied to the nucleus. These morphological and histochemical observations further add to the functional significance of these cells lining the central canal or the ventricles.

Friede (1961a) showed that the ependymal lining of certain parts of the ventricular system, such as the infundibulum, shows exceptional behavior; for example, little oxidative enzyme activity (Shimizu and Morikawa, 1957; Chason and Pearse, 1961; Chason *et al.*, 1963). Similar observations have been made on the cells lining the area postrema (Friede and Pax, 1961). It is possible that in some areas the ependymal cells, because of their location, may not be as actively functional as in other areas and hence may not need a full complement of oxidative enzymes. In the developing brain, SDH activity in the ependyma reaches adult levels within two weeks after birth in rats; may decrease 10–15% in old age in hamsters, mice, and cats; is not readily modified by various techniques that may modify CSF pressure or composition; and is particularly high in certain specialized rodents (Quay, 1960).

In addition to oxidative enzymes, the ependymal cells of the spinal cord and the glial cells show extremely strong phosphorylase activity in the nuclei and a mild to moderate reaction in the cytoplasm (Manocha *et al.*, 1967). The subependyma gives a strong reaction for the phosphorylases. This is in contrast to the lack of oxidative enzymes in the subependymal layer. A mild to moderate phosphorylase reaction in the ependymal cells and a strong reaction in the subependymal tissue for phosphorylases may suggest that the subependymal tissue acts as energy donor to the ependyma in the form of glucose-1-phosphate or glycogen. This is especially so in view if the lack of direct vascular supply to the ependyma. The subependyma also shows a strong reaction for monoamine oxidase and ATPase, which is hard to explain in a general context. According to Friede (1961a), the metabolic differentiation between the ependyma and subependyma "may be the cause of the greater vulnerability of the ependyma under various pathological conditions and also of the early postmortem changes of the ependyma."

MENINGES AND PERINEURAL EPITHELIUM

In the rhesus monkey, as in man, the meninges consist of dura mater, arachnoid, and pia mater (Figs. 249–253). All these layers cover the entire central nervous system, starting from the brain and ending at the filum terminale. The pia-arachnoid mater is made up of multiple layers of flat squamous cells with connective tissue and blood vessels running in between. The dura mater is mainly made up of connective tissue elements. In a series of studies, Shanthaveerappa and Bourne (1962a, 1962b, 1963a, 1963b, 1964, 1965a, 1966a; Shantha and Bourne, 1968) showed that the leptomeninges extend on into the peripheral (both autonomic and somatic) nervous system as perineural epithelium (PE) after covering the dorsal and ventral roots. It also forms the capsule of the dorsal root ganglion. A similar capsule of the epithelial cells has also been observed in the ganglion situated farther away in the distal part of sciatic nerves (Shanthaveerappa and Bourne, 1963c). These studies have also shown that all the encapsulated sensory and motor end-organs, including the motor end-plate, are covered by the extension of this PE cell layer. The epi- and perineural connective tissue of the peripheral nervous system is similar to the dura mater of the

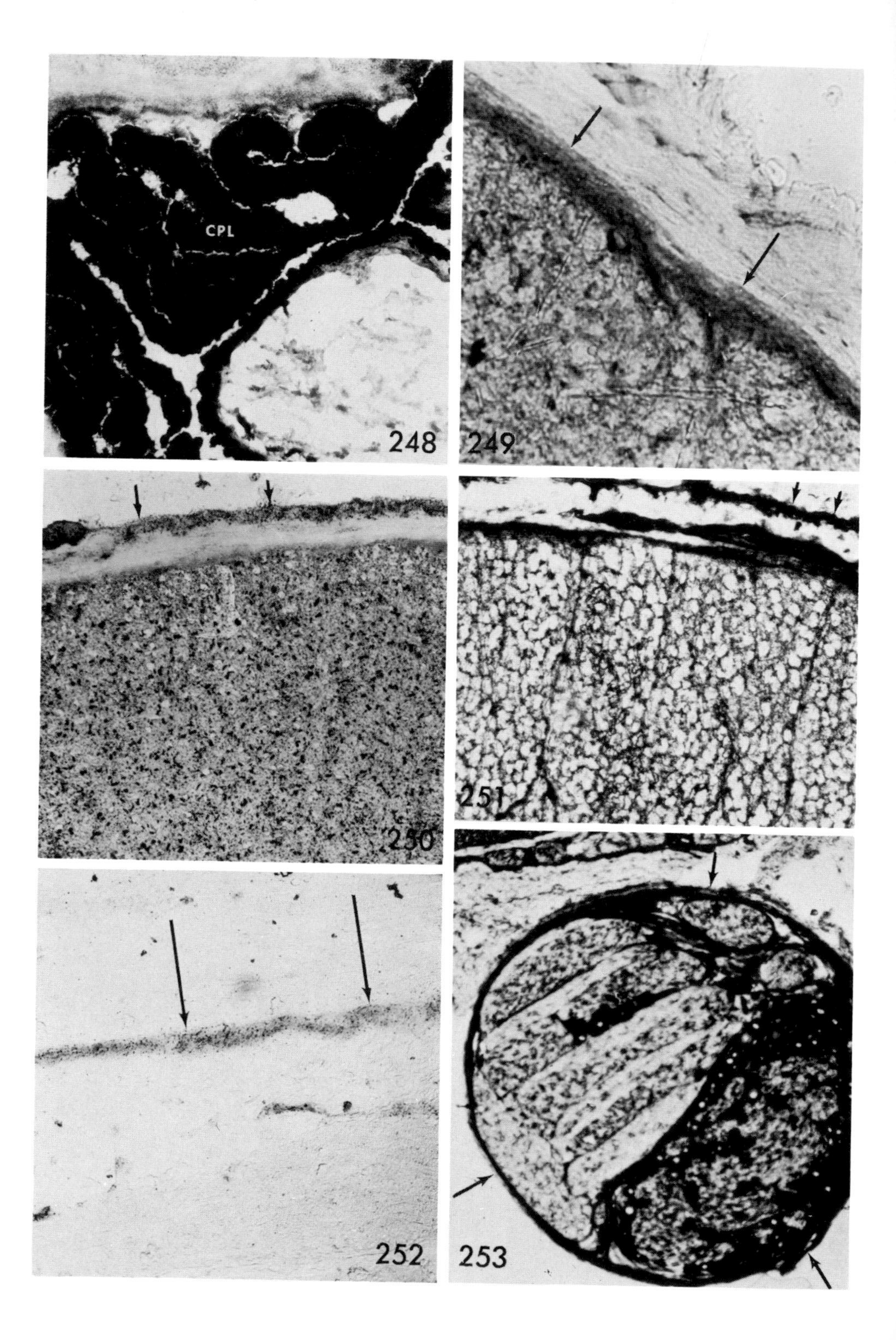

CPL
248
249
250
251
252
253

FIGURE 248. Choroid plexus showing strong AC activity throughout the cytoplasm of the choroid plexus lining cells and little activity in the stroma. ×148.

FIGURES 249–251. Leptomeninges (arrows) covering the spinal cord showing moderately strong LDH and AC (Figs. 249, 250) and strong ATPase activity (Fig. 251). Note also the variations in the enzyme activity in the white matter of the spinal cord. ×592, ×148, ×148.

FIGURES 252, 253. Perineural epithelial covering of the dorsal root ganglion cells showing moderate AK (Fig. 252) and strong ATPase (Fig. 253) activity. Note that some of the PE cells are peeling off from the inner surface (perineural septi), thus dividing the ganglion cells into compartments of different sizes in the ATPase preparation. ×148, ×37.

central nervous system and is in direct continuity with it. Thus the entire peripheral nervous system is isolated from the surrounding tissue fluids and plays a major role in nerve degeneration and regeneration (for details, see Shantha and Bourne, 1968).

Histochemically, the dura mater and the peri- and epineural connective tissue show very little enzyme activity. Only a few stromal cells, probably fibro-blasts, exhibit positive activity. On the other hand, perineural epithelial cells (Figs. 252–253) and leptomeninges (pia-arachnoid mater) give positive activity for a variety of enzymes tested. They show strong ATPase, moderately strong to strong LDH, moderate SDH and AC, mild SE and MAO, negligible AK, and negative AChE activity. The positive activity in all these enzymes is diffuse in character. Pepler (1960) observed strong AK activity in the arachnoidal cap cells of the arachnoidal granulations, arachnoid membrane, and in the arachnoidal cell rests. Similarly, Klika (1960) showed a strong AK reaction on the outer, periosteal surface of the dura mater cerebri and also in the inner, meningeal layer in the adult cat and pig, with a weaker reaction in the monkey. Intensely positive AK activity was also observed in the cell clusters on the subdural surface of the dura mater. The blood vessels, in the present study, are moderately to strongly positive in AK and ATPase. Sometimes thin AChE-positive areas are observed, which probably represent the AChE-positive nerve fibers supplying these membranes.

The presence of AC, AK, ATPase, TPPase, LDH, SDH, adenosine monophosphatase, glucose-6-phosphatase, and cytochrome oxidase in the leptomeninges surrounding the olfactory bulb and spinal cord has been described by Shanthaveerappa and Bourne (1965b) and Manocha *et al.* (1967). The PE cells surrounding the peripheral nerves including the ganglion cells, the sensory and motor end-organs like the Pacinian corpuscle, and the muscle spindle also show positive activity for these enzymes including β-glucuronidase (Shantha and Bourne, 1968; Shantha *et al.*, 1968; Nakajima *et al.*, 1968). Not only has AChE activity been observed in close proximity to the axonal membrane (Brzin *et al.*, 1966; Bloom and Barrnett, 1966; Schlaepfer and Torack, 1966), but also in the other membranes surrounding the nerve fibers, such as those of the myelin sheath. The PE cells also hydrolyze various high energy phosphates like adenosine-5′-monophosphate, adenosine-3′-monophosphate, adenosine-3′,5′-monophosphate, acetyl phosphate, cytidine mono- and triphosphate, guanosine triphosphate, inosine mono- and triphosphate, uridine tri- and monophosphate, di- and triphosphopyridine nucleotide, carbamyl phosphate, and thiamine pyrophosphate (Shantha *et al.*, 1968). The variability of the enzyme activity at different locations is another characteristic feature. For example, the degree of positive activity is much greater in the meninges, dorsal root ganglion coverings, and proximal parts of the cranial and spinal nerves which are close to the central nervous system than in the PE cells of the distal parts of the peripheral nervous system. This is probably due to constant bathing of these membranes by cerebrospinal fluid. As these membranes are traced back, there is probably less contact with cerebrospinal fluid, resulting in less activity. This is substantiated by the finding of more enzyme activity in PE cells covering nerve fasciculi in mesentery than in peripheral nerves because of the constant contact of the former nerves with the peritoneal fluids (Shantha, unpublished data). This is probably the reason why Adams (1965) found strong SDH activity in cranial nerves, whereas Shanthaveerappa and Bourne (1962b) found negligible activity in distal parts of peripheral nerves. It is also indicated that the enzyme content of these membranes is much higher in young animals than in old (Shantha, unpublished data). Both these membranes are also rich in PAS-positive material and give strong AK activity when the Gomori method is utilized and negligible to mild activity when the azo dye method is used. There is a possibility that there is more than one type of AK in tissue and that the azo dye method and Gomori method localize them differently.

The significance of these enzyme localizations and the role they play in protecting the entire peripheral and central nervous system become obvious. These studies provide ample evidence that the PE cell membrane has the potentiality of acting as a metabolically active diffusion barrier and maintaining a similar environment in both the central and the peripheral nervous systems. The central and peripheral nervous systems are thus enclosed in a mechanically as well as metabolically active barrier, which separates them from the surrounding tissues and body fluids.

As the dura mater and the epi- and perineural connective tissues are mostly made up of connective tissue elements lacking most of the enzymes, they cannot act as a metabolically active barrier. It appears to us that these connective tissue bands act as mechanical barriers for large, heavy, and solid molecules and may dampen the effects of sudden gushes of liquids on nervous tissue.

REFERENCES

Abe, T., Yamada, Y., and Hashimoto, P. H. Histochemical study of glucose-6-phosphate dehydrogenase in the brain of normal adult rat. *Med. J. Osaka Univ.* **14**:67–98 (1963).

Adams, C. W. M. (ed.) Histochemistry of the cells in the nervous system. In: *Neurohistochemistry,* pp. 253–331. Elsevier, Amsterdam, 1965.

Albert, Z., Orlowski, M., Rzucidlo, Z., and Orlowski, J. Studies on L-glutamyl transpeptidase activity and its histochemical localization in the central nervous system of man and different animal species. *Acta Histochem.* **25**:312–320 (1966).

Ariens-Kappers, J. Structural and functional changes in the telencephalic choroid plexus during human ontogenesis. *Ciba Found. Symp. Cerebrospinal Fluid* (1958).

Arioka, I., and Tanimukai, H. Histochemical studies on monoamine oxidase in the mid-brain of the mouse. *J. Neurochem.* **1**:311–315 (1957).

Barlow, R. M., d'Agostine, A. N., and Canalla, P. A. A morphological and histochemical study of the subcommissural organ of young and old sheep. *Z. Zellforsch.* **77**:299–315 (1967).

Becker, N. H., Goldfischer, S., Shin, Woo-Yung, and Novikoff, A. B. The localization of enzyme activities in the rat brain. *J. Biophys. Biochem. Cytol.* **8**:649–663 (1960).

Bloom, F. E., and Barrnett, R. J. Fine structural localization of acetylcholinesterase in electroplaque of the electric eel. *J. Cell Biol.* **29**:475–495 (1966).

Bourne, G. H. Histochemical demonstration of phosphatase in the central nervous system of the rat. *Exptl. Cell Res. Suppl.* **5**:101–117 (1958).

Brightman, M. W., and Palay, S. L. The fine structure of the ependyma in the brain of the rat. *J. Cell Biol.* **19**:415–439 (1963).

Brzin, M., Tennyson, V., and Daffy, P. E. Acetylcholinesterase in frog sympathetic and dorsal root ganglia. A study by electron microscope cytochemistry and microgasometric analysis with the magnetic diver. *J. Cell Biol.* **31**:215–242 (1966).

Burstone, M. S. Comparison of naphthol AS-Phosphates for demonstration of phosphatases. *J. Nat. Cancer Inst.* **20**:601–614 (1958).

Carpenter, S. J. An electron microscope study of the arboroid plexus of *Necturus maculosus. J. Comp. Neurol.* **127**:413–421 (1966).

Chason, J. L., and Pearse, A. G. E. Phenazine methosulphate and nicotinamide in the histochemical demonstration of dehydrogenases in rat brain. *J. Neurochem.* **6**:259–266 (1961).

Chason, J. L., Gonzalez, J. E., and Landers, J. W. Respiratory enzyme activity and distribution in the post mortem central nervous system. *J. Neuropathol. Exptl. Neurol.* **22**:248–254 (1963).

DeDuve, C. Lysosomes, a new group of cytoplasmic particles. In: *Subcellular Particles* (T. Hayashi, ed.), pp. 158–159. Ronald Press, New York, 1959.

DeLong, M., and Balogh, K. Glucose-6-phosphate dehydrogenase activity in the subcommissural organ of rats. A histochemical study. *Endocrinology* **76**:996–998 (1965).

Fisher, R. G., Copenhaver, J. H., Jr., and Maline, D. S. Effect of urea on enzyme activity of the choroid plexus. *Proc. Soc. Exptl. Biol. Med.* **101**:797–798 (1959).

Friede, R. L. Histochemical investigations on succinic dehydrogenase in the central nervous system. I. Distribution in the developing rat's brain. *J. Neurochem.* **4**:101–111 (1959).

Friede, R. L. Surface structures of the aqueduct and the ventricular wall; a morphologic, comparative and histochemical study. *J. Comp. Neurol.* **116**:229–297 (1961a).

Friede, R. L. *A Histochemical Atlas of Tissue Oxidation in the Brain Stem of the Cat.* Karger, Basel, and Stechert-Hafner, New York, 1961b.

Friede, R. L., and Knoller, M. A quantitative mapping of acid phosphatase in the brain of the rhesus monkey. *J. Neurochem.* **12**:441–450 (1965).

Friede, R. L., and Pax, R. H. Mitochondria and mitochondrial enzymes; a comparative study of localization in the cat's brain stem. *Histochemie* **2**:186–191 (1961).

Gomori, G. Microtechnical demonstration of phosphatase in tissue sections. *Proc. Soc. Exptl. Biol. N.Y.* **42**:23–26 (1939).

Herlin, L. On phosphate exchange in the central nervous system with special reference to metabolic activity in barriers. *Acta Physiol. Scand.* **37**:Suppl. 127, 1–86 (1956).

Iijima, K., Bourne, G. H., and Shantha, T. R. Histochemical studies on the distribution of enzymes of glycolytic pathways in the area postrema of the squirrel monkey. *Acta Histochem.* **27**:42–54 (1967).

Kaluza, J. S., Burstone, M. S. and Klatzo, I. Enzyme histochemistry of the chick choroid plexus. *Acta Neuropathol.* 3:480–489 (1964).

Kemeny, A., Kutas, F., Caspar, J., and Stützel, M. A study *in vitro* of respiration and glycolysis of the vascular plexus under different experimental conditions. *Biokhimiya* **26**:787–793 (1961).

Klika, E. Histochemistry of meninges in mammals. *Acta Histochem.* **10**:95–108 (1960).

Kopsch, F. Das Binnengerust in den Zellen einiger Organen des Menschen. *Z. Mikr. Anat. Forsch.* **5**:222–284 (1926).

Krnjevic, K., and Silver, A. Acetylcholinesterase in the developing fore brain. *J. Anat. (London)* **100**:63–89 (1966).

Lazarus, S. S., Wallace, B. J., and Volk, B. W. Neuronal enzyme alterations in Tay-Sachs disease. *Am. J. Pathol.* **41**:579–592 (1962).

Leduc, E. H., and Wislocki, G. B. The histochemical localization of acid and alkaline phosphatase, non-specific esterase, and succinic dehydrogenase in the structures comprising the hematoencephalic barrier of the rat. *J. Comp. Neurol.* **97**:241–279 (1952).

Manocha, S. L., Shantha, T. R., and Bourne, G. H. Histochemical studies on the spinal cord of squirrel monkey (*Saimiri sciureus*). *Exptl. Brain Res.* 3:25–39 (1967).

Masai, H., and Matano, S. Comparative neurological studies on respiratory enzymic activity in the central nervous system of submammals. II. Fishes and amphibia. *Yokohama Med. Bull.* **12**:271–276 (1961).

Maxwell, D. S., and Pearse, D. C. The electron microscopy of the choroid plexus. *J. Biophys. Biochem. Cytol.* **3**:467–476 (1956).

Miller, J. W., and Rogers, G. E. An electron microscopic study of the choroid plexus in the rabbit. *J. Biophys. Biochem. Cytol.* **2**:407–416 (1956).

Nachlas, M. M., Tsou, K., DeSouza, E., Cheng, C., and Seligman, A. M. Cytochemical demonstration of succinic dehydrogenase by the use of a new *p*-nitrophenyl substituted ditetrazole. *J. Histochem. Cytochem.* **5**:420–436 (1957).

Nakajima, Y., Shantha, T. R., and Bourne, G. H. Enzyme histochemical studies on the muscle spindle. *Histochemie* **16**:1–8 (1968).

Nandy, K., and Bourne, G. H. Histochemical studies on the ependyma lining the lateral ventricle of the rat with a note on its possible functional significance. *Ann. Histochim.* **9**:305–314 (1964).

Nandy, K., and Bourne, G. H. Histochemical studies on the ependyma lining the central canal of the spinal cord in the rat with a note on its functional significance. *Acta Anat.* **60**:539–550 (1965).

Novikoff, A. B. Approaches to the *in vivo* function of subcellular particles. In: *Subcellular Particles* (T. Hayashi, ed.), pp. 1–20. The Ronald Press, New York, 1959a.

Novikoff, A. B. Lysosomes and the physiology and pathology of cells. *Biol. Bull.* **117**:385 (1959b).

Pappenheimer, J. R., Heisery, S. R., and Jordan, E. F. Active transport of diodrast and phenolsulfonphthalein from cerebrospinal fluid to blood. *Am. J. Physiol.* **200**:1–10 (1961).

Paul, E. Histochemical studies on the choroid plexuses, the paraphysis cerebri and the ependyma of *Rana temporaria. Z. Zellforsch.* **91**:519–547 (1968).

Pepler, W. J. Alkaline phosphatase in the meninges and in meningiomas. *Nature* **186**:979 (1960).

Pighini, G. Chemische, und biochemische Untersuchungen über das Nervensystem unter normalen und pathologischen Bedingungen. *Biochem. Z.* **42**:124–136 (1912).

Pope, A., and Hess, H. H. Cytochemistry of neurones and neurolia. In: *Metabolism of the Nervous System* (D. Richter, ed.), pp. 72–86. Pergamon Press, London, 1957.

Quay, W. B. Experimental and comparative studies of succinic dehydrogenase activity in mammalian choroid plexuses, ependyma, and pineal organ. *Physiol. Zool.* **33**:206–212 (1960).

Rudolph, G., and Sotelo, C. Etude histochimique des enzymes non spécifiques et des déshydrogénase spécifiques danes les plexus choroïdes et l'ependyme chez le rat. *Ann. Histochim.* **4**:57–64 (1962).

Rutenburg, A. M., Wolman, M., and Seligman, A. M. Comparative distribution of succinic dehydrogenase in six mammals and modifications in the histochemical technique. *J. Histochem. Cytochem.* **1**:66–81 (1953).

Schaltenbrand, G. Plexus und Meningen. In: *Handbuch der Microskopischen Anatomie des Menschen,* Vol. IV, pp. 1–39. Springer-Verlag, Berlin, 1955.

Schlaepfer, W. W., and Torack, R. M. The ultrastructural localization of cholinesterase activity in the sciatic nerve of the rat. *J. Histochem. Cytochem.* **14**:369–378 (1966).

Scott, T. G. The distribution of 5′-nucleotidase in the brain of the mouse. *J. Comp. Neurol.* **129**:97–113 (1967).

Shantha, T. R., and Bourne, G. H. The perineural epithelium—a new concept. In: *The Structure and Function of the Nervous System* (G. H. Bourne, ed.), Vol. I. Academic Press, New York and London, 1968.

Shantha, T. R., and Manocha, S. L. Enzyme histochemistry of the choroid plexus in rat and squirrel monkey. *Histochemie* **14**:149–160 (1968).

Shantha, T. R., Golarz, M. N., and Bourne, G. H. Histological and histochemical observations on the capsule of the muscle spindle in normal and denervated muscle. *Acta Anat.* **69**:632–642 (1968).

Shanthaveerappa, T. R., and Bourne, G. H. A perineural epithelium. *Gen. Cell. Biol.* **14**:343–346 (1962a).

Shanthaveerappa, T. R., and Bourne, G. H. The 'perineural epithelium,' a metabolically active, continuous, protoplasmic cell barrier surrounding a peripheral nerve fasiculi. *J. Anat.* (*London*) **96**:527–537 (1962b).

Shanthaveerappa, T. R., and Bourne, G. H. Demonstration of perineural epithelium in vagus nerves. *Acta Anat.* **52**:95–100 (1963a).

Shanthaveerappa, T. R., and Bourne, G. H. New observations on the structure of the Pacinian corpsucle and its relation to the perineural epithelium of peripheral nerves. *Am. J. Anat.* **112**:97–109 (1963b).

Shanthaveerappa, T. R., and Bourne, G. H. An autonomic ganglion neuron in a small branch of the right sciatic nerve of the rat. *Nature* **198**:607–608 (1963c).

Shanthaveerappa, T. R., and Bourne, G. H. The perineural epithelium of synaptic nerves and ganglia and its relation to the pia-arachnoid of the central nervous system and the perineural epithelium of the peripheral nervous system. *Z. Zellforsch.* **61**:742–753 (1964).

Shanthaveerappa, T. R., and Bourne, G. H. Histological and histochemical studies of the choroid of the eye and its relation to the pia-arachnoid mater of the central nervous system and perineural epithelium of the peripheral nervous system. *Acta Anat.* **61**:379–398 (1965a).

Shanthaveerappa, T. R., and Bourne, G. H. Histochemical studies on the distribution of dephosphorylating and oxidative enzymes and esterases in olfactory bulb of the squirrel monkey. *J. Nat. Cancer Inst.* **35**:153–165 (1965b).

Shanthaveerappa, T. R., and Bourne, G. H. Perineural epithelium—a new concept in the integrity of the peripheral nervous system. *Science* **154**:1464–1467 (1966a).

Shanthaveerappa, T. R., and Bourne, G. H. Double central canal in spinal cord of rat. *Nature* **209**:729–731 (1966b).

Shimizu, N., and Morikawa, N. Histochemical studies of succinic dehydrogenase of the brain of mice, rats, guinea pigs, and rabbits. *J. Histochem. Cytochem.* **5**:334–345 (1957).

Shimizu, N., Morikawa, N., and Ishii, Y. Histochemical studies of succinic dehydrogenase and cytochrome oxidase of the rabbit brain, with special reference to the results in the paraventricular structures. *J. Comp. Neurol.* **108**:1–21 (1957).

Shuttleworth, E. C., Jr., and Allen, N. Acid hydrolases in pia-arachnoid and ependyma of the rat brain. *Neurology* **16**:979–985 (1966).

Tennyson, V. M., and Pappas, G. D. Electron microscopic studies of the developing telecephalic choroid plexus in normal and hydrocephalic rabbits. In: *Disorders of the Developing System* (W. Fields and M. Desmond, eds.), pp. 267–318, C. C. Thomas, Springfield, 1961.

Tennyson, V. M., and Pappas, G. D. Fine structure of developing telecephalic and myelencephalic choroid plexus in the rabbit. *J. Comp. Neurol.* **123**:379–411 (1964).

Thomas, E., and Pearse, A. G. E. The fine localization of dehydrogenases in the nervous system. *Histochemie* **2**:266–282 (1961).

Welch, K. Active transport of iodide by choroid plexus of the rabbit *in vitro. Am. J. Physiol.* **202**:757–760 (1962).

Wislocki, G. B., and Dempsey, E. W. The chemical cytology of the choroid plexus and blood brain barrier of the rhesus monkey (*Macaca mulatta*). *J. Comp. Neurol.* **88**:319–346 (1948).

Wislocki, G. B., and Ladman, A. J. The fine structure of the mammaliam choroid plexus. *Ciba Found. Symp. Cerebrospinal Fluid* **10**:55–79 (1958).

APPENDIX

LIST OF ENZYMES AND SUBSTRATES

AC	Acid phosphatase
ACh	Acetylcholine
AChE	Acetylcholinesterase (specific)
AD	Aldolase
ADH	Alcohol dehydrogenase
AK	Alkaline phosphatase
AMPase	Adenosinemonophosphatase (5′-nucleotidase)
3′, 5′-AMPase	Cyclic 3′, 5′-nucleotide phosphodiesterase
AP	Amylophosphorylase
ATPase	Adenosinetriphosphatase
BChE	Butyrylcholinesterase (nonspecific cholinesterase)
CYO	Cytochrome oxidase
DPN	Diphosphopyridine nucleotide
DPNH	Reduced diphosphopyridine nucleotide
G6P	Glucose-6-phosphatase
G6PD	Glucose-6-phosphate dehydrogenase
LDH	Lactic dehydrogenase
MAO	Monoamine oxidase
MDH	Malic dehydrogenase
NAD	Nicotinamide-adenine dinucleotide (DPN)
NADH	Reduced nicotinamide-adenine dinucleotide
NADP	Nicotinamide-adenine dinucleotide phosphate (TPN)
SDH	Succinic dehydrogenase
SE	Simple esterase
TPN	Triphosphopyridine nucleotide
TPNH	Reduced triphosphopyridine nucleotide
TPPase	Thiamine pyrophosphatase
UDPG	Uridine diphosphoglucose glycogen transferase

ABBREVIATIONS USED IN FIGURES

AAA	Area anterior amygdalae
AB	Nucleus ambiguus
AD	Nucleus anterior dorsalis thalami
ADH	Area dorsalis hypothalami
AH	Nucleus anterior hypothalami
ALH	Area lateralis hypothalami
AM	Nucleus anterior medialis thalami
AMH	Area medialis hypothalami
AN	Nucleus annularis
AO	Area olfactoria
AP	Area postrema
APH	Area posterior hypothalami
APL	Area praeoptica lateralis
APM	Area praeoptica medialis
APT	Area praetectalis
AS	Aquaeductus Sylvii
AST	Nucleus accumbens septi
AV	Nucleus anterior ventralis thalami
BAA	Nucleus basalis accessorius amygdalae
BAL	Nucleus basalis accessorius lateralis amygdalae
BAM	Nucleus basalis accessorius medialis amygdalae
BC	Brachium conjunctivum
BLA	Nucleus basalis lateralis amygdalae
BM	Nucleus basalis (Meynert)
BP	Brachium pontis
C	Canalis centralis
CA	Commissura anterior
CAM	Corpus amygdalae
CB	Cerebellum
CC	Corpus callosum
CD	Nucleus caudatus
CDC	Nucleus centralis densocellularis thalami
CDH	Commissura dorsalis hippocampi
CDS	Nucleus cochlearis dorsalis
CEA	Nucleus centralis amygdalae
CEL	Nucleus centralis lateralis thalami
CH	Commissura habenularis
CHO	Chiasma nervorum opticorum
CI	Capsula interna
CL	Claustrum
CLC	Nucleus centralis latocellularis thalami
CM	Nucleus centrum medianum thalami
CMM	Corpus mammillaris
CN	Nucleus cuneatus medialis
CNL	Nucleus cuneatus lateralis
COR	Corona radiata
CPB	Corpus pontobulbare

CPL	Plexus choroidae
CRF	Corpus restiforme
CSL	Nucleus centralis superior lateralis thalami
CSR	Colliculus superior
CT	Nucleus centralis thalami
CTA	Nucleus corticalis amygdalae
CVS	Nucleus cochlearis ventralis
D	Nucleus substantiae griseae dorsalis
DBC	Decussatio brachii conjunctivi
DG	Nucleus dorsalis tegmenti
DMH	Nucleus dorsomedialis hypothalami
DPY	Decussatio pyramidum
DR	Nucleus dorsalis raphae
DS	Nucleus dorsalis septi
DV	Nucleus dorsalis vagi
E	Ependyma
EM	Eminentia medialis
EP	Epiphysis
F	Fornix
FCT	Fibrae corpus trapezoidae
FD	Fascia dentata hippocampi
FH	Fimbria hippocampi
FI	Fossa interpeduncularis
FLM	Fasciculus longitudinalis medialis
FRG	Nucleus reticularis magnocellularis
FRM	Formatio reticularis myelencephali
FRP	Formatio reticularis pontis
FRS	Nucleus reticularis parvocellularis
FRTM	Formatio reticularis tegmenti mesencephali
FTP	Fibrae pontis transversae
G	Nucleus gracilis
GC	Substantia grisea centralis
GCP	Griseum centralis pontis
GH	Gyrus hippocampi
GL	Corpus geniculatum laterale
GLO	Corpus geniculatum laterale, pars oralis
GM	Corpus geniculatum mediale
GMM	Corpus geniculatum mediale, pars magnocellularis
GP	Globus pallidus
GPH	Griseum periventriculare hypothalami
GPO	Griseum pontis
GR	Granular layer of cerebellar cortex
HIP	Hippocampus
HL	Nucleus habenularis lateralis epithalami
HM	Nucleus habenularis medialis epithalami
I	Nucleus intermedius cerebelli (Nucleus emboliformis et globosus cerebelli)
IP	Nucleus interpeduncularis
IS	Nucleus interstitialis (Cajal)
ITA	Nucleus intercalatus amygdalae

ITP	Interthalamic peduncle
L	Nucleus lateralis cerebelli (nucleus dentatus cerebelli)
LA	Nucleus lateralis amygdalae
LCR	Locus coeruleus
LD	Nucleus lateralis dorsalis thalami
LGE	Globus pallidus, lamina medullaris externa
LGI	Globus pallidus, lamina medullaris interna
LM	Lemniscus medialis
LME	Lamina medullaris externa thalami
LMI	Lamina medullaris interna thalami
LMT	Nucleus limitans thalami
LP	Nucleus lateralis posterior thalami
LS	Nucleus lateralis septi
M	Nucleus medialis cerebelli (nucleus fastigii cerebelli)
MA	Nucleus medialis amygdalae
MD	Nucleus medialis dorsalis thalami
MI	Nucleus intermedius corpus mammillaris
ML	Nucleus lateralis corpus mammillaris
MM	Nucleus medialis corpus mammillaris
MO	Molecular layer of cerebellar cortex
MS	Nucleus medialis septi
MV	Nucleus tractus mesencephali n. trigemini
NAB	Nucleus n. abducentis
NAP	Nucleus ansa peduncularis
NCA	Nucleus commissurae anterioris
NCI	Nucleus colliculi inferioris
NCS	Nucleus centralis superior
NCT	Nucleus corpus trapezoidalis
NDK	Nucleus Darkschewitsch
NF	Nucleus n. facialis
NFDB	Nucleus fasciculi diagonalis band of Broca
NH	Nucleus n. hypoglossi
NI	Nucleus intercalatus
NIII	Nervus oculomotorius
NIV	Nervus trochlearis
NLL	Nucleus lemnisci lateralis
NME	Nucleus of the median eminence
NMV	Nucleus motorius n. trigemini
NOC	Nucleus centralis n. oculomotorii
NOD	Nucleus n. oculomotorii, pars dorsalis
NOV	Nucleus n. oculomotorii, pars ventralis
NR	Nucleus ruber
NST	Nucleus stria terminalis
NSTH	Nucleus subthalamicus
NSV	Nucleus tractus spinalis n. trigemini
NT	Nucleus n. trochlearis
NTS	Nucleus tractus solitarii
NVI	Nervus abducentis
NVII	Nervus facialis
NVIII	Nervus vestibularis

OI	Nucleus olivaris inferior
OID	Nucleus olivaris inferior dorsalis
OIM	Nucleus olivaris inferior medialis
OS	Nucleus olivaris superior
P	Purkinje cells
PBL	Nucleus parabrachialis lateralis
PBM	Nucleus parabrachialis medialis
PAS	Parasubiculum
PC	Nucleus paracentralis thalami
PCR	Pedunculus cerebri
PF	Nucleus parafascicularis thalami
PH	Nucleus paraventricularis hypothalami
PM	Nucleus praeopticus medianus
PP	Nucleus praepositus
PRS	Presubiculum
PS	Prosubiculum
PU	Pulvinar
PUT	Putamen
PV	Nucleus paraventricularis thalami
PVA	Nucleus principalis n. trigemini
PVI	Nucleus pulvinaris inferior thalami
PVL	Nucleus pulvinaris lateralis thalami
PVM	Nucleus pulvinaris medialis thalami
PY	Tractus pyramidalis
R	Nucleus reticularis thalami
RL	Nucleus reticularis lateralis
RN	Raphae nuclei
RP	Recess pineale
RTP	Nucleus reticularis tegmenti pontis
RU	Nucleus reuniens thalami
S	Substantia gelatinosa
SB	Subiculum
SG	Nucleus suprageniculatus
SGV	Nucleus substantiae griseae ventralis
SI	Substantia innominata
SM	Stria medullaris thalami
SMH	Nucleus supramammillaris hypothalami
SMT	Nucleus submedius thalami
SN	Substantia nigra
SOH	Nucleus supraopticus hypothalami
ST	Stria terminalis
TO	Tractus opticus
V	Nucleus substantiae griseae ventralis
VA	Nucleus ventralis anterior thalami
VI	Nucleus vestibularis inferior
VL	Nucleus vestibularis lateralis
VLM	Nucleus ventralis lateralis thalami, pars medialis
VLO	Nucleus ventralis lateralis thalami, pars oralis
VM	Nucleus vestibularis medialis
VMH	Nucleus ventromedialis hypothalami

VPI	Nucleus ventralis posterior inferior thalami
VPL	Nucleus ventralis posterior lateralis thalami
VPM	Nucleus ventralis posterior medialis thalami
VPMP	Nucleus ventralis posterior medialis thalami, pars parvocellularis
VS	Nucleus vestibularis superior
ZI	Zona incerta
II	Ventriculus lateralis
III	Ventriculus tertius
IV	Ventriculus quartus

AUTHOR INDEX

Numbers in italics refer to pages on which the complete references are cited.

C

D

G

H

L

M

T

U

V

W

SUBJECT INDEX

A

B

D

E

F

G

H

I

L

M

N

R

S

Z